Solar System Data

Body	Mass (kg)	Mean Radius (m)	Period (s)	Distance from Sun (m)
Mercury	3.18×10^{23}	2.43×10^{6}	7.60×10^{6}	5.79×10^{10}
Venus	4.88×10^{24}	6.06×10^{6}	1.94×10^{7}	1.08×10^{11}
Earth	5.98×10^{24}	6.37×10^{6}	3.156×10^{7}	1.496×10^{11}
Mars	6.42×10^{23}	3.37×10^{6}	5.94×10^{7}	2.28×10^{11}
Jupiter	1.90×10^{27}	6.99×10^{7}	3.74×10^{8}	7.78×10^{11}
Saturn	5.68×10^{26}	5.85×10^{7}	9.35×10^{8}	1.43×10^{12}
Uranus	8.68×10^{25}	2.33×10^{7}	2.64×10^{9}	2.87×10^{12}
Neptune	1.03×10^{26}	2.21×10^{7}	5.22×10^{9}	4.50×10^{12}
Pluto	$\approx 1.4 \times 10^{22}$	$\approx 1.5 \times 10^{6}$	7.82×10^{9}	5.91×10^{12}
Moon	7.36×10^{22}	1.74×10^{6}	—	—
Sun	1.991×10^{30}	6.96×10^{8}	—	—

Physical Data Often Used[a]

Average Earth-Moon distance	3.84×10^{8} m
Average Earth-Sun distance	1.496×10^{11} m
Average radius of the Earth	6.37×10^{6} m
Density of air (20°C and 1 atm)	1.20 kg/m^3
Density of water (20°C and 1 atm)	1.00×10^{3} kg/m^3
Free-fall acceleration	9.80 m/s^2
Mass of the Earth	5.98×10^{24} kg
Mass of the Moon	7.36×10^{22} kg
Mass of the Sun	1.99×10^{30} kg
Standard atmospheric pressure	1.013×10^{5} Pa

[a] These are the values of the constants as used in the text.

Some Prefixes for Powers of Ten

Power	Prefix	Abbreviation	Power	Prefix	Abbreviation
10^{-18}	atto	a	10^{1}	deka	da
10^{-15}	femto	f	10^{2}	hecto	h
10^{-12}	pico	p	10^{3}	kilo	k
10^{-9}	nano	n	10^{6}	mega	M
10^{-6}	micro	μ	10^{9}	giga	G
10^{-3}	milli	m	10^{12}	tera	T
10^{-2}	centi	c	10^{15}	peta	P
10^{-1}	deci	d	10^{18}	exa	E

Standard Abbreviations and Symbols of Units

Abbreviation	Unit	Abbreviation	Unit
A	ampere	in.	inch
Å	angstrom	J	joule
u	atomic mass unit	K	kelvin
atm	atmosphere	kcal	kilocalorie
Btu	British thermal unit	kg	kilogram
C	coulomb	kmol	kilomole
°C	degree Celsius	lb	pound
cal	calorie	m	meter
deg	degree (angle)	min	minute
eV	electron volt	N	newton
°F	degree Fahrenheit	Pa	pascal
F	farad	rev	revolution
ft	foot	s	second
G	gauss	T	tesla
g	gram	V	volt
H	henry	W	watt
h	hour	Wb	weber
hp	horsepower	μm	micrometer
Hz	hertz	Ω	ohm

Mathematical Symbols Used in the Text and Their Meaning

Symbol	Meaning
$=$	is equal to
$\equiv$	is defined as
$\neq$	is not equal to
$\propto$	is proportional to
$>$	is greater than
$<$	is less than
$\gg (\ll)$	is much greater (less) than
$\approx$	is approximately equal to
Δx	the change in x
$\sum_{i=1}^{N} x_i$	the sum of all quantities x_i from $i = 1$ to $i = N$
$\|x\|$	the magnitude of x (always a nonnegative quantity)
$\Delta x \rightarrow 0$	Δx approaches zero
$\dfrac{dx}{dt}$	the derivative of x with respect to t
$\dfrac{\partial x}{\partial t}$	the partial derivative of x with respect to t
$\int$	integral

PHYSICS

For Scientists & Engineers

| Fourth Edition |

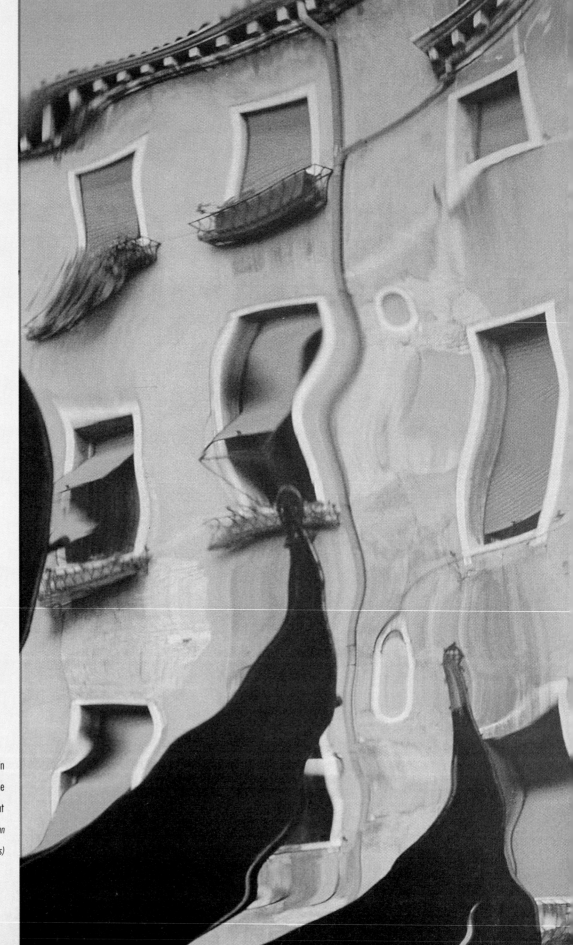

Photo of the image of an old building in Venice with gondolas in the foreground. The distorted image was formed by light reflected from a wavy water surface. *(John McDermott/Tony Stone Images)*

PART V

Light and Optics

"I procured me a Triangular glass-Prisme to try therewith the celebrated *Pheaenomena of Colours*. . . . I placed my Prisme at his entrance (the sunlight), that it might thereby be refracted to the opposite wall. It was a very pleasing divertisement to view the vivid and intense colours produced thereby; . . . I have often with admiration beheld that all the colours of the Prisme being made to converge, and thereby to be again mixed, as they were in the light before it was incident upon the Prisme, reproduced light, entirely and perfectly white, and not at all sensibly differing from the direct light of the sun. . . ."

ISAAC NEWTON

Scientists have long been intrigued by the nature of light, and philosophers have had endless arguments concerning the proper definition and perception of light. It is important to understand the nature of light because it is one of the basic ingredients of life on Earth. Plants convert light energy from the Sun to chemical energy through photosynthesis. Light is the principal means by which we are able to transmit and receive information from objects around us and throughout the Universe.

The nature and properties of light have been a subject of great interest and speculation since ancient times. The Greeks believed that light consisted of tiny particles (corpuscles) that were emitted by a light source and then stimulated the perception of vision upon striking the observer's eye. Newton used this corpuscular theory to explain the reflection and refraction of light. In 1670, one of Newton's contemporaries, the Dutch scientist Christian Huygens, was able to explain many properties of light by proposing that light is made up of waves. In 1801, Thomas Young showed that light beams can interfere with one another, giving strong support to the wave theory. In 1865, Maxwell developed a brilliant theory where he demonstrated that all electromagnetic waves travel at the same speed, c. (Chapter 34). By this time, the wave theory of light seemed to be on firm ground.

However, at the beginning of the 20th century, Albert Einstein returned to the corpuscular theory of light in order to explain the emission of electrons from a metal surface exposed to light (the photoelectric effect). We discuss these and other topics in the last part of this text, which is concerned with modern physics.

Today, scientists view light as having a dual nature. In some experiments, light displays particle-like behavior and in other experiments it exhibits wave properties.

In this part of the book, we concentrate on those aspects of light best understood through the wave model. First, we discuss the reflection of light at the boundary between two media and the refraction of light as it travels from one medium into another. Next we use these ideas to study the refraction of light as it passes through lenses and the reflection of light from surfaces. Then we describe how lenses and mirrors are used to view objects through such instruments as cameras, telescopes, and microscopes. Finally, we shall discuss the phenomena of diffraction, polarization, and interference as they apply to light.

CHAPTER 35

The Nature of Light and the Laws of Geometric Optics

This photograph taken in Salamanca, Spain, shows the reflection of the New Cathedral in the Tormes River. Are you able to distinguish the cathedral from its image? *(David Parker/Photo Researchers)*

35.1 THE NATURE OF LIGHT

Before the beginning of the 19th century, light was considered to be a stream of particles that were emitted by a light source and then stimulated the sense of sight upon entering the eye. The chief architect of this particle theory of light was, once again, Isaac Newton, who was able to explain on the basis of the particle theory some known experimental facts concerning the nature of light, namely, reflection and refraction.

Most scientists accepted Newton's particle theory of light. However, during his lifetime another theory was proposed—one that argued that light is some sort of wave motion. In 1678, a Dutch physicist and astronomer, Christian Huygens (1629–1695), showed that a wave theory of light could also explain the laws of reflection and refraction. The wave theory did not receive immediate acceptance for several reasons. First, all waves known at the time (sound, water, and so forth) traveled through some sort of medium. Because light could travel to us from the Sun through the vacuum of space, it could not possibly be a wave because wave travel requires a medium. Furthermore, it was argued that, if light were some form of wave, it would bend around obstacles; hence, we should be able to see around corners. It is now known that light does indeed bend around the edges of objects. This phenomenon, known as diffraction, is not easy to observe because light waves have such short wavelengths. Thus, although experimental evidence for the diffraction of light was discovered by Francesco Grimaldi (1618–1663) in approximately 1660, most scientists rejected the wave theory and adhered to Newton's particle theory for more than a century. This was, for the most part, due to Newton's great reputation as a scientist.

The first clear demonstration of the wave nature of light was provided in 1801 by Thomas Young (1773–1829), who showed that, under appropriate conditions, light rays interfere with each other. At that time, such behavior could not be explained by a particle theory because there is no conceivable way by which two or more particles could come together and cancel one another. Several years later, a French physicist, Augustin Fresnel (1788–1829), performed a number of experiments dealing with interference and diffraction. In 1850, Jean Foucault (1791–1868) provided further evidence of the inadequacy of the particle theory by showing that the speed of light in glasses and liquids is less than in air. According to the particle model, the speed of light would be higher in glasses and liquids than in air.

Additional developments during the 19th century led to the general acceptance of the wave theory of light, the most important being the work of Maxwell, who in 1873 asserted that light was a form of high-frequency electromagnetic wave (Chapter 34). As discussed in Chapter 34, Hertz provided experimental confirmation of Maxwell's theory by producing and detecting electromagnetic waves. Furthermore, Hertz and other investigators showed that *these waves exhibited reflection, refraction, and all the other characteristic properties of waves.*

Although the classical theory of electricity and magnetism was able to explain most known properties of light via the wave model, some subsequent experiments could not be explained by it. The most striking of these is the photoelectric effect, also discovered by Hertz: When light strikes a metal surface, electrons are sometimes ejected from the surface. As one example of the difficulties that arose, experiments showed that the kinetic energy of an ejected electron is independent of the light intensity. This was in contradiction of the wave theory, which held that a more intense beam of light should add more energy to the electron. An explanation of the photoelectric effect was proposed by Einstein in 1905 in a theory that used the concept of quantization developed by Max Planck (1858–1947) in 1900. The quantization model assumes that the energy of a light wave is present in bundles of energy called photons; hence, the energy is said to be quantized. (Any quantity that appears in discrete bundles is said to be quantized, just as charge and other properties are quantized.) According to Planck's theory, the energy of a photon is proportional to the frequency of the electromagnetic wave:

Energy of a photon

$$E = hf \tag{35.1}$$

where the constant of proportionality $h = 6.63 \times 10^{-34}$ J·s is Planck's constant. It is important to note that this theory retains some features of both the wave theory and the particle theory of light. As we discuss later, the photoelectric effect is the result of energy transfer from a single photon to an electron in the metal. That is, the electron interacts with one photon of light as if the electron had been struck by a particle. Yet this photon has wave-like characteristics because its energy is determined by the frequency (a wave-like quantity).

In view of these developments, light must be regarded as having a dual nature. That is, *in some cases light acts like a wave and in others it acts like a particle.* Classical electromagnetic wave theory provides an adequate explanation of light propagation and of the effects of interference, whereas the photoelectric effect and other experiments involving the interaction of light with matter are best explained by assuming that light is a particle. Light is light, to be sure. However, the question, "Is light a wave or a particle?" is an inappropriate one. In some experiments we measure its wave properties; in other experiments we measure its particle properties. In the next few chapters, we investigate the wave nature of light.

35.2 MEASUREMENTS OF THE SPEED OF LIGHT

Light travels at such a high speed ($c \approx 3 \times 10^8$ m/s) that early attempts to measure its speed were unsuccessful. Galileo attempted to measure the speed of light by positioning two observers in towers separated by approximately 5 mi. Each observer carried a shuttered lantern. One observer would open his lantern first, and then the other would open his lantern at the moment he saw the light from the first lantern. The speed could then be obtained, in principle, knowing the transit time of the light beams between lanterns. The results were inconclusive. Today, we realize (and as Galileo himself concluded) that it is impossible to measure the speed of light in this manner because the transit time is very small compared with the reaction time of the observers.

We now describe two methods for determining the speed of light.

Roemer's Method

The first successful estimate of the speed of light was made in 1675 by the Danish astronomer Ole Roemer (1644–1710). His technique involved astronomical observations of one of the moons of Jupiter, Io, which has a period of approximately 42.5 h; this was measured by observing the eclipse of Io as it passed behind Jupiter. The period of Jupiter is about 12 years, and so as the Earth moves through 180° around the Sun, Jupiter revolves through only 15° (Fig. 35.1).

Using the orbital motion of Io as a clock, its orbit around Jupiter would be expected to have a constant period over long time intervals. However, Roemer observed a systematic variation in Io's period during a year's time. He found that the periods were getting larger when the Earth was receding from Jupiter and smaller when the Earth was approaching Jupiter. For example, if Io had a constant period, Roemer should have seen an eclipse occurring at a particular instant and been able to predict when an eclipse should begin at a later time in the year. However, when he checked the second eclipse, he found that if the Earth was receding from Jupiter, the eclipse was late. In fact, if the interval between observations was three months, the delay was approximately 600 s. Roemer attributed this variation in period to the fact that the distance between the Earth and Jupiter was changing from one observation to the next. In three months (one quarter of the period of the Earth), the light from Jupiter has to travel an additional distance equal to the radius of the Earth's orbit.

Using Roemer's data, Huygens estimated the lower limit for the speed of light

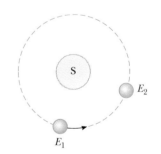

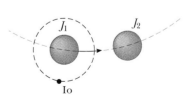

FIGURE 35.1 Roemer's method for measuring the speed of light. In the time it takes the Earth to travel 180° around the Sun (6 months), Jupiter travels only 15°.

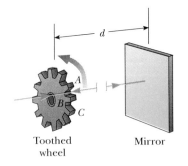

FIGURE 35.2 Fizeau's method for measuring the speed of light using a rotating toothed wheel.

to be approximately 2.3×10^8 m/s. This experiment is important historically because it demonstrated that light does have a finite speed and gave an estimate of this speed.

Fizeau's Technique

The first successful method for measuring the speed of light using purely terrestrial techniques was developed in 1849 by Armand H. L. Fizeau (1819–1896). Figure 35.2 represents a simplified diagram of his apparatus. The basic idea is to measure the total time it takes light to travel from some point to a distant mirror and back. If d is the distance between source and mirror and if the transit time for one round trip is t, then the speed of light is $c = 2d/t$. To measure the transit time, Fizeau used a rotating toothed wheel, which converts an otherwise continuous beam of light into a series of light pulses. Additionally, the rotation of the wheel controls what an observer at the light source sees. For example, if the light passing the opening at point A in Figure 35.2 should return at the instant that tooth B had rotated into position to cover the return path, the light would not reach the observer. At a faster rate of rotation, the opening at point C could move into position to allow the reflected beam to reach the observer. Knowing the distance d, the number of teeth in the wheel, and the angular speed of the wheel, Fizeau arrived at a value of $c = 3.1 \times 10^8$ m/s. Similar measurements made by subsequent investigators yielded more precise values for c, approximately 2.9977×10^8 m/s.

EXAMPLE 35.1 Measuring the Speed of Light with Fizeau's Toothed Wheel

Assume Fizeau's wheel has 360 teeth and is rotating at 27.5 rev/s when a burst of light passing through A in Figure 35.2 is blocked by tooth B on return. If the distance to the mirror is 7500 m, find the speed of light.

Solution Because the wheel has 360 teeth, it turns through an angle of 1/720 rev in the time that passes while the light makes its round trip. From the definition of angular speed, we see that the time is

$$t = \frac{\theta}{\omega} = \frac{(1/720) \text{ rev}}{27.5 \text{ rev/s}} = 5.05 \times 10^{-5} \text{ s}$$

Hence, the speed of light is

$$c = \frac{2d}{t} = \frac{2(7500 \text{ m})}{5.05 \times 10^{-5} \text{ s}} = 2.97 \times 10^8 \text{ m/s}$$

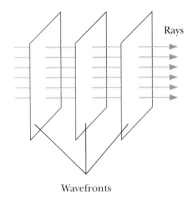

FIGURE 35.3 A plane wave propagating to the right. Note that the rays, which always point in the direction of wave motion, are straight lines perpendicular to the wave fronts.

35.3 THE RAY APPROXIMATION IN GEOMETRIC OPTICS

In studying geometric optics here and in Chapter 36, we use what is called the **ray approximation**. To understand this approximation, first note that the rays of a given wave are straight lines perpendicular to the wave fronts, as illustrated in Figure 35.3 for a plane wave. In the ray approximation, we assume that a wave moving through a medium travels in a straight line in the direction of its rays.

If the wave meets a barrier in which there is a circular opening whose diameter is large relative to the wavelength, as in Figure 35.4a, the wave emerging from the opening continues to move in a straight line (apart from some small edge effects); hence, the ray approximation continues to be valid. If, however, the diameter of the opening is of the order of the wavelength, as in Figure 35.4b, the waves spread out from the opening in all directions. We say that the outgoing wave is noticeably diffracted. Finally, if the opening is small relative to the wavelength, the opening can be approximated as a point source of waves (Fig. 34.4c). Thus, the effect of diffraction is more pronounced as the ratio d/λ approaches zero. Similar effects

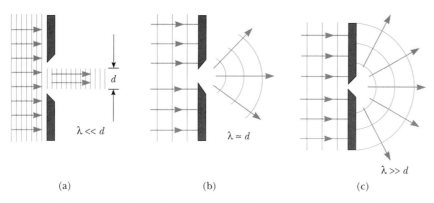

FIGURE 35.4 A plane wave of wavelength λ is incident on a barrier in which there is an opening of diameter d. (a) When $\lambda \ll d$, there is almost no observable diffraction and the ray approximation remains valid. (b) When $\lambda \approx d$, diffraction becomes significant. (c) When $\lambda \gg d$, the opening behaves as a point source emitting spherical waves.

are seen when waves encounter an opaque circular object of diameter d. In this case, when $\lambda \ll d$, the object casts a sharp shadow.

The ray approximation and the assumption that $\lambda \ll d$ are used here and in Chapter 36, both dealing with geometric optics. This approximation is very good for the study of mirrors, lenses, prisms, and associated optical instruments, such as telescopes, cameras, and eyeglasses. We return to the subject of diffraction (where $\lambda \approx d$) in Chapter 38.

35.4 REFLECTION AND REFRACTION

Reflection of Light

When a light ray traveling in a medium encounters a boundary leading into a second medium, part or all of the incident ray is reflected back into the first medium. Figure 35.5a shows several rays of a beam of light incident on a smooth, mirror-like, reflecting surface. The reflected rays are parallel to each other, as indicated in the figure. Reflection of light from such a smooth surface is called **specular reflection.** If the reflecting surface is rough, as in Figure 35.5b, the surface reflects the rays not as a parallel set but in various directions. Reflection from any rough surface is known as **diffuse reflection.** A surface behaves as a smooth surface as long as the surface variations are small compared with the wavelength of the incident light. Photographs of specular reflection and diffuse reflection using laser light are shown in Figures 35.5c and 35.5d. In this book, we concern ourselves only with specular reflection and use the term *reflection* to mean specular reflection.

Consider a light ray traveling in air and incident at an angle on a flat, smooth surface, as in Figure 35.6. The incident and reflected rays make angles θ_1 and θ_1', respectively, with a line drawn perpendicular to the surface at the point where the incident ray strikes. We call this line the normal to the surface. Experiments show that *the angle of reflection equals the angle of incidence:*

$$\theta_1' = \theta_1 \qquad (35.2)$$

By convention, the angles of incidence and reflection are measured from the normal to the surface rather than from the surface itself.

The pencil partially immersed in water appears bent because light is refracted as it travels across the boundary between water and air. (*Jim Lehman*)

Law of reflection

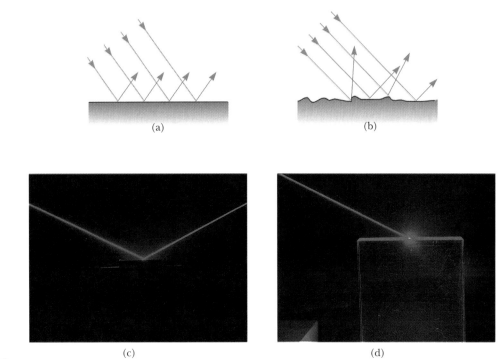

FIGURE 35.6 According to the law of reflection, $\theta_1 = \theta_1'$. The incident ray, the reflected ray, and the normal all lie in the same plane.

FIGURE 35.5 Schematic representation of (a) specular reflection, where the reflected rays are all parallel to each other, and (b) diffuse reflection, where the reflected rays travel in random directions. (c) and (d) Photographs of specular and diffuse reflection using laser light. *(Henry Leap and Jim Lehman)*

EXAMPLE 35.2 The Double-Reflected Light Ray

Two mirrors make an angle of 120° with each other, as in Figure 35.7. A ray is incident on mirror M_1 at an angle of 65° to the normal. Find the direction of the ray after it is reflected from mirror M_2.

Reasoning and Solution From the law of reflection, we know that the first reflected ray also makes an angle of 65° with the normal. Thus, this ray makes an angle of 90° − 65°, or 25°, with the horizontal. From the triangle made by the first reflected ray and the two mirrors, we see that the first reflected ray makes an angle of 35° with M_2 (since the sum of the interior angles of any triangle is 180°). This means that this ray makes an angle of 55° with the normal to M_2. Hence, from the law of reflection, the second reflected ray makes an angle of 55° with the normal to M_2. By comparing the direction of the ray incident on M_1 with its direction after reflecting from M_2, we see that the ray is reflected through 120°, which happens to correspond to the angle between the mirrors.

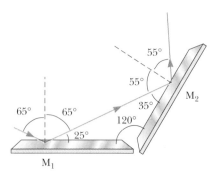

FIGURE 35.7 (Example 35.2) Mirrors M_1 and M_2 make an angle of 120° with each other.

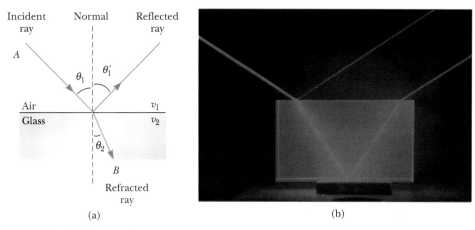

FIGURE 35.8 (a) A ray obliquely incident on an air-glass interface. The refracted ray is bent toward the normal because $v_2 < v_1$. All rays and the normal lie in the same plane. (b) Light incident on the Lucite block bends both when it enters the block and when it leaves the block. *(Henry Leap and Jim Lehman)*

Refraction of Light

When a ray of light traveling through a transparent medium encounters a boundary leading into another transparent medium, as in Figure 35.8, part of the ray is reflected and part enters the second medium. The ray that enters the second medium is bent at the boundary and is said to be refracted. *The incident ray, the reflected ray, the normal, and the refracted ray all lie in the same plane.* The angle of refraction, θ_2 in Figure 35.8, depends on the properties of the two media and on the angle of incidence through the relationship

$$\frac{\sin \theta_2}{\sin \theta_1} = \frac{v_2}{v_1} = \text{constant} \qquad (35.3)$$

where v_1 is the speed of light in medium 1 and v_2 is the speed of light in medium 2.

The experimental discovery of this relationship is usually credited to Willebrord Snell (1591–1627) and is therefore known as **Snell's law**.[1]

The path of a light ray through a refracting surface is reversible. For example, the ray in Figure 35.8 travels from point A to point B. If the ray originated at B, it would follow the same path to reach point A. In the latter case, however, the reflected ray would be in the glass.

When light moves from a material in which its speed is high to a material in which its speed is lower, the angle of refraction, θ_2, is less than the angle of incidence, as shown in Figure 35.9a. If the ray moves from a material in which it moves slowly to a material in which it moves more rapidly, it is bent away from the normal, as shown in Figure 35.9b.

The behavior of light as it passes from air into another substance and then re-emerges into air is often a source of confusion to students. Let us take a look at what happens and see why this behavior is so different from other occurrences in our daily lives. When light travels in air, its speed is equal to 3×10^8 m/s, but this speed is reduced to approximately 2×10^8 m/s when the light enters a block of

---

[1] This law was also deduced from the particle theory of light by René Descartes (1596–1650) and hence is known as *Descartes' law* in France.

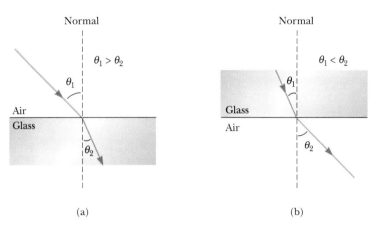

FIGURE 35.9 (a) When the light beam moves from air into glass, its path is bent toward the normal. (b) When the beam moves from glass into air, its path is bent away from the normal.

glass. When the light re-emerges into air, its speed instantaneously increases to its original value of 3×10^8 m/s. This is far different from what happens, for example, when a bullet is fired through a block of wood. In this case, the speed of the bullet is reduced as it moves through the wood because some of its original energy is used to tear apart the fibers of the wood. When the bullet enters the air once again, it emerges at the speed it had just before leaving the block of wood.

To see why light behaves as it does, consider Figure 35.10, which represents a beam of light entering a piece of glass from the left. Once inside the glass, the light may encounter an electron bound to an atom, indicated as point A in the figure. Let us assume that light is absorbed by the atom, which causes the electron to oscillate (a detail represented by the double-headed vertical arrows in the drawing). The oscillating electron then acts as an antenna and radiates the beam of light toward an atom at B, where the light is again absorbed by an atom. The details of these absorptions and radiations are best explained in terms of quantum mechanics, a subject we study in the extended version of this text. For now, it is sufficient to think of the process as one in which the light passes from one atom to another through the glass. (The situation is somewhat analogous to a relay race in which a baton is passed between runners on the same team.) Although light travels from one glass atom to another at 3×10^8 m/s, the absorption and radiation processes that take place in the glass cause the light speed to slow to 2×10^8 m/s. Once the light emerges into the air, the absorptions and radiations cease and its speed returns to the original value.

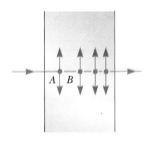

FIGURE 35.10 Light passing from one atom to another in a medium. The dots are electrons, and the vertical arrows represent their oscillations.

Index of Refraction

The speed of light in any material is less than the speed in vacuum except near very strong absorption bands.[2]

Index of refraction

$$n = \frac{\text{speed of light in vacuum}}{\text{speed of light in a medium}} = \frac{c}{v} \tag{35.4}$$

[2] There are two speeds for a wave, which may be the same or different. The group speed, or signal speed, for light cannot be greater than the speed of light in vacuum. The speed considered here is the phase speed, which may be greater than c, without violating special relativity.

From this definition, we see that the index of refraction is a dimensionless number usually greater than unity because v is usually less than c. Furthermore, n is equal to unity for vacuum. The indices of refraction for various substances are listed in Table 35.1.

As light travels from one medium to another, *its frequency does not change but its wavelength does*. To see why this is so, consider Figure 35.11. Wave fronts pass an observer at point A in medium 1 with a certain frequency and are incident on the boundary between medium 1 and medium 2. The frequency with which the wave fronts pass an observer at point B in medium 2 must equal the frequency at which they arrive at point A in medium 1. If this were not the case, either wave fronts would be piling up at the boundary or they would be destroyed or created at the boundary. Since there is no mechanism for this to occur, the frequency must be a constant as a light ray passes from one medium into another.

Therefore, because the relationship $v = f\lambda$ (Eq. 16.14) must be valid in both media and because $f_1 = f_2 = f$, we see that

$$v_1 = f\lambda_1 \quad \text{and} \quad v_2 = f\lambda_2 \tag{35.5}$$

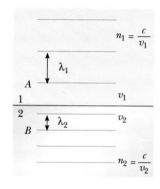

FIGURE 35.11 As a wave front moves from medium 1 to medium 2, its wavelength changes but its frequency remains constant.

Because $v_1 \neq v_2$, it follows that $\lambda_1 \neq \lambda_2$. A relationship between index of refraction and wavelength can be obtained by dividing the first of Equations 35.5 by the other and making use of the definition of the index of refraction given by Equation 35.4:

$$\frac{\lambda_1}{\lambda_2} = \frac{v_1}{v_2} = \frac{c/n_1}{c/n_2} = \frac{n_2}{n_1} \tag{35.6}$$

which gives

$$\lambda_1 n_1 = \lambda_2 n_2$$

If medium 1 is vacuum, or for all practical purposes air, then $n_1 = 1$. Hence, it follows from Equation 35.6 that the index of refraction of any medium can be

TABLE 35.1 Index of Refraction for Various Substances Measured with Light of Vacuum Wavelength $\lambda_0 = 589$ nm

Substance	Index of Refraction	Substance	Index of Refraction
Solids at 20°C		Liquids at 20°C	
Diamond (C)	2.419	Benzene	1.501
Fluorite (CaF_2)	1.434	Carbon disulfide	1.628
Silica (SiO_2)	1.458	Carbon tetrachloride	1.461
Glass, crown	1.52	Ethyl alcohol	1.361
Glass, flint	1.66	Glycerine	1.473
Ice (H_2O)	1.309	Water	1.333
Polystyrene	1.49	Gases at 0°C, 1 atm	
Sodium chloride (NaCl)	1.544		
Zircon	1.923	Air	1.000293
		Carbon dioxide	1.00045

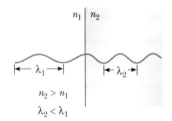

FIGURE 35.12 Schematic diagram of the reduction in wavelength when light travels from a medium of low index of refraction to one of higher index of refraction.

expressed as the ratio

$$n = \frac{\lambda_0}{\lambda_n} \tag{35.7}$$

where λ_0 is the wavelength of light in vacuum and λ_n is the wavelength in the medium whose index of refraction is n. A schematic representation of this reduction in wavelength is shown in Figure 35.12.

We are now in a position to express Snell's law of refraction (Eq. 35.3) in an alternative form. If we substitute Equation 35.6 into Equation 35.3, we get

$$n_1 \sin \theta_1 = n_2 \sin \theta_2 \tag{35.8}$$

This is the most widely used and practical form of Snell's law.

EXAMPLE 35.3 An Index of Refraction Measurement

A beam of light of wavelength 500 nm traveling in air is incident on a slab of transparent material. The incident beam makes an angle of 40.0° with the normal, and the refracted beam makes an angle of 26.0° with the normal. Find the index of refraction of the material.

Solution Snell's law of refraction (Eq. 35.8) with these data gives

$$n_1 \sin \theta_1 = n_2 \sin \theta_2$$

$$n_2 = \frac{n_1 \sin \theta_1}{\sin \theta_2} = (1.00) \frac{\sin 40.0°}{\sin 26.0°} = \frac{0.643}{0.438} = 1.47$$

If we compare this value with the data in Table 35.1, we see that the material may be silica.

Exercise What is the wavelength of light in the material?

Answer 374 nm.

EXAMPLE 35.4 Angle of Refraction for Glass

A light ray of wavelength 589 nm traveling through air is incident on a smooth, flat slab of crown glass at an angle of 30.0° to the normal, as sketched in Figure 35.13. Find the angle of refraction.

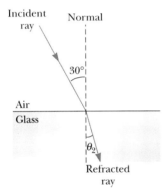

FIGURE 35.13 (Example 35.4) Refraction of light by glass.

Solution Snell's law given by Equation 35.8 can be rearranged as

$$\sin \theta_2 = \frac{n_1}{n_2} \sin \theta_1$$

From Table 35.1, we find that $n_1 = 1.00$ for air and $n_2 = 1.52$ for crown glass. Therefore,

$$\sin \theta_2 = \left(\frac{1.00}{1.52}\right)(\sin 30.0°) = 0.329$$

$$\theta_2 = \sin^{-1}(0.329) = 19.2°$$

The ray is bent toward the normal, as expected.

Exercise If the light ray moves from inside the glass toward the glass–air interface at an angle of 30.0° to the normal, determine the angle of refraction.

Answer 49.5° away from the normal.

EXAMPLE 35.5 The Speed of Light in Silica

Light of wavelength 589 nm in vacuum passes through a piece of silica ($n = 1.458$). (a) Find the speed of light in silica.

Solution The speed of light in silica can be easily obtained from Equation 35.4:

$$v = \frac{c}{n} = \frac{3.00 \times 10^8 \text{ m/s}}{1.458} = 2.06 \times 10^8 \text{ m/s}$$

(b) What is the wavelength of this light in silica?

Solution We use Equation 35.7 to calculate the wavelength in silica, noting that we are given the wavelength in vacuum to be $\lambda_0 = 589$ nm:

$$\lambda_n = \frac{\lambda_0}{n} = \frac{589 \text{ nm}}{1.458} = 404 \text{ nm}$$

Exercise Find the frequency of the light.

Answer 5.09×10^{14} Hz.

EXAMPLE 35.6 Light Passing Through a Slab

A light beam passes from medium 1 to medium 2 with the latter being a thick slab of material whose index of refraction is n_2 (Fig. 35.14). Show that the emerging beam is parallel to the incident beam.

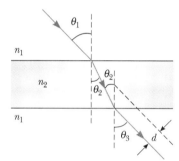

FIGURE 35.14 (Example 35.6) When light passes through a flat slab of material, the emerging beam is parallel to the incident beam, and therefore $\theta_1 = \theta_3$. The green dashed line represents the path the light would take if the slab were not there.

Solution First, let us apply Snell's law to the upper surface:

$$(1) \quad \sin \theta_2 = \frac{n_1}{n_2} \sin \theta_1$$

Applying Snell's law to the lower surface gives

$$(2) \quad \sin \theta_3 = \frac{n_2}{n_1} \sin \theta_2$$

Substituting (1) into (2) gives

$$\sin \theta_3 = \frac{n_2}{n_1}\left(\frac{n_1}{n_2} \sin \theta_1\right) = \sin \theta_1$$

That is, $\theta_3 = \theta_1$, and so the layer does not alter the direction of the beam as shown by the green dashed line. It does, however, produce a displacement of the beam. The same result is obtained when light passes through multiple layers of materials.

*35.5 DISPERSION AND PRISMS

An important property of the index of refraction is that, for a given material, the index varies with the wavelength of light passing through the material as Figure 35.15 shows. Since n is a function of wavelength, Snell's law indicates that light of different wavelengths is bent at different angles when incident on a refracting material. As we see from Figure 35.15, the index of refraction decreases with increasing wavelength. This means that blue light bends more than red light when passing into a refracting material. The various wavelengths are refracted at different wavelengths. This phenomenon is known as **dispersion**.

To understand the effects that dispersion can have on light, let us consider

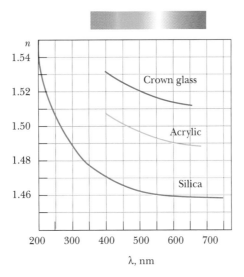

FIGURE 35.15 Variation of index of refraction with vacuum wavelength for three materials.

what happens when light strikes a prism, as in Figure 35.16. A ray of single-wavelength light incident on the prism from the left emerges bent away from its original direction of travel by an angle δ, called the **angle of deviation.** Now suppose a beam of white light (a combination of all visible wavelengths) is incident on a prism, as in Figure 35.17. The rays that emerge from the second face spread out in a series of colors known as a **spectrum.** These colors, in order of decreasing wavelength, are red, orange, yellow, green, blue, indigo, and violet. Newton showed that each color has a particular angle of deviation, that the spectrum cannot be further broken down, and that the colors can be recombined to form the original white light. Clearly, the angle of deviation, δ, depends on the wavelength of a given

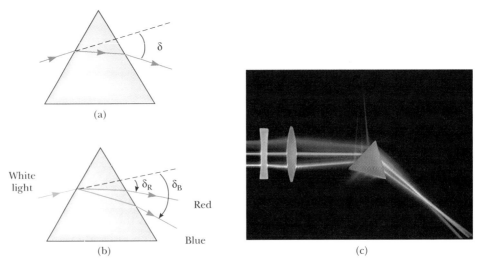

FIGURE 35.16 (a) A prism refracts single-wavelength light and deviates the light through an angle δ. (b) When light is incident on a prism, the blue light is bent more than the red. (c) Light of different colors passes through a prism and two lenses. As the light passes through the prism, different wavelengths are refracted at different angles. *(David Parker, SPL/Photo Researchers)*

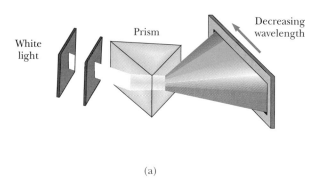 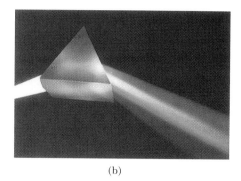

FIGURE 35.17 (a) Dispersion of white light by a prism. Because n varies with wavelength, the prism disperses the white light into its various spectral components. (b) Different colors are refracted at different angles because the index of refraction of the glass depends on wavelength. Violet light deviates the most; red light deviates the least. *(Photograph courtesy of Bausch and Lomb)*

color. Violet light deviates the most, red the least, and the remaining colors in the visible spectrum fall between these extremes.

A prism is often used in an instrument known as a **prism spectroscope,** the essential elements of which are shown in Figure 35.18. The instrument is commonly used to measure the wavelengths emitted by a light source. Light from the source is sent through a narrow, adjustable slit to produce a collimated beam. The light then passes through the prism and is dispersed into a spectrum. The refracted light is observed through a telescope. The experimenter sees an image of the slit through the eyepiece of the telescope. The telescope can be moved or the prism rotated in order to view the various images formed by different wavelengths at different angles of deviation.

All hot, low-pressure gases emit their own characteristic spectra. Thus, one use of a prism spectroscope is to identify gases. For example, sodium emits two wavelengths in the visible spectrum, which appear as two closely spaced yellow lines. Thus, a gas emitting these colors can be identified as having sodium as one of its constituents. Likewise, mercury vapor has its own characteristic spectrum, consisting of four prominent wavelengths — orange, green, blue, and violet lines — along with some wavelengths of lower intensity. The particular wavelengths emitted by a gas serve as "fingerprints" of that gas.

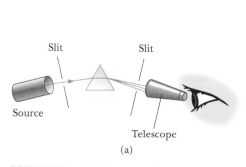 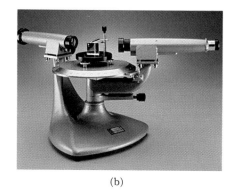

FIGURE 35.18 (a) Diagram of a prism spectrometer. The various colors in the spectrum are viewed through a telescope. (b) Photograph of a prism spectrometer. *(Courtesy of Central Scientific Co.)*

EXAMPLE 35.7 Measuring n Using a Prism

A prism is often used to measure the index of refraction of a transparent solid. Although we do not prove it here, the minimum angle of deviation, δ_{min}, occurs at the angle of incidence θ_1 where the refracted ray inside the prism makes the same angle α with the normal to the two prism faces,[3] as in Figure 35.19. Let us obtain an expression for the index of refraction of the prism material.

Using the geometry shown, we find that $\theta_2 = \Phi/2$, where Φ is the apex angle and

$$\theta_1 = \theta_2 + \alpha = \frac{\Phi}{2} + \frac{\delta_{min}}{2} = \frac{\Phi + \delta_{min}}{2}$$

From Snell's law of refraction,

$$\sin\theta_1 = n \sin\theta_2$$

$$\sin\left(\frac{\Phi + \delta_{min}}{2}\right) = n \sin(\Phi/2)$$

$$n = \frac{\sin\left(\dfrac{\Phi + \delta_{min}}{2}\right)}{\sin(\Phi/2)} \qquad (35.9)$$

[3] For details, see F. A. Jenkins and H. E. White, *Fundamentals of Optics*, New York, McGraw-Hill, 1976, Chapter 2.

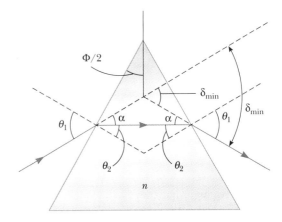

FIGURE 35.19 (Example 35.7) A light ray passing through a prism at the minimum angle of deviation, δ_{min}.

Hence, knowing the apex angle Φ of the prism and measuring δ_{min}, we can calculate the index of refraction of the prism material. Furthermore, we can use a hollow prism to determine the values of n for various liquids.

35.6 HUYGENS' PRINCIPLE

In this section, we develop the laws of reflection and refraction by using a method proposed by Huygens in 1678. As noted in Section 35.1, Huygens assumed that light is some form of wave motion rather than a stream of particles. He had no knowledge of the nature of light or of its electromagnetic character. Nevertheless, his simplified wave model is adequate for understanding many practical aspects of the propagation of light.

Huygens' principle is a construction for using knowledge of an earlier wave front to determine the position of a new wave front at some instant. In Huygens' construction,

Huygens' principle

> all points on a given wave front are taken as point sources for the production of spherical secondary waves, called wavelets, which propagate outward with speeds characteristic of waves in that medium. After some time has elapsed, the new position of the wave front is the surface tangent to the wavelets.

Figure 35.20 illustrates two simple examples of Huygens' construction. First, consider a plane wave moving through free space, as in Figure 35.20a. At $t = 0$, the wave front is indicated by the plane labeled AA'. In Huygens' construction, each

Christian Huygens
| 1629 – 1695 |

Christian Huygens was a Dutch physicist and astronomer who is best known for his contributions to the fields of dynamics and optics. As a physicist, his accomplishments include invention of the pendulum clock and the first exposition of a wave theory of light. As an astronomer, he was the first to recognize the rings around Saturn and to discover Titan, a satellite of Saturn.

Huygens was born in 1629 into a prominent family in The Hague. He was the son of Constantin Huygens, one of the most important figures of the Renaissance in Holland. Educated at the University of Leyden, Christian became a close friend of René Descartes, a frequent guest at the Huygens' home. Huygens published his first paper in 1651 on the subject of the quadrature of various curves.

Huygens' reputation in optics and dynamics spread throughout Europe, and in 1663 he was elected a charter member of the Royal Society. Louis XIV lured Huygens to France in 1666, in accordance with the king's policy of collecting scholars for the glory of his regime. While in France, Huygens became one of the founders of the French Academy of Science.

In 1673, in Paris, Huygens published *Horologium Oscillatorium*. In this work he described a solution to the problem of the compound pendulum, for which he calculated the equivalent simple pendulum length. In the same publication he also derived a formula for computing the period of oscillation of a simple pendulum and explained his laws of centrifugal force for uniform motion in a circle.

Huygens returned to Holland in 1681, constructed some lenses of large focal lengths, and invented the achromatic eyepiece for telescopes. Shortly after returning from a visit to England, where he met Newton, Huygens published his treatise on the wave theory of light. To Huygens, light was a vibratory motion in the ether, spreading out and producing the sensation of light when impinging on the eye. On the basis of this theory, he was able to deduce the laws of reflection and refraction and to explain the phenomenon of double refraction.

Huygens, second only to Newton among the greatest scientists in the second half of the 17th century, was the first to proceed in the field of dynamics beyond the point reached by Galileo and Descartes. It was Huygens who essentially solved the problem of centrifugal force. A solitary man, Huygens did not attract students or disciples and was very slow in publishing his findings. He died in 1695 after a long illness.

Photo credit: (Rijksmuseum voor de Geschiedenis der Natuurwetenschappen. Courtesy AIP Niels Bohr Library)

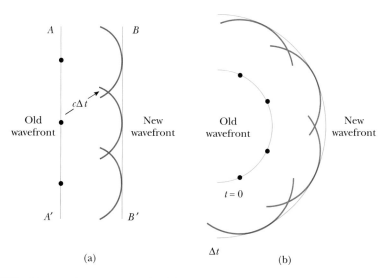

FIGURE 35.20 Huygens' construction for (a) a plane wave propagating to the right and (b) a spherical wave.

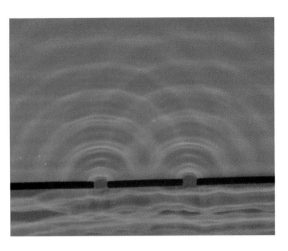

FIGURE 35.21 Water waves in a ripple tank are used to demonstrate Huygens' wavelets. A plane wave is incident on a barrier with two small openings. Each opening acts as a source of circular wavelets. *(Erich Schrempp/ Photo Researchers)*

point on this wave front is considered a point source for generating other waves. For clarity, only a few points on AA' are shown. With these points as sources for the wavelets, we draw circles each of radius $c\,\Delta t$, where c is the speed of light in free space and Δt is the time of propagation from one wave front to the next. The surface drawn tangent to these wavelets is the plane BB', which is parallel to AA'. In a similar manner, Figure 35.20b shows Huygens' construction for an outgoing spherical wave.

A convincing demonstration of Huygens' principle is obtained with water waves in a shallow tank (called a ripple tank), as in Figure 35.21. Plane waves produced below the slit emerge above the slit as two-dimensional circular waves propagating outward.

Huygens' Principle Applied to Reflection and Refraction

The laws of reflection and refraction were stated earlier in this chapter without proof. We now derive these laws using Huygens' principle.

Consider a plane wave incident on a reflecting surface as in Figure 35.22. At $t = 0$, suppose the wave front labeled I_1 is in contact with the surface at A. Point A is a source of Huygens' waves, and in a time interval Δt, these waves expand radially outward a distance $c\,\Delta t$ from A. In the same time interval, the incident wave travels a distance $c\,\Delta t$ (corresponding to the new wave front I_2), striking the surface at C.

From Huygens' construction, the reflected wave front R travels in the direction perpendicular to the wave front. From the construction in Figure 35.22, we see that $BC = AD = c\,\Delta t$, and the hypotenuse AC is common to both right triangles ADC and ABC. Hence, the two triangles are congruent, and $\theta_1 = \theta_1'$. Because the incident ray is perpendicular to AB, and the reflected ray is perpendicular to DC, the

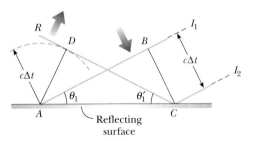

FIGURE 35.22 Huygens' construction for proving the law of reflection. Triangles ABC and ADC are congruent.

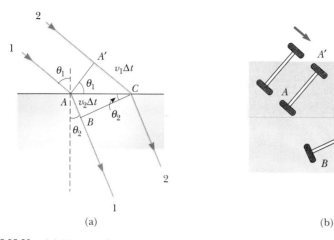

(a) (b)

FIGURE 35.23 (a) Huygens' construction for proving the law of refraction. (b) A mechanical analog of refraction.

angles that these rays make with the normal to the surface are also the angles of incidence and reflection. Hence, we have proved the law of reflection.

Now let us use Huygens' principle and Figure 35.23a to derive Snell's law of refraction. Note that in the time interval Δt, ray 1 moves from A to B and ray 2 moves from A' to C. The radius of the outgoing spherical wavelet centered at A is equal to $v_2 \Delta t$. The distance $A'C$ is equal to $v_1 \Delta t$. Geometric considerations show that angle $A'AC$ equals θ_1 and angle ACB equals θ_2. From triangles $AA'C$ and ACB, we find that

$$\sin \theta_1 = \frac{v_1 \Delta t}{AC} \quad \text{and} \quad \sin \theta_2 = \frac{v_2 \Delta t}{AC}$$

Dividing these two equations, we get

$$\frac{\sin \theta_1}{\sin \theta_2} = \frac{v_1}{v_2}$$

But from Equation 35.4 we know that $v_1 = c/n_1$ and $v_2 = c/n_2$. Therefore,

$$\frac{\sin \theta_1}{\sin \theta_2} = \frac{c/n_1}{c/n_2} = \frac{n_2}{n_1}$$

$$n_1 \sin \theta_1 = n_2 \sin \theta_2$$

which is Snell's law of refraction.

A mechanical analog of refraction is shown in Figure 35.23b. The wheels on a device such as a wagon change their direction as they move from a concrete surface to a grass surface.

35.7 TOTAL INTERNAL REFLECTION

An interesting effect called total internal reflection can occur when light attempts to move from a medium having a given index of refraction to one having a lower index of refraction. Consider a light beam traveling in medium 1 and meeting the boundary between medium 1 and medium 2, where n_1 is greater than n_2 (Fig. 35.24). Various possible directions of the beam are indicated by rays 1 through 5. The refracted rays are bent away from the normal because n_1 is greater than n_2.

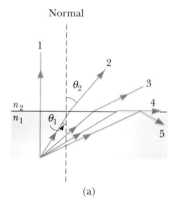

(a)

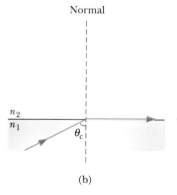

(b)

FIGURE 35.24 (a) Rays from a medium of index of refraction n_1 travel into a medium of index of refraction n_2, where $n_2 < n_1$. As the angle of incidence increases, the angle of refraction θ_2 increases until θ_2 is 90° (ray 4). For this and larger angles of incidence, total internal reflection occurs (ray 5). (b) The angle of incidence producing an angle of refraction equal to 90° is the critical angle, θ_c.

This photograph shows nonparallel light rays entering a glass prism. The bottom two rays undergo total internal reflection at the longest side of the prism. The top three rays are refracted at the longest side as they leave the prism. *(Henry Leap and Jim Lehman)*

(Remember that when light refracts at the interface between the two media, it is also partially reflected. For simplicity, we ignore these reflected rays here.) At some particular angle of incidence, θ_c, called the **critical angle,** the refracted light ray would move parallel to the boundary so that $\theta_2 = 90°$ (Fig. 35.24b). For angles of incidence greater than or equal to θ_c, the beam is entirely reflected at the boundary as shown by ray 5 in Figure 35.24a. This ray is reflected at the boundary as though it had struck a perfectly reflecting surface. This ray, and all those like it, obey the law of reflection; that is, the angle of incidence equals the angle of reflection.

We can use Snell's law of refraction to find the critical angle. When $\theta_1 = \theta_c$, $\theta_2 = 90°$ and Snell's law (Eq. 35.8) gives

$$n_1 \sin \theta_c = n_2 \sin 90° = n_2$$

$$\sin \theta_c = \frac{n_2}{n_1} \quad \text{(for } n_1 > n_2\text{)} \tag{35.10}$$

In this equation n_1 is always greater than n_2. That is,

> total internal reflection occurs only when light attempts to move from a medium of given index of refraction to a medium of lower index of refraction.

If n_1 were less than n_2, Equation 35.10 would give $\sin \theta_c > 1$, which is meaningless because the sine of an angle can never be greater than unity.

The critical angle for total internal reflection will be small when n_1 is considerably larger than n_2. Examples of this combination are diamond ($n = 2.42$ and $\theta_c = 24°$) and crown glass ($n = 1.52$ and $\theta_c = 41°$). This property combined with proper faceting causes diamonds and crystal glass to sparkle.

A prism and the phenomenon of total internal reflection can be used to alter the direction of travel of a propagation beam. Two such possibilities are illustrated in Figure 35.25. In one case the light beam is deflected through 90° (Fig. 35.25a), and in the second case the path of the beam is reversed (Fig. 35.25b). A common

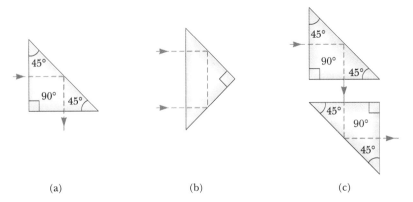

FIGURE 35.25 Internal reflection in a prism. (a) The ray is deviated by 90°. (b) The direction of the ray is reversed. (c) Two prisms used as a periscope.

application of total internal reflection is in a submarine periscope. In this device, two prisms are arranged as in Figure 35.25c so that an incident beam of light follows the path shown and one is able to "see around corners."

EXAMPLE 35.8 A View from the Fish's Eye

(a) Find the critical angle for a water–air boundary if the index of refraction of water is 1.33.

Solution Applying Equation 35.10, we find the critical angle to be

$$\sin \theta_c = \frac{n_2}{n_1} = \frac{1}{1.33} = 0.752$$

$$\theta_c = \boxed{48.8°}$$

(b) Use the results of part (a) to predict what a fish sees when it looks upward toward the water surface at an angle of 40°, 49°, and 60° (Fig. 35.26).

Reasoning Examine Figure 35.24a. Note that because the path of a light ray is reversible, light traveling from medium 2 into medium 1 follows the paths shown in Figure 35.24a, but in the *opposite* direction. This situation is illustrated in Figure 35.26 for the case of a fish looking upward toward the water surface. A fish can see out of the water if it looks toward the surface at an angle less than the critical angle. Thus, at 40°, the fish can see into the air above the water. At an angle of 49°, the critical angle for water, the light that reaches the fish

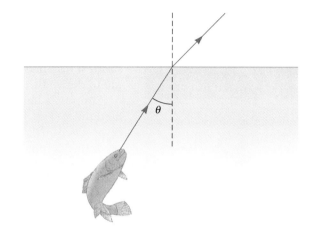

FIGURE 35.26 (Example 35.8).

has to skim along the water surface before being refracted to the fish's eye. At angles greater than the critical angle, the light reaching the fish comes via internal reflection at the surface. Thus, at 60°, the fish sees a reflection of some object on the bottom of the pool.

Fiber Optics

Another interesting application of total internal reflection is the use of glass or transparent plastic rods to "pipe" light from one place to another. As indicated in Figure 35.27, light is confined to traveling within the rods, even around gentle curves, as the result of successive internal reflections. Such a "light pipe" is flexible if thin fibers are used rather than thick rods. If a bundle of parallel fibers is used to construct an optical transmission line, images can be transferred from one point to another.

This technique is used in a sizable industry known as fiber optics. There is very little light intensity lost in these fibers as a result of reflections on the sides. Any loss in intensity is due essentially to reflections from the two ends and absorption by the fiber material. These devices are particularly useful when an image of an object that is at an inaccessible location is to be viewed. For example, physicians often use this technique to examine internal organs of the body. The field of fiber optics is finding increasing use in telecommunications, since the fibers can carry a much higher volume of telephone calls or other forms of communication than electrical wires.

FIGURE 35.27 Light travels in a curved transparent rod by multiple internal reflections.

Strands of glass optical fibers are used to carry voice, video, and data signals in telecommunication networks. Typical fibers have diameters of 60 μm. (© Richard Megna 1983, Fundamental Photographs)

*35.8 FERMAT'S PRINCIPLE

A general principle that can be used to determine light paths was developed by Pierre de Fermat (1601–1665). **Fermat's principle** states that

> when a light ray travels between any two points P and Q, its path is the one that requires the least time, or constant time.[4]

An obvious consequence of this principle is that when the rays travel in a single, homogeneous medium, the paths are straight lines because a straight line is the shortest distance between two points. Let us illustrate how to use Fermat's principle to derive the law of refraction.

Suppose a light ray is to travel from P to Q, where P is in medium 1 and Q is in medium 2 (Fig. 35.28). The points P and Q are at perpendicular distances a and b, respectively, from the interface. The speed of light is c/n_1 in medium 1 and c/n_2 in medium 2. Using the geometry of Figure 35.28, we see that the time it takes the ray to travel from P to Q is

$$t = \frac{r_1}{v_1} + \frac{r_2}{v_2} = \frac{\sqrt{a^2 + x^2}}{c/n_1} + \frac{\sqrt{b^2 + (d-x)^2}}{c/n_2}$$

We obtain the least time, or the minimum value of t, by taking the derivative of t with respect to x (the variable) and setting the derivative equal to zero:

$$\frac{dt}{dx} = \frac{n_1}{c} \frac{d}{dx} \sqrt{a^2 + x^2} + \frac{n_2}{c} \frac{d}{dx} \sqrt{b^2 + (d-x)^2}$$

$$= \frac{n_1}{c}\left(\frac{1}{2}\right)\frac{2x}{(a^2+x^2)^{1/2}} + \frac{n_2}{c}\left(\frac{1}{2}\right)\frac{2(d-x)(-1)}{[b^2+(d-x)^2]^{1/2}}$$

$$= \frac{n_1 x}{c(a^2+x^2)^{1/2}} - \frac{n_2(d-x)}{c[b^2+(d-x)^2]^{1/2}} = 0$$

From Figure 35.28 and recognizing that in this equation,

$$\sin\theta_1 = \frac{x}{(a^2+x^2)^{1/2}} \qquad \sin\theta_2 = \frac{d-x}{[b^2+(d-x)^2]^{1/2}}$$

we find that

$$n_1 \sin\theta_1 = n_2 \sin\theta_2$$

which is Snell's law of refraction.

It is a simple matter to use a similar procedure to derive the law of reflection. The calculation is left for you to carry out (Problem 47).

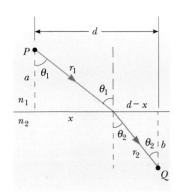

FIGURE 35.28 Geometry for deriving the law of refraction using Fermat's principle.

SUMMARY

In geometric optics, we use the **ray approximation,** which assumes that a wave travels through a medium in straight lines in the direction of the rays.

The basic laws of geometric optics are the laws of reflection and refraction for

---

[4] More generally, Fermat's principle requires only that the time be an extremum with respect to small variations in path.

light rays. The **law of reflection** states that the angle of reflection, θ_1', equals the angle of incidence, θ_1. The **law of refraction**, or **Snell's law**, states that

$$\frac{\sin \theta_2}{\sin \theta_1} = \frac{v_2}{v_1} = \text{constant} \tag{35.3}$$

where θ_2 is the angle of refraction.

The **index of refraction** of a medium, n, is defined by the ratio

$$n \equiv \frac{c}{v} \tag{35.4}$$

where c is the speed of light in a vacuum and v is the speed of light in the medium. In general, n varies with wavelength and is given by

$$n = \frac{\lambda_0}{\lambda_n} \tag{35.7}$$

where λ_0 is the vacuum wavelength and λ_n is the wavelength in the medium.

An alternate form of Snell's law of refraction is

$$n_1 \sin \theta_1 = n_2 \sin \theta_2 \tag{35.8}$$

where n_1 and n_2 are the indices of refraction in the two media. The incident ray, the reflected ray, the refracted ray, and the normal to the surface all lie in the same plane.

Huygens' principle states that all points on a wave front can be taken as point sources for the production of secondary wavelets. At some later time, the new position of the wave front is the surface tangent to these wavelets.

Total internal reflection can occur when light travels from a medium of high index of refraction to one of lower index of refraction. The minimum angle of incidence, θ_c, for which total reflection occurs at an interface is given by

$$\sin \theta_c = \frac{n_2}{n_1} \quad \text{(where } n_1 > n_2\text{)} \tag{35.10}$$

Fermat's principle states that when a light ray travels between two points, its path is the one that requires the least time (or more generally an extremum, which may be a maximum).

QUESTIONS

1. Light of wavelength λ is incident on a slit of width d. Under what conditions is the ray approximation valid? Under what circumstances does the slit produce enough diffraction to make the ray approximation invalid?
2. Sound waves have much in common with light waves, including the properties of reflection and refraction. Give examples of these phenomena for sound waves.
3. Does a light ray traveling from one medium into another always bend toward the normal as in Figure 35.8? Explain.
4. As light travels from one medium to another, does the wavelength of the light change? Does the frequency change? Does the speed change? Explain.
5. A laser beam passing through a nonhomogeneous sugar solution follows a curved path. Explain.
6. A laser beam ($\lambda = 632.8$ nm) is incident on a piece of Lucite as in Figure 35.29. Part of the beam is reflected and part is refracted. What information can you get from this photograph?
7. Suppose blue light were used instead of red light in the experiment shown in Figure 35.29. Would the refracted beam be bent at a larger or smaller angle?
8. The level of water in a clear, colorless glass is easily observed with the naked eye. The level of liquid helium in a clear glass vessel is extremely difficult to see with the naked eye. Explain.

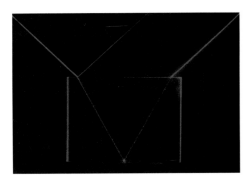

FIGURE 35.29 (Questions 6 and 7) Light from a helium–neon laser beam ($\lambda = 632.8$ nm, red light) is incident on a block of Lucite. The photograph shows both reflected and refracted rays. Can you identify the incident, reflected, and refracted rays? From this photograph, estimate the index of refraction of Lucite at this wavelength. *(Henry Leap and Jim Lehman)*

9. Describe an experiment in which internal reflection is used to determine the index of refraction of a medium.
10. Why does a diamond show flashes of color when observed under white light?
11. Explain why a diamond sparkles more than a glass crystal of the same shape and size.
12. Explain why an oar in the water appears bent.
13. Redesign the periscope of Figure 35.25c so that it can show you where you have been rather than where you are going.
14. Under certain circumstances, sound can be heard over extremely long distances. This frequently happens over a body of water, where the air near the water surface is cooler than the air higher up. Explain how the refraction of sound waves in such a situation could increase the distance over which the sound can be heard.
15. Why do astronomers looking at distant galaxies talk about looking backward in time?
16. A solar eclipse occurs when the Moon gets between the Earth and the Sun. Use a diagram to show why some areas of the Earth see a total eclipse, other areas see a partial eclipse, and most areas see no eclipse.
17. Some department stores have their windows slanted slightly inward at the bottom. This is to decrease the glare from streetlights or the Sun, which would make it difficult for shoppers to see the display inside. Sketch a light ray reflecting off such a window to show how this technique works.
18. When two colors of light (X and Y) are sent through a glass prism, X is bent more than Y. Which color travels more slowly in the prism?
19. Figure 35.30 represents sunlight striking a drop of water in the atmosphere. Use the laws of refraction and reflection and the fact that sunlight consists of a wide range of wavelengths to discuss the formation of rainbows.

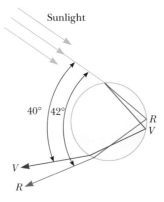

FIGURE 35.30 (Question 19) Refraction of sunlight by a spherical raindrop.

20. Why does the arc of a rainbow appear with red on top and violet on the bottom?
21. How is it possible that a complete circle of rainbow can sometimes be seen from an airplane?
22. You can make a corner reflector by placing three flat mirrors in the corner of a room where the ceiling meets the walls. Show that no matter where you are in the room, you can see yourself reflected in the mirrors—upside down.
23. Several corner reflectors were left on the Moon's Sea of Tranquility by the astronauts of Apollo 11. How can scientists utilize a laser beam sent from Earth even today to determine the precise distance from the Earth to the Moon?
24. Under what conditions is a mirage formed? On a hot day, what are we seeing when we observe "water on the road"?

PROBLEMS

Section 35.2 Measurements of the Speed of Light

1. During the Apollo XI Moon landing, a highly reflecting screen was erected on the Moon's surface. The speed of light is found by measuring the time it takes a laser beam to travel from Earth, reflect from the screen, and return to Earth. If this interval is measured to be 2.51 s, what is the measured speed of

□ indicates problems that have full solutions available in the Student Solutions Manual and Study Guide.

light? Take the center-to-center distance from Earth to Moon to be 3.84×10^8 m, and do not neglect the sizes of the Earth and Moon.

2. As a result of his observations, Roemer concluded that the time interval between eclipses of Io by Jupiter increased by 22 min during a 6-month period as the Earth moved from a point in its orbit where its motion is toward Jupiter to a diametrically opposite point where it moves away from Jupiter. Using 1.5×10^8 km as the average radius of the Earth's orbit around the Sun, calculate the speed of light from these data.

3. Figure P35.3 shows an apparatus used to measure the speed distribution of gas molecules. It consists of two slotted rotating disks separated by a distance s, with the slots displaced by the angle θ. Suppose the speed of light is measured by sending a light beam toward the right disk of this apparatus. (a) Show that a light beam will be seen in the detector (that is, will make it through both slots) only if its speed is given by $c = \omega s/\theta$, where ω is the angular speed of the disks and θ is measured in radians. (b) What is the measured speed of light if the distance between the two slotted rotating disks is 2.5 m, the slot in the second disk is displaced 0.017° from the slot in the first disk, and the disks are rotating at 5555 rev/s?

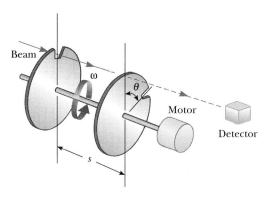

FIGURE P35.3

4. Albert Michelson used an improved version of the technique developed by Fizeau to measure the speed of light. In one of Michelson's experiments, the toothed wheel was replaced by a wheel with 32 identical mirrors mounted on its perimeter, with the plane of each mirror perpendicular to a radius of the wheel. The total light path was 8 mi (obtained by multiple reflections of a light beam within an evacuated tube 1 mi long). For what minimum angular speed of the mirror would Michelson have calculated the speed of light to be 2.998×10^8 m/s?

5. In an experiment to measure the speed of light using the apparatus of Fizeau (Fig. 35.2), the distance between light source and mirror was 11.45 km and the wheel had 720 notches. The experimentally determined value of c was 2.998×10^8 m/s. Calculate the minimum angular speed of the wheel for this experiment.

6. If the Fizeau experiment is performed such that the round-trip distance for the light is 40 m, find the two lowest speeds of rotation that allow the light to pass through the notches. Assume that the wheel has 360 teeth and that the speed of light is 3×10^8 m/s. Repeat for a round-trip distance of 4000 m.

7. Use Roemer's value of 22 min discussed in Problem 2 and the presently accepted value of the speed of light in vacuum to find an average value for the distance between the Earth and the Sun.

Section 35.4 Reflection and Refraction

(*Note:* In this section if an index of refraction value is not given, refer to Table 35.1.)

8. A coin is on the bottom of a swimming pool 1.00 m deep. What is the apparent depth of the coin, seen from above the water surface?

9. The wavelength of red helium–neon laser light in air is 632.8 nm. (a) What is its frequency? (b) What is its wavelength in glass that has an index of refraction of 1.50? (c) What is its speed in the glass?

10. A narrow beam of sodium yellow light is incident from air onto a smooth water surface at an angle $\theta_1 = 35.0°$. Determine the angle of refraction θ_2 and the wavelength of the light in water.

11. An underwater scuba diver sees the Sun at an apparent angle of 45° from the vertical. Where is the Sun?

12. A light ray in air is incident on a water surface at an angle of 30.0° with respect to the normal to the surface. What is the angle of the refracted ray relative to this normal?

13. A ray of light in air is incident on a planar surface of silica. The refracted ray makes an angle of 37.0° with the normal. Calculate the angle of incidence.

14. A light ray initially in water enters a transparent substance at an angle of incidence of 37.0°, and the transmitted ray is refracted at an angle of 25.0°. Calculate the speed of light in the transparent substance.

15. A ray of light strikes a flat block of glass ($n = 1.50$) of thickness 2.00 cm at an angle of 30.0° with the normal. Trace the light beam through the glass, and find the angles of incidence and refraction at each surface.

16. Find the speed of light in (a) flint glass, (b) water, and (c) zircon.

17. Light of wavelength 436 nm in air enters a fishbowl filled with water, then exits through the crown glass wall of the container. What is the wavelength of the light (a) in the water and (b) in the glass?

18. The reflecting surfaces of two intersecting plane

mirrors are at an angle of θ ($0° < \theta < 90°$), as in Figure P35.18. If a light ray strikes the horizontal mirror, show that the emerging ray will intersect the incident ray at an angle of $\beta = 180° - 2\theta$.

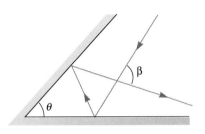

FIGURE P35.18

19. A 1.00-cm-thick by 4.00-cm-long glass plate is made up of two fused prisms. The top prism has an index of refraction of 1.486 for blue light and 1.472 for red light. The bottom prism has an index of refraction of 1.878 for blue light and 1.862 for red light. A ray consisting of red and blue light is incident at 50.0° on the top face as shown in Figure P35.19. Determine the exit angles for both rays that pass through the prisms and the angle between them.

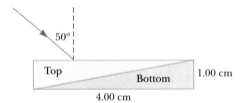

FIGURE P35.19

20. A glass block having $n = 1.52$ and surrounded by air measures 10.0 cm × 10.0 cm. For an angle of incidence of 45.0°, what is the maximum distance x shown in Figure P35.20 so that the ray will emerge from the opposite side?

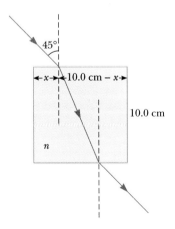

FIGURE P35.20

*Section 35.5 Dispersion and Prisms

21. A ray of light strikes the midpoint of one face of an equiangular glass prism ($n = 1.50$) at an angle of incidence of 30.0°. Trace the path of the light ray through the glass and find the angles of incidence and refraction at each surface.

22. A light ray enters the atmosphere of a planet along a radius and then descends to the surface 20.0 km below. The index of refraction where the light enters the atmosphere is 1.000, and it increases uniformly to the surface, where it has a value of 1.005. (a) How long does it take the ray to traverse this path? (b) Compare this with the time needed to cover the same distance in a vacuum.

22A. A light ray enters the atmosphere of a planet along a radius and then descends to the surface a distance h below. The index of refraction where the light enters the atmosphere is 1.000, and it increases uniformly to the surface, where it has a value of n. (a) How long does it take the ray to traverse this path? (b) Compare this with the time needed to cover the same distance in a vacuum.

23. A narrow white light ray is incident on a block of silica at an angle of 30.0°. Find the angular width of the light ray inside the silica.

23A. A narrow white light ray is incident on a block of fused silica at an angle θ. Find the angular width of the light ray inside the silica.

24. A certain kind of glass has an index of refraction of 1.6500 for blue light (430 nm) and an index of 1.6150 for red light (680 nm). If a beam containing these two colors is incident at an angle of 30° on a piece of this glass, what is the angle between the two beams inside the glass?

25. Show that if the apex angle Φ of a prism is small, an approximate value for the angle of minimum deviation is $\delta_{min} = (n - 1)\Phi$.

26. The index of refraction for red light in water is 1.331, and that for blue light is 1.340. If a ray of white light enters the water at an angle of incidence of 83.00°, what are the underwater angles of refraction for the blue and red components of the light?

27. An experimental apparatus includes a prism made of sodium chloride. The angle of minimum deviation for light of wavelength 589 nm is to be 10.0°. What is the required apex angle of the prism?

28. Light of wavelength 700 nm is incident on the face of a silica prism at an angle of 75° (with respect to the normal to the surface). The apex angle of the prism is 60°. Use the value of n from Figure 35.15 and calculate the angle (a) of refraction at this first surface,

(b) of incidence at the second surface, (c) of refraction at the second surface, and (d) between the incident and emerging rays.

29. A prism that has an apex angle of 50.0° is made of cubic zirconia, with $n = 2.20$. What is its angle of minimum deviation?

30. A triangular glass prism with apex angle 60.0° has an index of refraction $n = 1.50$. (a) What is the smallest angle of incidence θ_1 for which a light ray can emerge from the other side? (See Figure 35.19.) (b) For what angle of incidence θ_1 does the light ray leave at the same angle θ_1?

31. The index of refraction for violet light in silica flint glass is 1.66, and that for red light is 1.62. What is the average angular dispersion of visible light passing through a prism of apex angle 60.0° if the angle of incidence is 50.0° (Fig. P35.31)?

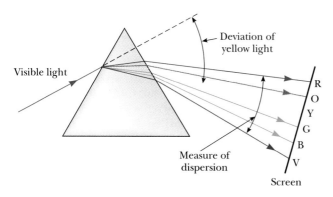

FIGURE P35.31

Section 35.7 Total Internal Reflection

32. A large Lucite cube ($n = 1.59$) has a small air bubble (a defect in the casting process) below one surface. When a penny (diameter 1.90 cm) is placed directly over the bubble on the outside of the cube, the bubble cannot be seen by looking down into the cube at any angle. However, when a dime (radius 1.75 cm) is placed directly over it, the bubble can be seen by looking down into the cube. What is the range of the possible depths of the air bubble beneath the surface?

33. A fiber optic cable ($n = 1.50$) is submerged in water ($n = 1.33$). What is the critical angle for light to stay inside the cable?

34. A glass cube is placed on a newspaper, which rests on a table. A person reads the news through the vertical side of the cube. Determine the maximum index of refraction of the cube.

35. For 589-nm light, calculate the critical angle for the following materials surrounded by air: (a) diamond, (b) flint glass, and (c) ice.

36. Repeat Problem 35 when the materials are surrounded by water.

37. Consider a common mirage formed by super-heated air just above the roadway. If an observer viewing from 2.00 m above the road (where $n = 1.0003$) sees water up the road at $\theta_1 = 88.8°$, find the index of refraction of the air just above the road surface. (*Hint:* Treat this as a problem in total internal reflection.)

38. Traveling inside a diamond, a light ray is incident on the interface between diamond and air. What is the critical angle for total internal reflection? Use Table 35.1. (The smallness of θ_c for diamond means that light is easily "trapped" within a diamond and eventually emerges from the many cut faces; this makes a diamond more brilliant than stones with smaller n and larger θ_c.)

39. An optical fiber is made of a clear plastic for which the index of refraction is 1.50. For what angles with the surface does light remain contained within the fiber?

ADDITIONAL PROBLEMS

40. A 4.00-m-long pole stands vertically in a river having a depth of 2.00 m. When the Sun is 40.0° above the horizontal, determine the length of the pole's shadow on the bottom of the river. Take the index of refraction for water to be 1.33.

40A. A pole of length L stands vertically in a river having a depth d. When the Sun is at an angle θ above the horizontal, determine the length of the pole's shadow on the bottom of the river. Take the index of refraction for water to be n.

41. A specimen of glass has an index of refraction of 1.61 for the wavelength corresponding to the prominent bright line in the sodium spectrum. If an equiangular prism is made from this glass, what angle of incidence results in minimum deviation of the sodium line?

42. A coin is at the bottom of a 6.00-cm-deep beaker. The beaker is filled to the top with 3.50 cm of water (index of refraction = 1.33) covered by 2.50 cm of ether (index of refraction = 1.36). How deep does the coin appear to be when seen from the top of the beaker?

43. A small underwater pool light is 1.00 m below the surface. The light emerging from the water forms a circle on the water surface. What is the diameter of this circle?

44. When the Sun is directly overhead, a narrow shaft of light enters a cathedral through a small hole in the ceiling and forms a spot on the floor 10.0 m below. (a) At what speed (in centimeters per minute) does the spot move across the (flat) floor? (b) If a mirror is placed on the floor to intercept the light, at what speed does the reflected spot move across the ceiling?

45. A drinking glass is 4.00 cm wide at the bottom, as shown in Figure P35.45. When an observer's eye is placed as shown, the observer sees the edge of the bottom of the glass. When this glass is filled with water, the observer sees the center of the bottom of the glass. Find the height of the glass.

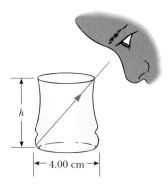

FIGURE P35.45

46. A material having an index of refraction $n = 2.0$ is surrounded by a vacuum and is in the shape of a quarter circle of radius $R = 10$ cm (Fig. P35.46). A light ray parallel to the base of the material is incident from the left at a distance of $L = 5.0$ cm above the base and emerges out of the material at the angle θ. Determine the value of θ.

46A. A material having an index of refraction n is surrounded by a vacuum and is in the shape of a quarter circle of radius R (Fig. P35.46). A light ray parallel to the base of the material is incident from the left at a distance of L above the base and emerges out of the material at the angle θ. Determine an expression for θ in terms of L, n, and R.

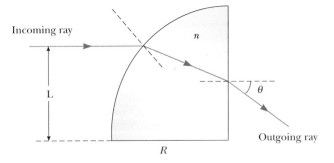

FIGURE P35.46

47. Derive the law of reflection (Eq. 35.2) from Fermat's principle of least time. (See the procedure outlined in Section 35.8 for the derivation of the law of refraction from Fermat's principle.)

48. The angle between the two mirrors in Figure P35.48 is a right angle. The beam of light in the vertical plane P strikes mirror 1 as shown. (a) Determine the distance the reflected light beam travels before striking mirror 2. (b) In what direction does the light beam travel after being reflected from mirror 2?

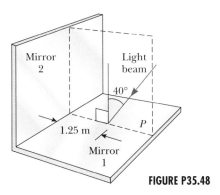

FIGURE P35.48

49. A light ray of wavelength 589 nm is incident at an angle θ on the top surface of a block of polystyrene, as shown in Figure P35.49. (a) Find the maximum value of θ for which the refracted ray undergoes total internal reflection at the left vertical face of the block. Repeat the calculation for the case in which the polystyrene block is immersed in (b) water and (c) carbon disulfide.

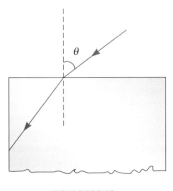

FIGURE P35.49

50. A cylindrical material of radius $R = 2.00$ m has a mirrored surface on its right half, as in Figure P35.50. A light ray traveling in air is incident on the left side of the cylinder. If the incident light ray and exiting light ray are parallel and $d = 2.00$ m, determine the index of refraction of the material.

51. A hiker stands on a mountain peak near sunset and observes a rainbow caused by water droplets in the air about 8 km away. The valley is 2 km below the mountain peak and entirely flat. What fraction of the complete circular arc of the rainbow is visible to the hiker? See Question 19.

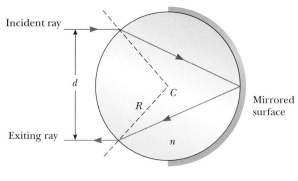

FIGURE P35.50

52. A fish is at a depth d under water. Show that when viewed from an angle of incidence θ_1, the apparent depth z of the fish is

$$z = \frac{3d \cos \theta_1}{\sqrt{7 + 9 \cos^2 \theta_1}}$$

53. A light ray is incident on a prism and refracted at the first surface as shown in Figure P35.53. Let Φ represent the apex angle of the prism and n its index of refraction. Find in terms of n and Φ the smallest allowed value of the angle of incidence at the first surface for which the refracted ray does not undergo internal reflection at the second surface.

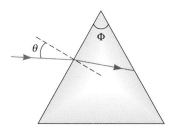

FIGURE P35.53

54. The prism shown in Figure P35.54 has an index of refraction of 1.55. Light is incident at an angle of 20°. Determine the angle θ at which the light emerges.

55. A laser beam strikes one end of a slab of material, as shown in Figure P35.55. The index of refraction of the slab is 1.48. Determine the number of internal reflections of the beam before it emerges from the opposite end of the slab.

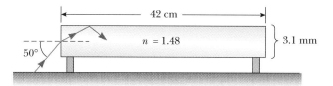

FIGURE P35.55

56. A. H. Pfund's method for measuring the index of refraction of glass is illustrated in Figure P35.56. One face of a slab of thickness t is painted white, and a small hole scraped clear at point P serves as a source of diverging rays when the slab is illuminated from below. Ray PBB' strikes the clear surface at the critical angle and is totally reflected, as are rays such as PCC'. Rays such as PAA' emerge from the clear surface. On the painted surface there appears a dark circle of diameter d, surrounded by an illuminated region, or halo. (a) Derive a formula for n in terms of the measured quantities d and t. (b) What is the diameter of the dark circle if $n = 1.52$ for a slab 0.600 cm thick? (c) If white light is used, the critical angle depends on color caused by dispersion. Is the inner edge of the white halo tinged with red light or violet light? Explain.

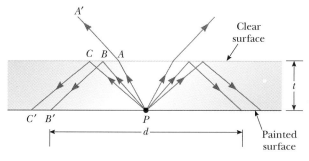

FIGURE P35.56

57. The light beam in Figure P35.57 strikes surface 2 at the critical angle. Determine the angle of incidence θ_1.

58. Students allow a narrow beam of laser light to strike a water surface. They arrange to measure the angle of refraction for selected angles of incidence and

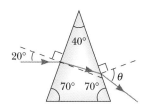

FIGURE P35.54

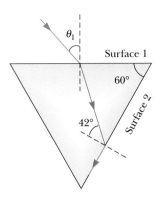

FIGURE P35.57

record the data shown in the accompanying table. Use the data to verify Snell's law of refraction by plotting the sine of the angle of incidence versus the sine of the angle of refraction. Use the resulting plot to deduce the index of refraction of water.

Angle of Incidence (degrees)	Angle of Refraction (degrees)
10.0	7.5
20.0	15.1
30.0	22.3
40.0	28.7
50.0	35.2
60.0	40.3
70.0	45.3
80.0	47.7

59. A light ray traveling in air is incident on one face of a right-angle prism of index of refraction $n = 1.5$ as in Figure P35.59, and the ray follows the path shown in the figure. If $\theta = 60°$ and the base of the prism is mirrored, determine the angle ϕ made by the outgoing ray with the normal to the right face of the prism.

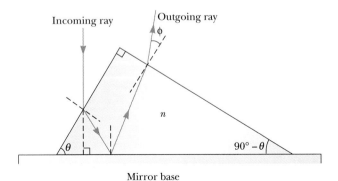

FIGURE P35.59

59A. A light ray traveling in air is incident normally on the base of a right-angle prism of index of refraction n. The ray follows the path shown in Figure P35.59. If the base of the prism is mirrored, determine the angle ϕ made by the outgoing ray with the normal to the right face of the prism in terms of n and the angle θ.

60. A piece of wire is bent through an angle θ. The bent wire is partially submerged in benzene (index of refraction = 1.50) so that looking along the "dry" part of the wire, it appears to be straight and makes a 30.0° angle with the horizontal. Determine the value of θ.

61. A light ray enters a rectangular block of plastic at an angle of $\theta_1 = 45°$ and emerges at an angle of $\theta_2 = 76°$, as in Figure P35.61. (a) Determine the index of refraction for the plastic. (b) If the light ray enters the plastic at a point $L = 50$ cm from the bottom edge, how long does it take the light ray to travel through the plastic?

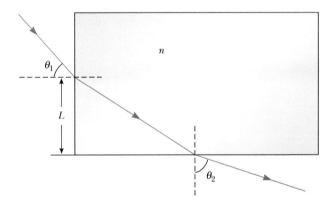

FIGURE P35.61

SPREADSHEET PROBLEM

S1. Spreadsheet 35.1 calculates the angle of refraction and angle of deviation as light beams of different incidence angles travel through a prism of apex angle Φ. Input parameters are the index of refraction n and the apex angle Φ (Fig. PS35.1). As the angle of incidence θ_1 increases from zero, the angle of deviation δ decreases, reaches a minimum value, and then increases. At minimum deviation

$$n \sin \frac{\Phi}{2} = \sin \frac{\Phi + \delta_{min}}{2} = \sin \theta$$

This equation is often used to experimentally find n of any transparent prism. (a) For crown glass ($n = 1.52$ and $\Phi = 60°$), find the minimum angle of de-

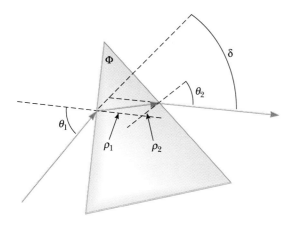

FIGURE PS35.1

iation from the graph. Using the spreadsheet, verify these equations. At the angle of minimum deviation, how is angle ρ_1 related to ρ_2? (b) Because the index of refraction depends slightly on wavelength, vary n between 1.50 and 1.55, and note how the minimum angle of deviation changes with wavelength. (c) Choose several different values for Φ, and note the minimum angles of deviation for each.

"Most mirrors reverse left and right. This one reverses top and bottom."

Chapter 36

Geometric Optics

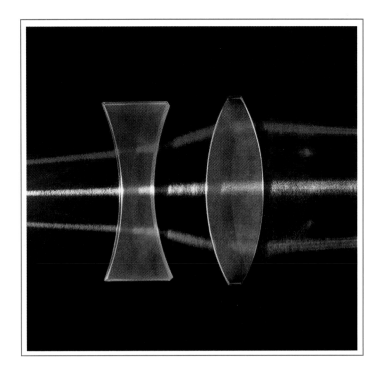

Nearly parallel light rays are incident on a diverging (concave) lens at the left. The red and blue rays refracted by the diverging lens are bent away from the horizontal, while the same rays are bent toward the horizontal as they pass through the converging lens at the right. The net effect is to produce nearly parallel rays at the right. *(Richard Megna/ Fundamental Photographs)*

This chapter is concerned with the images formed when spherical waves fall on flat and spherical surfaces. We find that images can be formed by reflection or by refraction and that mirrors and lenses work because of this reflection and refraction. Such devices, commonly used in optical instruments and systems, are described in detail. In this chapter, we continue to use the ray approximation and to assume that light travels in straight lines, both steps valid because here we are studying the field called *geometric optics*. The field of wave optics is discussed in subsequent chapters.

36.1 IMAGES FORMED BY FLAT MIRRORS

In this chapter we discuss the manner in which optical instruments such as lenses and mirrors form images. We begin this investigation by considering the simplest possible mirror, the flat mirror.

Consider a point source of light placed at O in Figure 36.1, a distance p in front of a flat mirror. The distance p is called the **object distance**. Light rays leave the source and are reflected from the mirror. After reflection, the rays diverge (spread apart), but they appear to the viewer to come from a point *I* located behind the mirror. Point *I* is called the **image** of the object at *O*. Regardless of the system under study, images are always formed in the same way. *Images are formed either at the point where rays of light actually intersect or at the point from which they appear to originate.* Since the rays in Figure 36.1 appear to originate at *I*, which is a distance q behind the mirror, this is the location of the image. The distance q is called the **image distance.**

Images are classified as real or virtual. A **real image** is *one in which rays converge at the image point;* a **virtual image** is *one in which the light rays do not converge to the image point but appear to emanate from that point.* The image formed by the mirror in Figure 36.1 is virtual. The images seen in flat mirrors *are always virtual.* Real images can usually be displayed on a screen (as at a movie), but virtual images cannot be displayed on a screen.

We examine the properties of the images formed by flat mirrors by using the simple geometric techniques shown in Figure 36.2. In order to find out where an image is formed, it is always necessary to follow at least two rays of light as they reflect from the mirror. One of those rays starts at *P*, follows a horizontal path to the mirror, and reflects back on itself. The second ray follows the oblique path *PR* and reflects as shown. An observer to the left of the mirror would trace the two reflected rays back to the point from which they appear to have originated, that is, point *P′*. A continuation of this process for points on the object other than *P* would result in a virtual image (drawn as a yellow arrow) to the right of the mirror. Since triangles *PQR* and *P′QR* are congruent, *PQ = P′Q*. Hence, we conclude that *the image formed by an object placed in front of a flat mirror is as far behind the mirror as the object is in front of the mirror.*

Geometry also shows that the object height, h, equals the image height, h'. Let us define **lateral magnification,** *M*, as follows:

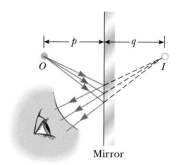

FIGURE 36.1 An image formed by reflection from a flat mirror. The image point, *I*, is located behind the mirror at a distance q, which is equal to the object distance, p.

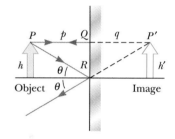

FIGURE 36.2 Geometric construction used to locate the image of an object placed in front of a flat mirror. Because the triangles *PQR* and *P′QR* are congruent, $p = q$ and $h = h'$.

$$M \equiv \frac{\text{image height}}{\text{object height}} = \frac{h'}{h} \quad (36.1)$$

This is a general definition of the lateral magnification of any type of mirror. $M = 1$ for a flat mirror because $h' = h$ in this case.

The image formed by a flat mirror has one more important property: it reverses forward and back, but it does not reverse left and right or top and bottom. If you stand in front of the mirror and raise your right hand, the image raises the hand on your right, which you interpret to be the left hand of the image because the image is facing the opposite direction from you.

Thus, we conclude that the image formed by a flat mirror has the following properties:

- The image is as far behind the mirror as the object is in front.
- The image is unmagnified, virtual, and upright. (By upright we mean that, if the object arrow points upward as in Figure 36.2, so does the image arrow.)
- The image has forward-back reversal.

CONCEPTUAL EXAMPLE 36.1   Multiple Images Formed by Two Mirrors

Two flat mirrors are at right angles to each other, as in Figure 36.3, and an object is placed at point O. In this situation, multiple images are formed. Locate the positions of these images.

Reasoning   The image of the object is at I_1 in mirror 1 and at I_2 in mirror 2. In addition, a third image is formed at I_3, which is the image of I_1 in mirror 2 or, equivalently, the image of I_2 in mirror 1. That is, the image at I_1 (or I_2) serves as the object for I_3. Note that in order to form this image at I_3, the rays reflect twice after leaving the object at O.

Exercise   Sketch the rays corresponding to viewing the images at I_1 and I_2 and show that the light is reflected only once in these cases.

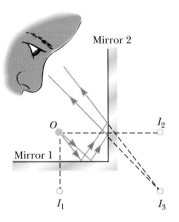

FIGURE 36.3   (Conceptual Example 36.1) When an object is placed in front of two mutually perpendicular mirrors as shown, three images are formed.

CONCEPTUAL EXAMPLE 36.2   The Levitated Professor

The professor in the box shown in Figure 36.4 appears to be balancing himself on a few fingers with both of his feet elevated from the floor. The professor can maintain this position for a long time, and he appears to defy gravity. How do you suppose this illusion was created?

Reasoning   This is one example of an optical illusion, used by magicians, that makes use of a mirror. The box that the professor is standing in is a cubical frame that contains a flat vertical mirror through a diagonal plane. The professor straddles the mirror so that one foot is in front of the mirror which you see, and one foot is behind the mirror which you cannot see. When he raises the foot that you see in front of the mirror, the reflection of this foot also rises, so he appears to float in air.

FIGURE 36.4   (Conceptual Example 36.2) An optical illusion. *(Henry Leap and Jim Lehman)*

CONCEPTUAL EXAMPLE 36.3   The Tilting Rearview Mirror

Most rearview mirrors on cars have a day setting and a night setting. The night setting greatly diminishes the intensity of the image so that lights from trailing vehicles do not blind the driver. How does such a mirror work?

Reasoning   Consider Figure 36.5, which represents a cross-sectional view of the mirror for the two settings. The mirror is a wedge of glass with a reflecting mirror on the back side. When the mirror is in the day setting, as in Figure 36.5a, the light from an object behind the car strikes the mirror at point 1. Most of the light enters the wedge, is refracted, and reflects from the back of the mirror to return to the front surface, where it is refracted again as it re-enters the air as ray B (for *bright*). In addition, a small portion of the light is reflected at the front surface, as indicated by ray D (for *dim*). This dim reflected light is responsible for the image observed when the mirror is in the night setting, as in Figure 36.5b. In this case, the wedge is rotated so that the path followed by the bright light (ray B) does not lead to the eye. Instead, the dim light reflected from the front surface travels to the eye, and the brightness of trailing headlights does not become a hazard.

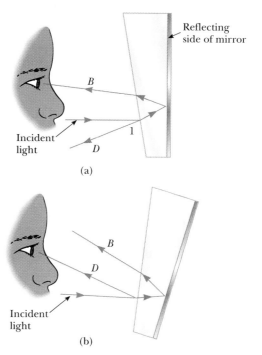

FIGURE 36.5 (Conceptual Example 36.3) A cross-sectional view of a rearview mirror. (a) The day setting forms a bright image, B. (b) The night setting forms a dim image, D.

36.2 IMAGES FORMED BY SPHERICAL MIRRORS

Concave Mirrors

A **spherical mirror**, as its name implies, has the shape of a segment of a sphere. Figure 36.6a shows the cross-section of a spherical mirror with its surface represented by the solid curved black line. (The blue band represents the area behind the mirror.) Such a mirror, in which light is reflected from the inner, concave surface, is called a **concave mirror**. The mirror has a radius of curvature R, and its center of curvature is located at point C. Point V is the center of the spherical

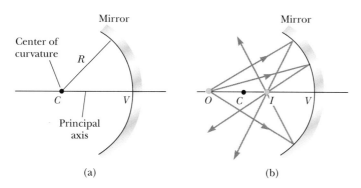

FIGURE 36.6 (a) A concave mirror of radius R whose center of curvature is at C located on the principal axis. (b) A point object placed at O in front of a concave spherical mirror of radius R, where O is any point on the principal axis that is farther than R from the mirror surface, forms a real image at I. If the rays diverge from O at small angles, they all reflect through the same image point.

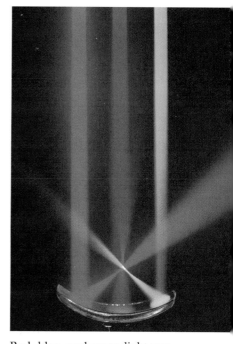

Red, blue, and green light rays are reflected by a parabolic mirror. Note that the focal point where the three colors meet is white light. *(Ken Kay/Fundamental Photographs)*

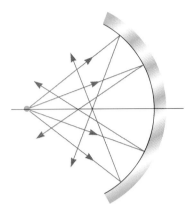

FIGURE 36.7 Rays at large angles from the horizontal axis reflect from a spherical concave mirror to intersect the principal axis at different points, resulting in a blurred image. This is called *spherical aberration.*

segment, and a line drawn from C to V is called the **principal axis** of the optical system.

Now consider a point source of light placed at point O in Figure 36.6b, located on the principal axis to the left of point C. Several diverging rays originating at O are shown. After reflecting from the mirror, these rays converge (come together) at the image point I. The rays then continue to diverge from I as if there were an object there. As a result, a real image is formed.

In what follows, we assume that all rays that diverge from the object make a small angle with the principal axis. Such rays are called **paraxial rays.** All such rays reflect through the image point, as in Figure 36.6b. Rays that are far from the principal axis, as in Figure 36.7, converge to other points on the principal axis, producing a blurred image. This effect, called **spherical aberration,** is present to some extent for any spherical mirror and is discussed in Section 36.5.

We can use the geometry shown in Figure 36.8 to calculate the image distance q from a knowledge of the object distance p and the mirror radius of curvature, R. By convention, these distances are measured from point V. Figure 36.8 shows two rays of light leaving the tip of the object. One of these rays passes through the center of curvature, C, of the mirror, falls normal to the surface of the mirror, and reflects back on itself. The second ray strikes the mirror at the center, point V, and reflects as shown, obeying the law of reflection. The image of the tip of the arrow is located at the point where these two rays intersect. From the largest right triangle in Figure 36.8 whose base is OV, we see that $\tan\theta = h/p$, while the blue-shaded right triangle gives $\tan\theta = -h'/q$. The negative sign is introduced because the image is inverted, and so h' is taken to be negative. Thus, from Equation 36.1 and these results, we find that the magnification of the mirror is

$$M = \frac{h'}{h} = -\frac{q}{p} \qquad (36.2)$$

We also note from two other triangles in Figure 36.8 that

$$\tan\alpha = \frac{h}{p - R} \quad \text{and} \quad \tan\alpha = -\frac{h'}{R - q}$$

from which we find that

$$\frac{h'}{h} = -\frac{R - q}{p - R} \qquad (36.3)$$

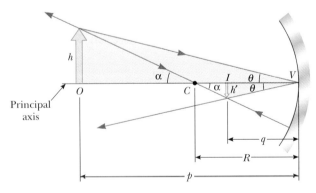

FIGURE 36.8 The image formed by a spherical concave mirror where the object O lies outside the center of curvature, C.

If we compare Equations 36.2 and 36.3, we see that

$$\frac{R-q}{p-R} = \frac{q}{p}$$

Simple algebra reduces this to

$$\frac{1}{p} + \frac{1}{q} = \frac{2}{R} \tag{36.4}$$

This expression is called the **mirror equation**. It is applicable only to paraxial rays.

If the object is very far from the mirror, that is, if the object distance, p, is large enough compared with R that p can be said to approach infinity, then $1/p \approx 0$, and we see from Equation 36.4 that $q \approx R/2$. That is, when the object is very far from the mirror, *the image point is halfway between the center of curvature and the center of the mirror*, as in Figure 36.9a. The rays are essentially parallel in this figure because the source is assumed to be very far from the mirror. We call the image point in this special case the **focal point**, F, and the image distance the **focal length**, f, where

$$f = \frac{R}{2} \tag{36.5}$$

Focal length

The mirror equation can therefore be expressed in terms of the focal length:

$$\frac{1}{p} + \frac{1}{q} = \frac{1}{f} \tag{36.6}$$

Mirror equation

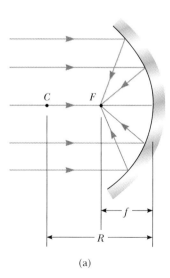
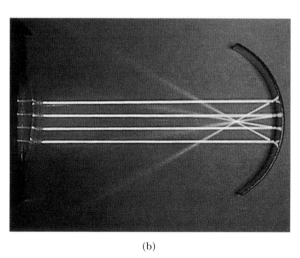

(a) (b)

FIGURE 36.9 (a) Light rays from a distant object ($p \approx \infty$) reflect from a concave mirror through the focal point, F. In this case, the image distance $q = R/2 = f$, where f is the focal length of the mirror. (b) Photograph of the reflection of parallel rays from a concave mirror. *(Henry Leap and Jim Lehman)*

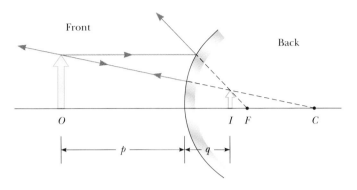

FIGURE 36.10 Formation of an image by a spherical convex mirror. The image formed by the real object is virtual and upright.

Convex cylindrical mirror: reflection of parallel lines. The image of any object in front of the mirror is virtual, erect, and diminished in size. *(© Richard Megna 1990, Fundamental Photographs)*

Convex Mirrors

Figure 36.10 shows the formation of an image by a **convex mirror,** that is, one silvered so that light is reflected from the outer, convex surface. This is sometimes called a **diverging mirror** because the rays from any point on a real object diverge after reflection as though they were coming from some point behind the mirror. The image in Figure 36.10 is virtual because the reflected rays only appear to originate at the image point. Furthermore, the image is always upright and smaller than the object, as shown in the figure.

We do not derive any equations for convex spherical mirrors because we can use Equations 36.2, 36.4, and 36.6 for either concave or convex mirrors if we adhere to the following procedure. Let us refer to the region in which light rays move as the *front side* of the mirror and the other side, where virtual images are formed, as the *back side*. For example, in Figures 36.7 and 36.9, the side to the left of the mirrors is the front side and the side to the right of the mirrors is the back side. Figure 36.11 is helpful for understanding the rules for object and image distances, and Table 36.1 summarizes the sign conventions for all the necessary quantities.

Ray Diagrams for Mirrors

The position and size of images formed by mirrors can be conveniently determined by using *ray diagrams*. These graphical constructions tell us the total nature of the image and can be used to check results calculated from the mirror and

FIGURE 36.11 A diagram for describing the signs of p and q for convex and concave mirrors.

TABLE 36.1 Sign Convention for Mirrors

p is $+$ if the object is in front of the mirror (real object).
p is $-$ if the object is in back of the mirror (virtual object).

q is $+$ if the image is in front of the mirror (real image).
q is $-$ if the image is in back of the mirror (virtual image).

Both f and R are $+$ if the center of curvature is in front of the mirror (concave mirror).
Both f and R are $-$ if the center of curvature is in back of the mirror (convex mirror).

If M is positive, the image is upright.
If M is negative, the image is inverted.

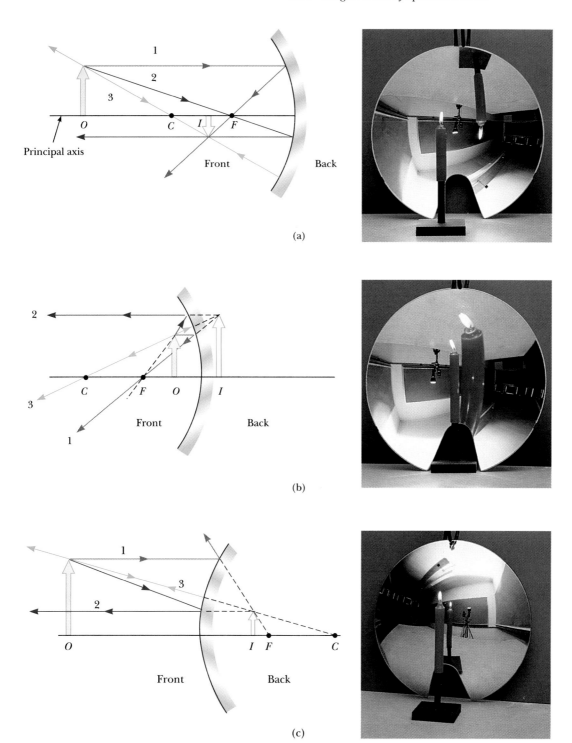

FIGURE 36.12 Ray diagrams for spherical mirrors, and corresponding photographs of the images of candles. (a) When the object is located so that the center of curvature lies between the object and a concave mirror surface, the image is real, inverted, and reduced in size. (b) When the object is located between the focal point and a concave mirror surface, the image is virtual, upright, and enlarged. (c) When the object is in front of a spherical convex mirror, the image is virtual, upright, and reduced in size.

magnification equations. In these diagrams, we need to know the position of the object and the location of the center of curvature. In order to locate the image, three rays are then constructed, as shown by the various examples in Figure 36.12. These rays all start from any object point (although in our examples we always choose the tip of the arrow for simplicity) and are drawn as follows:

- Ray 1 is drawn from the top of the object parallel to the optical axis and is reflected back through the focal point, F.
- Ray 2 is drawn from the top of the object through the focal point. Thus, it is reflected parallel to the optic axis.
- Ray 3 is drawn from the top of the object through the center of curvature, C, and is reflected back on itself.

The intersection of any two of these rays locates the image. The third ray serves as a check on your construction. The image point obtained in this fashion must always agree with the value of q calculated from the mirror equation.

With concave mirrors, note what happens as the object is moved closer to the mirror. The real, inverted image in Figure 36.12a moves to the left as the object approaches the focal point. When the object is at the focal point, the image is infinitely far to the left. However, when the object lies between the focal point and the mirror surface, as in Figure 36.12b, the image is virtual and upright. Finally, for the convex mirror shown in Figure 36.12c, the image of a real object is always virtual and upright. In this case, as the object distance increases, the virtual image decreases in size and approaches the focal point as p approaches infinity. You should construct other diagrams to verify how the image position varies with object position.

EXAMPLE 36.4 The Image for a Concave Mirror

Assume that a certain concave spherical mirror has a focal length of 10.0 cm. Find the location of the image for object distances of (a) 25.0 cm, (b) 10.0 cm, and (c) 5.00 cm. Describe the image in each case.

Solution (a) For an object distance of 25.0 cm, we find the image distance using the mirror equation:

$$\frac{1}{p} + \frac{1}{q} = \frac{1}{f}$$

$$\frac{1}{25.0 \text{ cm}} + \frac{1}{q} = \frac{1}{10.0 \text{ cm}}$$

$$q = 16.7 \text{ cm}$$

The magnification is given by Equation 36.2:

$$M = -\frac{q}{p} = -\frac{16.7 \text{ cm}}{25.0 \text{ cm}} = -0.668$$

This value of M means that the image is smaller than the object. The negative sign means that the image is inverted. Finally, because q is positive, the image is located on the front side of the mirror and is real. This situation is pictured in Figure 36.12a.

(b) When the object distance is 10.0 cm, the object is located at the focal point. Substituting the values $p = 10.0$ cm and $f = 10.0$ cm into the mirror equation, we find

$$\frac{1}{10.0 \text{ cm}} + \frac{1}{q} = \frac{1}{10.0 \text{ cm}}$$

$$q = \infty$$

Thus, we see that rays of light originating from an object located at the focal point of a mirror are reflected so that the image is formed at an infinite distance from the mirror; that is, the rays travel parallel to one another after reflection.

(c) When the object is at the position $p = 5.00$ cm, it lies between the focal point and the mirror surface. In this case, the mirror equation gives

$$\frac{1}{5.00 \text{ cm}} + \frac{1}{q} = \frac{1}{10.0 \text{ cm}}$$

$$q = -10.0 \text{ cm}$$

That is, the image is virtual because it is located behind the mirror. The magnification is

$$M = -\frac{q}{p} = -\left(\frac{-10.0 \text{ cm}}{5.00 \text{ cm}}\right) = 2.00$$

From this, we see that the image is twice as large as the object and the positive sign for M indicates that the image is upright (Fig. 36.12b). The negative value of q means that the image is behind the mirror and is virtual.

Note the characteristics of the images formed by a concave spherical mirror. When the focal point lies between the object and mirror surface, the image is inverted and real; with the object at the focal point, the image is formed at infinity; with the object between the focal point and mirror surface, the image is upright and virtual.

Exercise If the object distance is 20.0 cm, find the image distance and the magnification of the mirror.

Answer $q = 20.0$ cm, $M = -1.00$.

EXAMPLE 36.5 The Image for a Convex Mirror

An object 3.00 cm high is placed 20.0 cm from a convex mirror having a focal length of 8.00 cm. Find (a) the position of the final image and (b) the magnification.

Solution (a) Because the mirror is convex, its focal length is negative. To find the image position, we use the mirror equation:

$$\frac{1}{p} + \frac{1}{q} = \frac{1}{f} = -\frac{1}{8.00 \text{ cm}}$$

$$\frac{1}{q} = -\frac{1}{8.00 \text{ cm}} - \frac{1}{20.0 \text{ cm}}$$

$$q = -5.71 \text{ cm}$$

The negative value of q indicates that the image is virtual, or behind the mirror, as in Figure 36.12c.

(b) The magnification is

$$M = -\frac{q}{p} = -\left(\frac{-5.71 \text{ cm}}{20.0 \text{ cm}}\right) = 0.286$$

The image is about 30% of the size of the object and upright because M is positive.

Exercise Find the height of the image.

Answer 0.857 cm.

36.3 IMAGES FORMED BY REFRACTION

In this section, we describe how images are formed by the refraction of rays at a spherical surface of a transparent material. Consider two transparent media with indices of refraction n_1 and n_2, where the boundary between the two media is a spherical surface of radius R (Fig. 36.13). We assume that the object at point O is in the medium whose index of refraction is n_1. Furthermore, of all the paraxial rays leaving O, let us consider only those that make a small angle with the axis and with each other. As we shall see, all such rays originating at the object point are refracted at the spherical surface and focus at a single point I, the image point.

Let us proceed by considering the geometric construction in Figure 36.14,

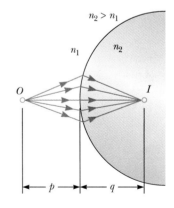

FIGURE 36.13 An image formed by refraction at a spherical surface. Rays making small angles with the optic axis diverge from a point object at O and pass through the image point, I.

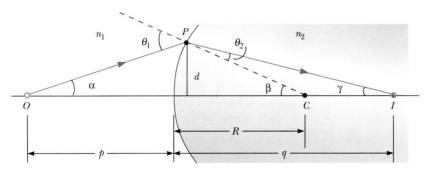

FIGURE 36.14 Geometry used to derive Equation 36.8.

which shows a single ray leaving point O and focusing at point I. Snell's law applied to this refracted ray gives

$$n_1 \sin \theta_1 = n_2 \sin \theta_2$$

Because the angles θ_1 and θ_2 are assumed small, we can use the small-angle approximation $\sin \theta \approx \theta$ (angles in radians). Therefore, Snell's law becomes

$$n_1 \theta_1 = n_2 \theta_2$$

Now we use the fact that an exterior angle of any triangle equals the sum of the two opposite interior angles. Applying this to the triangles OPC and PIC in Figure 36.14 gives

$$\theta_1 = \alpha + \beta$$
$$\beta = \theta_2 + \gamma$$

If we combine the last three expressions, and eliminate θ_1 and θ_2, we find

$$n_1 \alpha + n_2 \gamma = (n_2 - n_1)\beta \tag{36.7}$$

Again, in the small angle approximation, $\tan \theta \approx \theta$, and so we can write the approximate relationships

$$\alpha \cong \frac{d}{p} \qquad \beta \cong \frac{d}{R} \qquad \gamma \cong \frac{d}{q}$$

where d is the distance shown in Figure 36.14. We substitute these expressions into Equation 36.7 and divide through by d to give

$$\frac{n_1}{p} + \frac{n_2}{q} = \frac{n_2 - n_1}{R} \tag{36.8}$$

For a fixed object distance p, the image distance q is independent of the angle that the ray makes with the axis. This result tells us that all paraxial rays focus at the same point I.

As with mirrors, we must use a sign convention if we are to apply this equation to a variety of circumstances. First, note that real images are formed on the side of the surface that is opposite the side from which the light comes, in contrast to mirrors, where real images are formed on the same side of the reflecting surface. Therefore, *the sign convention for spherical refracting surfaces is similar to the convention for mirrors, recognizing the change in sides of the surface for real and virtual images.* For example, in Figure 36.14, p, q, and R are all positive.

The sign convention for spherical refracting surfaces is summarized in Table 36.2. (The same sign convention is used for thin lenses, which we discuss in the

TABLE 36.2 Sign Convention for Refracting Surfaces

p is + if the object is in front of the surface (real object).
p is − if the object is in back of the surface (virtual object).

q is + if the image is in back of the surface (real image).
q is − if the image is in front of the surface (virtual image).

R is + if the center of curvature is in back of the surface.
R is − if the center of curvature is in front of the surface.

next section.) As with mirrors, we assume that the front of the refracting surface is the side from which the light approaches the surface.

Flat Refracting Surfaces

If the refracting surface is flat, then R approaches infinity and Equation 36.8 reduces to

$$\frac{n_1}{p} = -\frac{n_2}{q}$$

$$q = -\frac{n_2}{n_1} p \qquad (36.9)$$

From Equation 36.9 we see that the sign of q is opposite that of p. Thus, *the image formed by a flat refracting surface is on the same side of the surface as the object.* This is illustrated in Figure 36.15 for the situation in which n_1 is greater than n_2, where a virtual image is formed between the object and the surface. If n_1 is less than n_2, the image is still virtual but is formed to the left of the object.

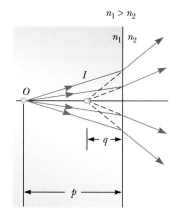

FIGURE 36.15 The image formed by a flat refracting surface is virtual; that is, it forms to the left of the refracting surface. All rays are assumed to be paraxial.

CONCEPTUAL EXAMPLE 36.6 Let's Go Scuba Diving

It is well known that objects viewed underwater with the naked eye appear blurred and out of focus. However, a scuba diver using goggles has a clear view of underwater objects. Explain how this works, using the fact that the indices of refraction of the cornea, water, and air are 1.376, 1.333, and 1.00029, respectively.

Reasoning Because the cornea and water have almost identical indices of refraction, very little refraction occurs when underwater objects are viewed with the naked eye. In this case, light from the object focuses behind the retina and causes a blurred image. When goggles are used, the air space between the eye and goggle surface provides the normal amount of refraction at the eye-air interface, and the light from the object is focused on the retina.

EXAMPLE 36.7 Gaze into the Crystal Ball

A coin 2.00 cm in diameter is embedded in a solid glass ball of radius 30.0 cm (Fig. 36.16). The index of refraction of the ball is $n_1 = 1.5$, and the coin is 20.0 cm from the surface. Find the position of the image.

Solution Because $n_1 > n_2$, where $n_2 = 1.00$ is the index of refraction for air, the rays originating from the object are refracted away from the normal at the surface and diverge outward. Hence, the image is formed in the glass and is virtual. Applying Equation 36.8, we get

$$\frac{n_1}{p} + \frac{n_2}{q} = \frac{n_2 - n_1}{R}$$

$$\frac{1.50}{20.0 \text{ cm}} + \frac{1}{q} = \frac{1.00 - 1.50}{-30.0 \text{ cm}}$$

$$q = \boxed{-17.1 \text{ cm}}$$

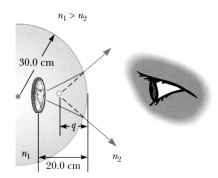

FIGURE 36.16 (Example 36.7) A coin embedded in a glass ball forms a virtual image between the coin and the glass surface. All rays are assumed to be paraxial.

The negative sign indicates that the image is in the same medium as the object (the side of incident light), in agreement with our ray diagram. Being in the same medium as the object, the image must be virtual.

EXAMPLE 36.8 The One That Got Away

A small fish is swimming at a depth d below the surface of a pond (Fig. 36.17). What is the apparent depth of the fish as viewed from directly overhead?

Solution In this example, the refracting surface is flat, and so R is infinite. Hence, we can use Equation 36.9 to determine the location of the image. Using the facts that $n_1 = 1.33$ for water and $p = d$ gives

$$q = -\frac{n_2}{n_1}p = -\frac{1}{1.33}d = \boxed{-0.750d}$$

Again, since q is negative, the image is virtual, as indicated in Figure 36.17. The apparent depth is three-fourths the actual depth.

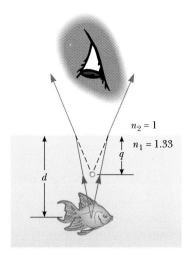

FIGURE 36.17 (Example 36.8) The apparent depth, q, of the fish is less than the true depth, d. All rays are assumed to be paraxial.

36.4 THIN LENSES

Lenses are commonly used to form images by refraction in optical instruments, such as cameras, telescopes, and microscopes. The essential idea in locating the final image of a lens is to *use the image formed by one refracting surface as the object for the second surface.*

Consider a lens having an index of refraction n and two spherical surfaces of radii of curvature R_1 and R_2, as in Figure 36.18. An object is placed at point O at a distance p_1 in front of surface 1. For this example, p_1 has been chosen so as to produce a virtual image I_1 to the left of the lens. This image is then used as the object for surface 2, which results in a real image I_2.

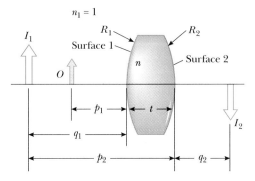

FIGURE 36.18 To locate the image of a lens, the image at I_1 formed by the first surface is used as the object for the second surface. The final image is at I_2.

Using Equation 36.8 and assuming $n_1 = 1$ because the lens is surrounded by air, we find that the image formed by surface 1 satisfies the equation

$$(1) \quad \frac{1}{p_1} + \frac{n}{q_1} = \frac{n-1}{R_1}$$

Now we apply Equation 36.8 to surface 2, taking $n_1 = n$ and $n_2 = 1$. That is, light approaches surface 2 as if it had come from I_1. Taking p_2 as the object distance and q_2 as the image distance for surface 2 gives

$$(2) \quad \frac{n}{p_2} + \frac{1}{q_2} = \frac{1-n}{R_2}$$

But $p_2 = -q_1 + t$, where t is the thickness of the lens. (Remember q_1 is a negative number and p_2 must be positive by our sign convention.) For a thin lens, we can neglect t. In this approximation and from Figure 36.18, we see that $p_2 = -q_1$. Hence, (2) becomes

$$(3) \quad -\frac{n}{q_1} + \frac{1}{q_2} = \frac{1-n}{R_2}$$

Adding (1) and (3), we find that

$$(4) \quad \frac{1}{p_1} + \frac{1}{q_2} = (n-1)\left(\frac{1}{R_1} - \frac{1}{R_2}\right)$$

For the thin lens, we can omit the subscripts on p_1 and q_2 in (4) and call the object distance p and the image distance q, as in Figure 36.19. Hence, we can write (4) in the form

$$\frac{1}{p} + \frac{1}{q} = (n-1)\left(\frac{1}{R_1} - \frac{1}{R_2}\right) \quad (36.10)$$

This expression relates the image distance q of the image formed by a thin lens to the object distance p and to the thin lens properties (index of refraction and radii of curvature). It is valid only for paraxial rays and only when the lens thickness is small relative to R_1 and R_2.

We now define the focal length f of a thin lens as the image distance that corresponds to an infinite object distance, as we did with mirrors. According to this definition and from Equation 36.10, we see that as $p \to \infty$, $q \to f$; therefore, the inverse of the focal length for a thin lens is

$$\frac{1}{f} = (n-1)\left(\frac{1}{R_1} - \frac{1}{R_2}\right) \quad (36.11)$$

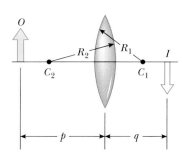

FIGURE 36.19 The biconvex lens.

Lens makers' equation

Equation 36.11 is called the **lens makers' equation** because it enables f to be calculated from the known properties of the lens. It can also be used to determine the values of R_1 and R_2 needed for a given index of refraction and desired focal length.

Using Equation 36.11, we can write Equation 36.10 in an alternate form identical to Equation 36.6 for mirrors:

$$\frac{1}{p} + \frac{1}{q} = \frac{1}{f} \quad (36.12)$$

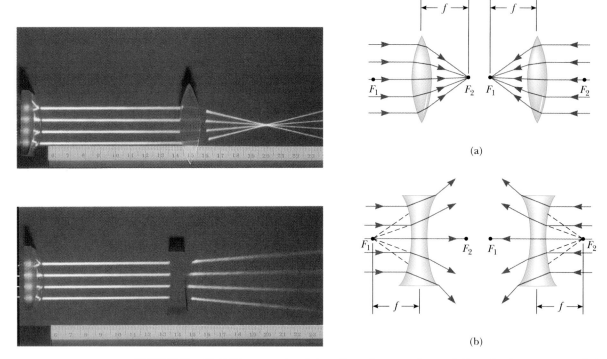

FIGURE 36.20 *(Left)* Photographs of the effect of converging and diverging lenses on parallel rays. *(Henry Leap and Jim Lehman)* *(Right)* The object and image focal points of (a) the biconvex lens and (b) the biconcave lens.

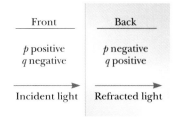

FIGURE 36.21 A diagram for obtaining the signs of p and q for a thin lens or a refracting surface.

A thin lens has two focal points, corresponding to incident parallel light rays traveling from the left or right. This is illustrated in Figure 36.20 for a biconvex lens (converging, positive f) and a biconcave lens (diverging, negative f). Focal point F_1 is sometimes called the *object focal point*, and F_2 is called the *image focal point*.

Figure 36.21 is useful for obtaining the signs of p and q, and Table 36.3 gives the complete sign conventions for lenses. Note that the sign conventions for thin lenses are the same as for refracting surfaces. Applying these rules to a converging lens, we see that when $p > f$, the quantities p, q, and R_1 are positive and R_2 is negative. Therefore, when a converging lens forms a real image from a real object, p, q, and f are all positive. For a diverging lens, p and R_2 are positive, q and R_1 are negative, and so f is negative for a diverging lens.

Sketches of various lens shapes are shown in Figure 36.22. In general, note that a converging lens is thicker at the center than at the edge, whereas a diverging lens is thinner at the center than at the edge.

TABLE 36.3 Sign Convention for Thin Lenses

p is $+$ if the object is in front of the lens.
p is $-$ if the object is in back of the lens.

q is $+$ if the image is in back of the lens.
q is $-$ if the image is in front of the lens.

R_1 and R_2 are $+$ if the center of curvature is in back of the lens.
R_1 and R_2 are $-$ if the center of curvature is in front of the lens.

Consider a single thin lens illuminated by a real object (that the object is real means that $p > 0$). As with mirrors, the lateral magnification of a thin lens is defined as the ratio of the image height h' to the object height h:

$$M = \frac{h'}{h} = -\frac{q}{p}$$

From this expression, it follows that when M is positive, the image is upright and on the same side of the lens as the object. When M is negative, the image is inverted and on the side of the lens opposite the object.

Ray Diagrams for Thin Lenses

Ray diagrams are very convenient for locating the image formed by a thin lens or a system of lenses. Such constructions also help clarify the sign conventions that have been discussed. Figure 36.23 illustrates this method for three single-lens situations. To locate the image of a converging lens (Figs. 36.23a and 36.23b), the following three rays are drawn from the top of the object:

- Ray 1 is drawn parallel to the optic axis. After being refracted by the lens, this ray passes through (or appears to come from) one of the focal points.
- Ray 2 is drawn through the center of the lens. This ray continues in a straight line.
- Ray 3 is drawn through the object focal point and emerges from the lens parallel to the optic axis.

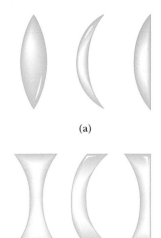

FIGURE 36.22 Various lens shapes: (a) Converging lenses have a positive focal length and are thickest at the middle. From left to right are biconvex, convex-concave, and plano-convex lenses. (b) Diverging lenses have a negative focal length and are thickest at the edges. From left to right are biconcave, convex-concave, and plano-concave lenses.

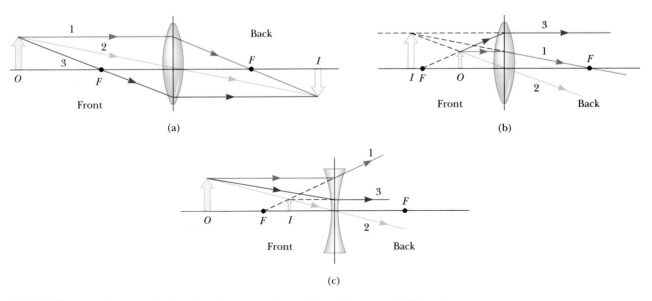

FIGURE 36.23 Ray diagrams for locating the image formed by a thin lens. (a) The object is located to the left of the object focal point of a converging lens. (b) The object is located between the object focal point and a converging lens. (c) The object is located to the left of the object focal point of a diverging lens.

A similar construction is used to locate the image of a diverging lens, as shown in Figure 36.23c.

For the converging lens in Figure 36.23a, where the object is to the left of the object focal point ($p > f$), the image is real and inverted. When the real object is between the object focal point and the lens ($p < f$), as in Figure 36.23b, the image is virtual and erect. Finally, for a diverging lens (Fig. 36.23c), the image is always virtual and upright. These geometric constructions are reasonably accurate only if the distance between the rays and the principal axis is small compared with the radii of the lens surfaces.

EXAMPLE 36.9 An Image Formed by a Diverging Lens

A diverging lens has a focal length of -20.0 cm. An object 2.00 cm in height is placed 30.0 cm in front of the lens. Locate the position of the image.

Solution Using the thin lens equation (Eq. 36.12) with $p = 30.0$ cm and $f = -20.0$ cm, we get

$$\frac{1}{p} + \frac{1}{q} = \frac{1}{f}$$

$$\frac{1}{30.0 \text{ cm}} + \frac{1}{q} = -\frac{1}{20.0 \text{ cm}}$$

$$q = -12.0 \text{ cm}$$

The negative sign tells us that the image is virtual, as indicated in Figure 36.23c.

Exercise Find the magnification and the height of the image.

Answer $M = 0.400$, $h' = 0.800$ cm.

EXAMPLE 36.10 An Image Formed by a Converging Lens

A converging lens of focal length 10.0 cm forms an image of an object placed (a) 30.0 cm, (b) 10.0 cm, and (c) 5.00 cm from the lens. Find the image distance and describe the image in each case.

Solution (a) The thin lens equation, Equation 36.12, can be used to find the image distance:

$$\frac{1}{p} + \frac{1}{q} = \frac{1}{f}$$

$$\frac{1}{30.0 \text{ cm}} + \frac{1}{q} = \frac{1}{10.0 \text{ cm}}$$

$$q = 15.0 \text{ cm}$$

The positive sign tells us that the image is real. The magnification is

$$M = -\frac{q}{p} = -\frac{15.0 \text{ cm}}{30.0 \text{ cm}} = -0.500$$

Thus, the image is reduced in size by one half, and the negative sign for M tells us that the image is inverted. The situation is like that pictured in Figure 36.23a.

(b) No calculation is necessary for this case because we know that, when the object is placed at the focal point, the image is formed at infinity. This is readily verified by substituting $p = 10.0$ cm into the lens equation.

(c) We now move inside the focal point, to an object distance of 5.00 cm. In this case, the thin lens equation gives

$$\frac{1}{5.00 \text{ cm}} + \frac{1}{q} = \frac{1}{10.0 \text{ cm}}$$

$$q = -10.0 \text{ cm}$$

$$M = -\frac{q}{p} = -\left(\frac{-10.0 \text{ cm}}{5.00 \text{ cm}}\right) = 2.00$$

The negative image distance tells us that the image is virtual. The image is enlarged, and the positive sign for M tells us that the image is upright, as in Figure 36.23b.

There are two general cases for a converging lens. When the object distance is greater than the focal length ($p > f$), the image is real and inverted. When the object is between the object focal point and lens ($p < f$), the image is virtual, upright, and enlarged.

EXAMPLE 36.11 A Lens Under Water

A converging glass lens ($n = 1.52$) has a focal length of 40.0 cm in air. Find its focal length when it is immersed in water, which has an index of refraction of 1.33.

Solution We can use the lens makers' formula (Eq. 36.11) in both cases, noting that R_1 and R_2 remain the same in air and water:

$$\frac{1}{f_{air}} = (n - 1)\left(\frac{1}{R_1} - \frac{1}{R_2}\right)$$

$$\frac{1}{f_{water}} = (n' - 1)\left(\frac{1}{R_1} - \frac{1}{R_2}\right)$$

where n' is the index of refraction of glass relative to water. That is, $n' = 1.52/1.33 = 1.14$. Dividing the two equations gives

$$\frac{f_{water}}{f_{air}} = \frac{n - 1}{n' - 1} = \frac{1.52 - 1}{1.14 - 1} = 3.71$$

Since $f_{air} = 40.0$ cm, we find that

$$f_{water} = 3.71 f_{air} = 3.71(40.0 \text{ cm}) = \boxed{148 \text{ cm}}$$

In fact, the focal length of any glass lens is increased by the factor $(n - 1)/(n' - 1)$ when immersed in water.

Combination of Thin Lenses

If two thin lenses are used to form an image, the system can be treated in the following manner. First, the image of the first lens is located as if the second lens were not present. The light then approaches the second lens as if it had come from the image formed by the first lens. In other words, the image of the first lens is treated as the object of the second lens. The image of the second lens is the final image of the system. If the image formed by the first lens lies to the right of the second lens, then that image is treated as a virtual object for the second lens (that is, p negative). The same procedure can be extended to a system of three or more lenses. The overall magnification of a system of thin lenses equals the product of the magnification of the separate lenses.

Now suppose two thin lenses of focal lengths f_1 and f_2 are placed in contact with each other. If p is the object distance for the combination, then application of the thin lens equation to the first lens gives

$$\frac{1}{p} + \frac{1}{q_1} = \frac{1}{f_1}$$

where q_1 is the image distance for the first lens. Treating this image as the object for the second lens, we see that the object distance for the second lens must be $-q_1$. Therefore, for the second lens

$$-\frac{1}{q_1} + \frac{1}{q} = \frac{1}{f_2}$$

where q is the final image distance from the second lens. Adding these equations eliminates q_1 and gives

$$\frac{1}{p} + \frac{1}{q} = \frac{1}{f_1} + \frac{1}{f_2}$$

$$\frac{1}{f} = \frac{1}{f_1} + \frac{1}{f_2} \qquad (36.13)$$

If the two thin lenses are in contact with one another, then q is also the distance of the final image from the first lens. Therefore, *two thin lenses in contact are equivalent to a single thin lens whose focal length is given by Equation 36.13.*

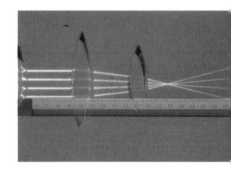

Light from a distant object brought into focus by two converging lenses. Can you estimate the overall focal length of this combination from the photograph? *(Henry Leap and Jim Lehman)*

Focal length of two thin lenses in contact

EXAMPLE 36.12 Where Is the Final Image?

Two thin converging lenses of focal lengths 10.0 cm and 20.0 cm are separated by 20.0 cm, as in Figure 36.24. An object is placed 15.0 cm to the left of the first lens. Find the position of the final image and the magnification of the system.

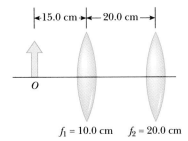

FIGURE 36.24 (Example 36.12) A combination of two converging lenses.

Solution First, we find the image position for the first lens while neglecting the second lens:

$$\frac{1}{p_1} + \frac{1}{q_1} = \frac{1}{15.0 \text{ cm}} + \frac{1}{q_1} = \frac{1}{10.0 \text{ cm}}$$

$$q_1 = 30.0 \text{ cm}$$

where q_1 is measured from the first lens.

Because q_1 is greater than the separation between the two lenses, the image of the first lens lies 10.0 cm to the right of the second lens. We take this as the object distance for the second lens. That is, we apply the thin lens equation to the second lens with $p_2 = -10.0$ cm, where distances are now measured from the second lens, whose focal length is 20.0 cm:

$$\frac{1}{p_2} + \frac{1}{q_2} = \frac{1}{f_2}$$

$$\frac{1}{-10.0 \text{ cm}} + \frac{1}{q_2} = \frac{1}{20.0 \text{ cm}}$$

$$q_2 = 6.67 \text{ cm}$$

That is, the final image lies 6.67 cm to the right of the second lens.

The magnification of each lens separately is given by

$$M_1 = \frac{-q_1}{p_1} = -\frac{30.0 \text{ cm}}{15.0 \text{ cm}} = -2.00$$

$$M_2 = \frac{-q_2}{p_2} = -\frac{6.67 \text{ cm}}{-10.0 \text{ cm}} = 0.667$$

The total magnification M of the two lenses is the product $M_1 M_2 = (-2.00)(0.667) = -1.33$. Hence, the final image is real, inverted, and enlarged.

*36.5 LENS ABERRATIONS

One of the basic problems of lenses and lens systems is imperfect images, largely the result of defects in the shape and form of the lenses. The simple theory of mirrors and lenses assumes that rays make small angles with the optic axis. In this simple model, all rays leaving a point source focus at a single point, producing a sharp image. Clearly, this is not always true. When the approximations used in this theory do not hold, imperfect images are formed.

If one wishes to perform a precise analysis of image formation, it is necessary to trace each ray using Snell's law at each refracting surface. This procedure shows that the rays from a point object do not focus at a single point. That is, there is no single point image; instead, the image is blurred. The departures of real (imperfect) images from the ideal image predicted by the simple theory are called **aberrations**.

Spherical Aberrations

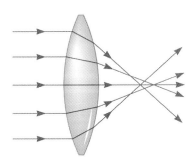

FIGURE 36.25 Spherical aberration caused by a converging lens. Does a diverging lens cause spherical aberration?

Spherical aberrations result from the fact that the focal points of light rays far from the optic axis of a spherical lens (or mirror) are different from the focal points of rays of the same wavelength passing near the center. Figure 36.25 illustrates spherical aberration for parallel rays passing through a converging lens. Rays near the middle of the lens are imaged farther from the lens than rays at the edges. Hence, there is no single focal length for a lens.

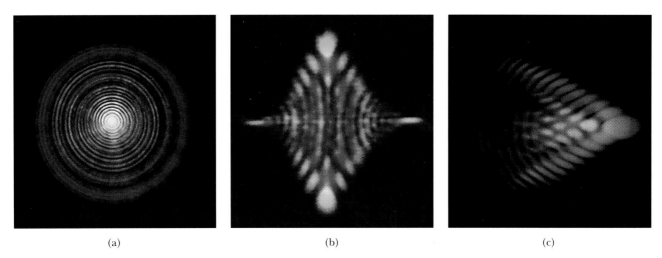

(a)   (b)   (c)

Lenses can produce various forms of aberrations, as shown by the blurred images of a point source in these photos. (a) Spherical aberration occurs when light passing through the lens at different distances from the optical axis is focused at different points. (b) Astigmatism is an aberration that occurs for objects not located on the optical axis of the lens. (c) Coma. This aberration occurs as light passing through the lens far from the optical axis and light passing near the center of the lens focus at different parts of the focal plane. *(Photographs by Norman Goldberg)*

Many cameras are equipped with a variable aperture to control the light intensity and reduce spherical aberration when possible. (A variable aperture is used to control the amount of light transmitted through the lens.) Sharper images are produced as the aperture size is reduced, because for small apertures only the central portion of the lens is exposed to the incident light. At the same time, however, less light is imaged. To compensate for this lower light intensity, a longer exposure time is used on the photographic film. A good example is the sharp image produced by a "pin-hole" camera, whose aperture size is approximately 1 mm.

In the case of mirrors used for very distant objects, spherical aberrations can be eliminated, or at least minimized, by using a parabolic surface rather than a spherical surface. Parabolic surfaces are not often used, however, because those with high-quality optics are very expensive to make. Parallel light rays incident on such a surface focus at a common point. Parabolic reflecting surfaces are used in many astronomical telescopes in order to enhance the image quality. They are also used in flashlights, where a nearby parallel light beam is produced from a small lamp placed at the focus of the surface.

Chromatic Aberrations

The fact that different wavelengths of light refracted by a lens focus at different points gives rise to chromatic aberrations. In Chapter 35, we described how the index of refraction of a material varies with wavelength. When white light passes through a lens, it is found, for example, that violet light rays are refracted more than red light rays (Fig. 36.26). From this we see that the focal length is larger for red light than for violet light. Other wavelengths (not shown in Fig. 36.26) have intermediate focal points. The chromatic aberration for a diverging lens is opposite that for a converging lens. Chromatic aberration can be greatly reduced by

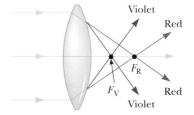

FIGURE 36.26 Chromatic aberration caused by a converging lens. Rays of different wavelengths focus at different points.

using a combination of a converging and diverging lens made from two different types of glass.

*36.6 THE CAMERA

The photographic **camera** is a simple optical instrument whose essential features are shown in Figure 36.27. It consists of a light-tight box, a converging lens that produces a real image, and a film behind the lens to receive the image. Focusing is accomplished by varying the distance between lens and film with an adjustable bellows in older style cameras or some other mechanical arrangement in modern cameras. For proper focusing, or sharp images, the lens-to-film distance will depend on the object distance as well as on the focal length of the lens. The shutter, located behind the lens, is a mechanical device that is opened for selected time intervals. With this arrangement, one can photograph moving objects by using short exposure times or dark scenes (low light levels) by using long exposure times. If this arrangement were not available, it would be impossible to take stop-action photographs. For example, a rapidly moving vehicle could move far enough in the time that the shutter was open to produce a blurred image. Another major cause of blurred image is the movement of the camera while the shutter is open. For this reason, short exposure times or a tripod should be used, even for stationary objects. Typical shutter speeds are 1/30, 1/60, 1/125, and 1/250 s. A stationary object is normally shot with a shutter speed of 1/60 s.

More expensive cameras also have an aperture of adjustable diameter either behind or between the lenses to further control the intensity of the light reaching the film. When an aperture of small diameter is used, only light from the central portion of the lens reaches the film and so the aberration is reduced somewhat.

The brightness of the image depends on the focal length and diameter D of the lens. Clearly, the light intensity I is proportional to the area of the lens. Since the area is proportional to D^2, we conclude that $I \propto D^2$. Furthermore, the intensity is a measure of the energy received by the film per unit area of the image. Since the area of the image is proportional to $(q)^2$, and $q \approx f$ (for objects with $p \gg f$), we conclude that the intensity is also proportional to $1/f^2$, so that $I \propto D^2/f^2$. The ratio f/D is defined to be the **f-number** of a lens:

$$f\text{-number} \equiv \frac{f}{D} \qquad (36.14)$$

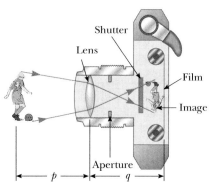

FIGURE 36.27 Cross-sectional view of a simple camera. Note that in reality, $p \gg q$.

Hence, the intensity of light incident on the film can be expressed as

$$I \propto \frac{1}{(f/D)^2} \propto \frac{1}{(f\text{-number})^2} \qquad (36.15)$$

The *f*-number is a measure of the "light-concentrating" power and determines the "speed" of the lens. A "fast" lens has a small *f*-number. Fast lenses, with an *f*-number as low as approximately 1.4, are more expensive because it is more difficult to keep aberrations acceptably small. Camera lenses are often marked with various *f*-numbers such as $f/2.8, f/4, f/5.6, f/8, f/11, f/16$. These settings are obtained by adjusting the aperture, which effectively changes *D*. The smallest *f*-number corresponds to the aperture wide open and the full lens area in use. Simple cameras for routine snapshots usually have a fixed focal length and fixed aperture size, with an *f*-number of about $f/11$.

EXAMPLE 36.13 Finding the Correct Exposure Time

The lens of a certain 35-mm camera (where 35 mm is the width of the film strip) has a focal length of 55 mm and a speed of $f/1.8$. The correct exposure time for this speed under certain conditions is known to be $(1/500)$ s. (a) Determine the diameter of the lens.

Solution From Equation 36.14, we find that

$$D = \frac{f}{f\text{-number}} = \frac{55 \text{ mm}}{1.8} = \boxed{31 \text{ mm}}$$

(b) Calculate the correct exposure time if the *f*-number is changed to $f/4$ under the same lighting conditions.

Reasoning and Solution The total light energy received by each part of the image is proportional to the product of the flux and the exposure time. If I is the light intensity reaching the film, then in a time t, the energy received by the film is It. Comparing the two situations, we require that $I_1 t_1 = I_2 t_2$, where t_1 is the correct exposure time for $f/1.8$ and t_2 is the correct exposure time for some other *f*-number. Using this result, together with Equation 36.15, we find that

$$\frac{t_1}{(f_1\text{-number})^2} = \frac{t_2}{(f_2\text{-number})^2}$$

$$t_2 = \left(\frac{f_2\text{-number}}{f_1\text{-number}}\right)^2 t_1 = \left(\frac{4}{1.8}\right)^2 \left(\frac{1}{500} \text{ s}\right) \approx \boxed{\frac{1}{100} \text{ s}}$$

That is, as the aperture is reduced in size, the exposure time must increase.

*36.7 THE EYE

The eye is an extremely complex part of the body, and because of its complexity, certain defects often arise that can cause impaired vision. In this section we describe the parts of the eye, their purpose, and some of the corrections that can be made when the eye does not function properly.

Like a camera, a normal eye focuses light and produces a sharp image. However, the mechanisms by which the eye controls the amount of light admitted and adjusts to produce correctly focused images are far more complex, intricate, and effective than those in even the most sophisticated camera. In all respects, the eye is an architectural wonder.

Figure 36.28 shows the essential parts of the eye. The front is covered by a transparent membrane called the *cornea*, behind which are a clear liquid region (the *aqueous humor*), a variable aperture (the *iris* and *pupil*), and the *crystalline lens*. Most of the refraction occurs in the cornea because the liquid medium surrounding the lens has an average index of refraction close to that of the lens. The iris, which is the colored portion of the eye, is a muscular diaphragm that controls pupil size. The iris regulates the amount of light entering the eye by dilating the

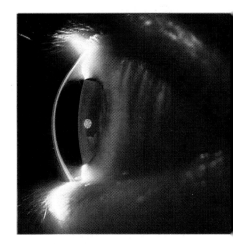

Close-up photograph of the cornea of the human eye. *(Lennart Nilsson,* Behold Man, *Little Brown and Company)*

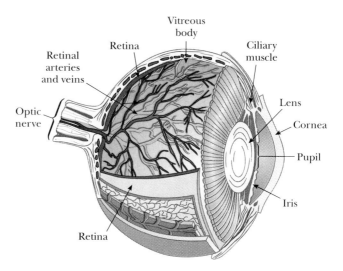

FIGURE 36.28 Essential parts of the eye. Note the similarity between the eye and the simple camera. Can you correlate the parts of the eye with those of the camera?

pupil in light of low intensity and contracting the pupil in high-intensity light. The f-number range of the eye is from about $f/2.8$ to $f/16$.

Light entering the eye is focused by the cornea-lens system onto the back surface of the eye, called the *retina*, which consists of millions of sensitive structures called *rods* and *cones*. When stimulated by light, these receptors send impulses via the optic nerve to the brain, where an image is perceived. By this process, a distinct image of an object is observed when the image falls on the retina.

The eye focuses on a given object by varying the shape of the pliable crystalline lens through an amazing process called **accommodation.** An important component in accommodation is the ciliary muscle, which is attached to the lens. When the eye is focused on distant objects, the ciliary muscle is relaxed. For an object distance of infinity, the focal length of the eye (the distance between the lens and the retina) is about 1.7 cm. The eye focuses on nearby objects by tensing the ciliary muscle. This action effectively decreases the focal length by slightly decreasing the radius of curvature of the lens, which allows the image to be focused on the retina. This lens adjustment takes place so swiftly that we are not even aware of the change. Again in this respect, even the finest electronic camera is a toy compared with the eye. It is evident that there is a limit to accommodation because objects that are very close to the eye produce blurred images. The **near point** represents the closest distance for which the lens of the relaxed eye can focus light on the retina. This distance usually increases with age and has an average value of 25 cm. Typically, at age ten the near point of the eye is about 18 cm. This increases to about 25 cm at age 20, to 50 cm at age 40, and to 500 cm or greater at age 60. The **far point** of the eye represents the largest distance for which the lens of the relaxed eye can focus light on the retina. A person with normal vision is able to see very distant objects, such as the Moon, and thus has a far point near infinity.

Conditions of the Eye

The eye may have several abnormalities, which can often be corrected with eyeglasses, contact lenses, or surgery.

When the relaxed (unaccommodated) eye produces an image of a distant object behind the retina, as in Figure 36.29a, the condition is known as **farsighted-**

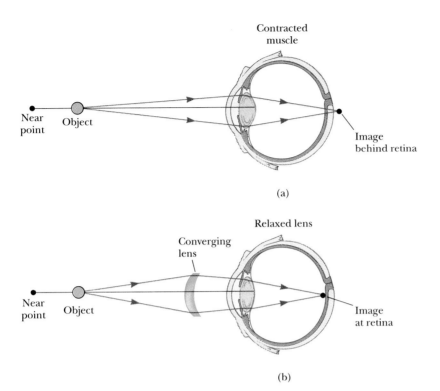

FIGURE 36.29 (a) When a farsighted person looks at an object inside the near point, the image is formed behind the retina, resulting in blurred vision. The eye muscle contracts to try to bring the object into focus. (b) The condition of farsightedness can be corrected with a converging lens.

ness (or *hyperopia*). A farsighted person can usually see faraway objects clearly but cannot focus on nearby objects. Although the near point of a normal eye is approximately 25 cm, the near point of a farsighted person is much farther than this. The eye of a farsighted person tries to focus on objects closer than the near point by accommodation, that is, by shortening its focal length. However, because the focal length of the eye is longer than normal, the light from a nearby object forms a sharp image behind the retina, causing a blurred image. The condition can be corrected by placing a converging lens in front of the eye, as in Figure 36.29b. The lens refracts the incoming rays more toward the principal axis before entering the eye, allowing them to converge and focus on the retina.

Nearsightedness (or *myopia*) is a condition in which a person is able to focus on nearby objects but cannot see faraway objects clearly. In most cases, nearsightedness is due to an eye whose lens is too far from the retina. The far point of the nearsightedness is not at infinity, and may be as close as a few meters. The maximum focal length of the nearsighted eye is insufficient to produce a clearly formed image on the retina. In this case, rays from a distant object are focused in front of the retina, causing blurred vision (Fig. 36.30a). Nearsightedness can be corrected with a diverging lens, as in Figure 36.30b. The lens refracts the rays away from the principal axis before entering the eye, allowing them to focus on the retina.

Beginning with middle age, most people lose some of their accommodation power as the ciliary muscle weakens and the lens hardens. This causes an individual to become farsighted. Fortunately, the condition can be corrected with converging lenses.

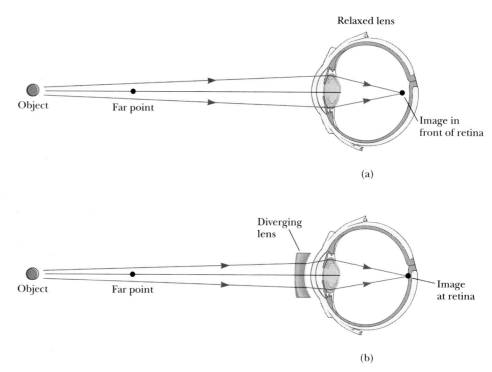

FIGURE 36.30 (a) When a nearsighted person looks at a distant object that lies beyond the far point, the image is formed in front of the retina, resulting in blurred vision. (b) The condition of nearsightedness can be corrected with a diverging lens.

A person may also have an eye defect known as **astigmatism,** in which light from a point source produces a line image on the retina. This condition arises either when the cornea or the lens or both are not perfectly spherical. Astigmatism can be corrected with lenses having different curvatures in two mutually perpendicular directions.

Optometrists and ophthalmologists usually prescribe lenses measured in **diopters.**

> The **power**, P, of a lens in diopters equals the inverse of the focal length in meters, that is, $P = 1/f$.

For example, a converging lens whose focal length is $+20$ cm has a power of $+5.0$ diopters, and a diverging lens whose focal length is -40 cm has a power of -2.5 diopters.

EXAMPLE 36.14 A Case of Nearsightedness

A particular nearsighted person is unable to see objects clearly when they are beyond 2.5 m (the far point of the eye). What should the focal length of the lens prescribed to correct this problem be?

Solution The purpose of the lens in this instance is to "move" an object from infinity to a distance where it can be seen clearly. This is accomplished by having the lens produce an image at the far point of the eye. From the thin lens equation, we have

$$\frac{1}{p} + \frac{1}{q} = \frac{1}{\infty} - \frac{1}{2.5 \text{ m}} = \frac{1}{f}$$

$$f = -2.5 \text{ m}$$

Why did we use a negative sign for the image distance? As you should have suspected, the lens must be a diverging lens (negative focal length) to correct nearsightedness.

Exercise What is the power of this lens?

Answer -0.40 diopters.

*36.8 THE SIMPLE MAGNIFIER

The simple magnifier consists of a single converging lens. As the name implies, this device is used to increase the apparent size of an object. Suppose an object is viewed at some distance p from the eye, as in Figure 36.31. Clearly, the size of the image formed at the retina depends on the angle θ subtended by the object at the eye. As the object moves closer to the eye, θ increases and a larger image is observed.[1] However, an average normal eye is unable to focus on an object closer than about 25 cm, the near point (Fig. 36.32a). Try it! Therefore, θ is maximum at the near point.

To further increase the apparent angular size of an object, a converging lens can be placed in front of the eye with the object located at point O, just inside the focal point of the lens, as in Figure 36.32b. At this location, the lens forms a virtual, upright, and enlarged image, as shown. Clearly, the lens increases the angular size of the object. We define the **angular magnification,** m, as the ratio of the angle subtended by an object with a lens in use (angle θ in Fig. 36.32b) to that subtended by the object placed at the near point with no lens (angle θ_0 in Fig. 36.32a):

$$m \equiv \frac{\theta}{\theta_0} \tag{36.16}$$

A magnifying glass is useful when viewing fine details on a map. *(Jim Lehman)*

Angular magnification with the object at the near point

The angular magnification is a maximum when the image is at the near point of the eye, that is, when $q = -25$ cm. The object distance corresponding to this image distance can be calculated from the thin lens formula:

$$\frac{1}{p} + \frac{1}{-25 \text{ cm}} = \frac{1}{f}$$

$$p = \frac{25f}{25 + f}$$

where f is the focal length of the magnifier in centimeters. If we make the small angle approximations

$$\theta_0 \approx \frac{h}{25} \quad \text{and} \quad \theta \approx \frac{h}{p} \tag{36.17}$$

Equation 36.16 becomes

$$m = \frac{\theta}{\theta_0} = \frac{h/p}{h/25} = \frac{25}{p} = \frac{25}{25f/(25+f)}$$

$$m = 1 + \frac{25 \text{ cm}}{f} \tag{36.18}$$

FIGURE 36.31 The size of the image formed on the retina depends on the angle θ subtended at the eye.

---

[1] Regular eyeglasses give some magnification because the lenses are not located at the lens of the eye. On the other hand, contact lenses minimize this effect because of their close proximity to the lens of the eye.

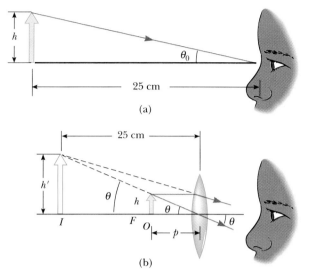

FIGURE 36.32 (a) An object placed at the near point of the eye ($p = 25$ cm) subtends an angle θ_0 at the eye, where $\theta_0 \approx h/25$. (b) An object placed near the focal point of a converging lens produces a magnified image, which subtends an angle $\theta \approx h'/25$ at the eye.

Although the eye can focus on an image formed anywhere between the near point and infinity, it is most relaxed when the image is at infinity (Section 36.7). For the image formed by the magnifying lens to appear at infinity, the object has to be at the focal point of the lens. In this case, Equations in 36.17 become

$$\theta_0 \approx \frac{h}{25} \quad \text{and} \quad \theta \approx \frac{h}{f}$$

and the magnification is

$$m = \frac{\theta}{\theta_0} = \frac{25 \text{ cm}}{f} \tag{36.19}$$

With a single lens, it is possible to obtain angular magnifications up to about 4 without serious aberrations. Magnifications up to about 20 can be achieved by using one or two additional lenses to correct for aberrations.

EXAMPLE 36.15 Maximum Magnification of a Lens

What is the maximum magnification of a lens having a focal length of 10 cm, and what is the magnification of this lens when the eye is relaxed?

Solution The maximum magnification occurs when the image is located at the near point of the eye. Under these circumstances, Equation 36.18 gives

$$m = 1 + \frac{25 \text{ cm}}{f} = 1 + \frac{25 \text{ cm}}{10 \text{ cm}} = 3.5$$

When the eye is relaxed, the image is at infinity. In this case, we use Equation 36.19:

$$m = \frac{25 \text{ cm}}{f} = \frac{25 \text{ cm}}{10 \text{ cm}} = 2.5$$

*36.9 THE COMPOUND MICROSCOPE

A simple magnifier provides only limited assistance in inspecting the minute details of an object. Greater magnification can be achieved by combining two lenses in a device called a compound microscope, a schematic diagram of which is shown in Figure 36.33a. It consists of an objective lens that has a very short focal length $f_o < 1$ cm, and an eyepiece lens having a focal length, f_e, of a few centimeters. The two lenses are separated by a distance L, where L is much greater than either f_o or f_e. The object, which is placed just to the left of the focal point of the objective,

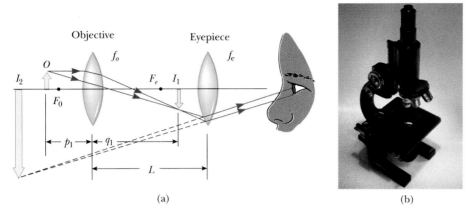

FIGURE 36.33 (a) Diagram of a compound microscope, which consists of an objective lens and an eyepiece lens. (b) A compound microscope. The three-objective turret allows the user to switch to several different powers of magnification. Combinations of eyepieces with different focal lengths and different objectives can produce a wide range of magnifications. *(Henry Leap and Jim Lehman)*

forms a real, inverted image at I_1, which is at or close to the focal point of the eyepiece. The eyepiece, which serves as a simple magnifier, produces at I_2 an image of the image at I_1, and this image at I_2 is virtual and inverted. The lateral magnification, M_1, of the first image is $-q_1/p_1$. Note from Figure 36.33a that q_1 is approximately equal to L, and recall that the object is very close to the focal point of the objective; thus, $p_1 \approx f_0$. This gives a magnification for the objective of

$$M_1 \approx -\frac{L}{f_0}$$

The angular magnification of the eyepiece for an object (corresponding to the image at I_1) placed at the focal point of the eyepiece is found from Equation 36.19 to be

$$m_e = \frac{25 \text{ cm}}{f_e}$$

The overall magnification of the compound microscope is defined as the product of the lateral and angular magnifications:

$$M = M_1 m_e = -\frac{L}{f_0}\left(\frac{25 \text{ cm}}{f_e}\right) \tag{36.20}$$

The negative sign indicates that the image is inverted.

The microscope has extended our vision to the previously unknown details of incredibly small objects. The capabilities of this instrument have steadily increased with improved techniques in the precision grinding of lenses. A question that is often asked about microscopes is, "If you were extremely patient and careful, would it be possible to construct a microscope that would enable you to see an atom?" The answer to this question is no, as long as light is used to illuminate the object. The reason is that, in order to be seen, the object under a microscope must be at least as large as a wavelength of light. An atom is many times smaller than the wavelengths of visible light, and so its mysteries have to be probed using other types of "microscopes."

The wavelength dependence of the "seeing" ability of a wave can be illustrated by water waves in a bathtub. Suppose you vibrate your hand in the water until waves having a wavelength of about 15 cm are moving along the surface. If you fix a small object, such as a toothpick, in the path of the waves, the waves are not disturbed appreciably by the toothpick but instead continue along their path, oblivious of the small object. Now suppose you fix a larger object, such as a toy sailboat, in the path of the waves. In this case, the waves are considerably "disturbed" by the object. In the first case, the toothpick is smaller than the wavelength of the waves, and as a result the waves do not "see" the toothpick. (The intensity of the scattered waves is low.) In the second case, the toy sailboat is about the same size as the wavelength of the waves and, hence, creates a disturbance. That is, the object acts as the source of scattered waves that appear to come from it. Light waves behave in this same general way. The ability of an optical microscope to view an object depends on the size of the object relative to the wavelength of the light used to observe it. Hence, we will never be able to observe atoms or molecules with a light microscope, because their dimensions are small (≈ 0.1 nm) relative to the wavelength of the light (≈ 500 nm).

*36.10 THE TELESCOPE

There are two fundamentally different types of **telescopes**, both designed to aid in viewing distant objects, such as the planets in our solar system. The **refracting telescope** uses a combination of lenses to form an image, and the **reflecting telescope** uses a curved mirror and a lens.

The telescope sketched in Figure 36.34a is a refracting telescope. The two lenses are arranged so that the objective forms a real, inverted image of the distant object very near the focal point of the eyepiece. Furthermore, the image at I_1 is formed at the focal point of the objective because the object is essentially at infinity. Hence, the two lenses are separated by a distance $f_o + f_e$, which corresponds to the length of the telescope's tube. The eyepiece finally forms, at I_2, an enlarged, inverted image of the image at I_1.

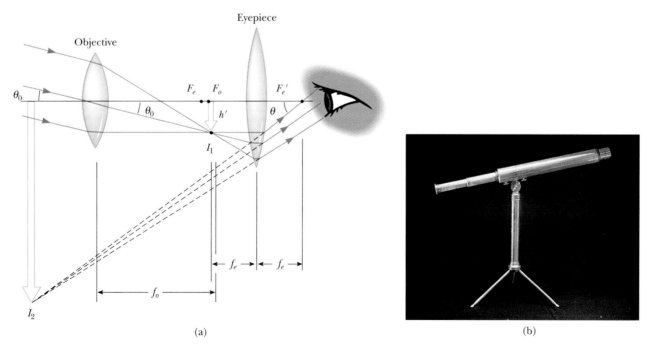

FIGURE 36.34 (a) Diagram of a refracting telescope, with the object at infinity. (b) Photograph of a refracting telescope. *(Henry Leap and Jim Lehman)*

Optical Instruments

Eyeglasses, mirrors, microscopes, and telescopes are examples of optical instruments without which our lives would be dramatically different. This simulator enables you to construct a multitude of optical instruments and investigate how their performance is affected when you modify parameters such as focal length, wavelength, and index of refraction. The simulations you create can include lenses, prisms, mirrors, and light sources of various wavelengths. Furthermore, the model you create will automatically show a new image as you move or modify any optical component. The adventuresome student might even invent a new optical instrument.

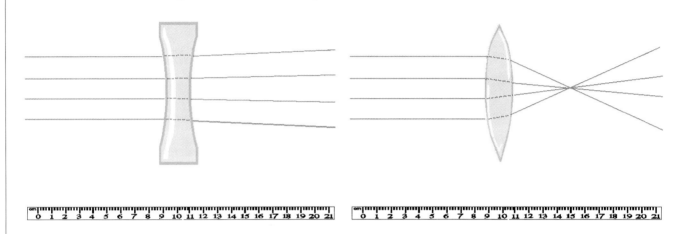

The angular magnification of the telescope is given by θ/θ_0, where θ_0 is the angle subtended by the object at the objective and θ is the angle subtended by the final image at the viewer's eye. From the triangles in Figure 36.34a, and for small angles, we have

$$\theta \approx -\frac{h'}{f_e} \quad \text{and} \quad \theta_0 \approx \frac{h'}{f_0}$$

Hence, the angular magnification of the telescope can be expressed as

$$m = \frac{\theta}{\theta_0} = \frac{-h'/f_e}{h'/f_0} = -\frac{f_0}{f_e} \qquad (36.21)$$

The minus sign indicates that the image is inverted. This expression says that the angular magnification of a telescope equals the ratio of the objective focal length to the eyepiece focal length. Here again, the magnification is the ratio of the angular size seen with the telescope to the angular size seen with the unaided eye.

In some applications, such as observing nearby objects like the Sun, Moon, or planets, magnification is important. However, stars are so far away that they always appear as small points of light regardless of how much magnification is used. Large research telescopes used to study very distant objects must have a large diameter in order to gather as much light as possible. It is difficult and expensive to manufacture large lenses for refracting telescopes. Another difficulty with large lenses is that their large weight leads to sagging, which is an additional source of aberration. These problems can be partially overcome by replacing the objective lens with a reflecting, concave mirror. Figure 36.35 shows the design for a typical reflecting

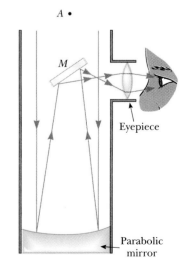

FIGURE 36.35 A reflecting telescope with a Newtonian focus.

telescope. Incoming light rays pass down the barrel of the telescope and are reflected by a parabolic mirror at the base. These rays converge toward point *A* in the figure, where an image would be formed. However, before this image is formed, a small flat mirror *M* reflects the light toward an opening in the side of the tube that passes into an eyepiece. This particular design is said to have a Newtonian focus because it was Newton who developed it. Note that the light never passes through glass in the reflecting telescope (except through the small eyepiece). As a result, problems associated with chromatic aberration are virtually eliminated.

The largest telescope in the world is the 6-m-diameter reflecting telescope on Mount Pastukhov in the Caucasus, Soviet Union. The largest reflecting telescope in the United States is the 5-m-diameter instrument on Mount Palomar in California. In contrast, the largest refracting telescope in the world, which is located at the Yerkes Observatory in Williams Bay, Wisconsin, has a diameter of only 1 m.

SUMMARY

The **lateral magnification** M of a mirror or lens is defined as the ratio of the image height h' to the object height h:

$$M = \frac{h'}{h} \tag{36.1}$$

In the paraxial ray approximation, the object distance p and image distance q for a spherical mirror of radius R are related by the **mirror equation**

$$\frac{1}{p} + \frac{1}{q} = \frac{2}{R} = \frac{1}{f} \tag{36.4, 36.6}$$

where $f = R/2$ is the **focal length** of the mirror.

An image can be formed by refraction from a spherical surface of radius R. The object and image distances for refraction from such a surface are related by

$$\frac{n_1}{p} + \frac{n_2}{q} = \frac{n_2 - n_1}{R} \tag{36.8}$$

where the light is incident in the medium of index of refraction n_1 and is refracted in the medium whose index of refraction is n_2.

The inverse of the **focal length** f of a thin lens in air is

$$\frac{1}{f} = (n-1)\left(\frac{1}{R_1} - \frac{1}{R_2}\right) \tag{36.11}$$

Converging lenses have positive focal lengths, and **diverging lenses** have negative focal lengths.

For a thin lens, and in the paraxial ray approximation, the object and image distances are related by

$$\frac{1}{p} + \frac{1}{q} = \frac{1}{f} \tag{36.12}$$

QUESTIONS

1. When you look in a mirror, your image is reversed left to right but not top to bottom. Explain.
2. Using a simple ray diagram, as in Figure 36.2, show that a flat mirror whose top is at eye level need not be as long as your height for you to see your entire body.
3. Consider a concave spherical mirror with a real object. Is the image always inverted? Is the image always real? Give conditions for your answers.
4. Repeat the previous question for a convex spherical mirror.
5. Why does a clear stream always appear to be shallower than it actually is?
6. Consider the image formed by a thin converging lens. Under what conditions is the image (a) inverted, (b) upright, (c) real, (d) virtual, (e) larger than the object, and (f) smaller than the object?
7. Repeat Question 6 for a thin diverging lens.
8. If a cylinder of solid glass or clear plastic is placed above the words LEAD OXIDE and viewed from the side as shown in Figure 36.36, the LEAD appears inverted but the OXIDE does not. Explain.

FIGURE 36.36 (Question 8) *(Henry Leap and Jim Lehman)*

9. Describe two types of aberration common in a spherical lens.
10. Explain why a mirror cannot give rise to chromatic aberration.
11. What is the magnification of a flat mirror? What is its focal length?
12. Why do some emergency vehicles have the symbol ƎƆИAJUᗺMA written on the front?
13. Explain why a fish in a spherical goldfish bowl appears larger than it really is.
14. Lenses used in eyeglasses, whether converging or diverging, are always designed such that the middle of the lens curves away from the eye, like the center lenses of Figure 36.22a and 36.22b. Why?
15. A mirage is formed when the air gets gradually cooler as the height above the ground increases. What might happen if the air grows gradually warmer as the height is increased? This often happens over bodies of water or snow-covered ground: The effect is called *looming*.
16. Consider a spherical concave mirror, with the object located to the left of the mirror beyond the focal point. Using ray diagrams, show that the image moves to the left as the object approaches the focal point.
17. In a Jules Verne novel, a piece of ice is shaped to form a magnifying lens to focus sunlight to start a fire. Is this possible?
18. The *f*-number of a camera is the focal length of the lens divided by its aperture (or diameter). How can the *f*-number of the lens be changed? How does changing this number affect the required exposure time?
19. A solar furnace can be constructed by using a concave mirror to reflect and focus sunlight into a furnace enclosure. What factors in the design of the reflecting mirror would guarantee very high temperatures?
20. One method for determining the position of an image, either real or virtual, is by means of *parallax*. If a finger or other object is placed at the position of the image, as in Figure 36.37, and the finger and image are viewed simultaneously (the image through the lens if it is virtual), the finger and image have the same parallax; that is, if it is viewed from different positions, the image will appear to move along with the finger. Use this method to locate the image formed by a lens. Explain why the method works.

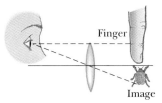

FIGURE 36.37 (Question 20)

FIGURE 36.38 (Question 21) *(M. C. Escher/Cordon Art-Baarn-Holland. All rights reserved.)*

21. Figure 36.38 shows a lithograph by M. C. Escher titled *Hand with Reflection Sphere (Self-Portrait in Spherical Mirror)*. Escher had this to say about the work: "The picture shows a spherical mirror, resting on a left hand. But as a print is the reverse of the original drawing on stone, it was my right hand that you see depicted. (Being left-handed, I needed my left hand to make the drawing.) Such a globe reflection collects almost one's whole surroundings in one disk-shaped image. The whole room, four walls, the floor, and the ceiling, everything, albeit distorted, is compressed into that one small circle. Your own head, or more exactly the point between your eyes, is the absolute center. No matter how you turn or twist yourself, you can't get out of that central point. You are immovably the focus, the unshakable core, of your world." Comment on the accuracy of Escher's description.

PROBLEMS

Review Problem

One of the surfaces of a lens is flat; the other one is convex and has a radius of curvature $R = 10.0$ cm. The speed of light in the lens material is $v = 1.50 \times 10^8$ m/s. Find (a) the index of refraction of the lens material, (b) the angle of total internal reflection, (c) the power of the lens, (d) the focal length of the lens, (e) the object distance if the magnification $M = 2.00$, (f) the image distance if an object is 1.00 m from the screen and an enlarged image appears on the screen, (g) the magnification M' in case (f), and (h) the angular magnification of the lens.

Section 36.1 Images Formed by Flat Mirrors

1. In a physics laboratory experiment, a torque is applied to a small-diameter wire suspended vertically under tensile stress. The small angle through which the wire turns as a consequence of the net torque is measured by attaching a small mirror to the wire and reflecting a beam of light off the mirror and onto a circular scale. Such an arrangement is known as an optical lever and is shown from a top view in Figure P36.1. Show that when the mirror turns through an angle θ, the reflected beam is rotated by an angle 2θ.

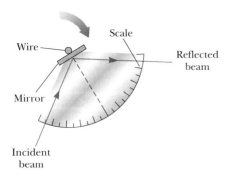

FIGURE P36.1

2. A light ray enters a small hole in a rectangular box that has three mirrored sides, as in Figure P36.2. (a) For what angle θ does the light ray emerge from the hole, assuming that it hits each mirror only once? (b) Show that your answer does not depend on the location of the hole.

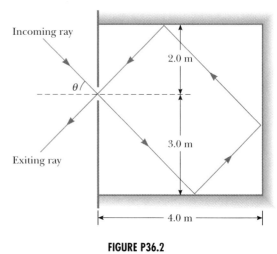

FIGURE P36.2

3. Determine the minimum height of a vertical flat mirror in which a person 5'10" in height can see his or her full image. (A ray diagram would be helpful.)

3A. Determine the minimum height of a vertical flat mirror in which a person of height h can see his or her full image. (A ray diagram would be helpful.)

4. Two flat mirrors have their reflecting surfaces facing one another, with the edge of one mirror in contact with an edge of the other, so that the angle between the mirrors is α. When an object is placed between the mirrors, a number of images are formed. In general, if the angle α is such that $n\alpha = 360°$, where n is an integer, the number of images formed is $n - 1$. Graphically, find all the image positions for

☐ indicates problems that have full solutions available in the Student Solutions Manual and Study Guide.

the case $n = 6$ when a point object is between the mirrors (but not on the angle bisector).

Section 36.2 Images Formed by Spherical Mirrors

5. A concave mirror has a focal length of 40.0 cm. Determine the object position for which the resulting image is upright and four times the size of the object.

6. A rectangle 10.0 cm × 20.0 cm is placed so that its right edge is 40.0 cm to the left of a concave spherical mirror, as in Figure P36.6. The radius of curvature of the mirror is 20.0 cm. (a) Draw the image seen through this mirror. (b) What is the area of the image?

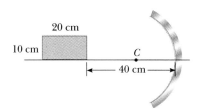

FIGURE P36.6

7. A concave mirror has a radius of curvature of 60 cm. Calculate the image position and magnification of an object placed in front of the mirror at distances of (a) 90 cm and (b) 20 cm. (c) Draw ray diagrams to obtain the image in each case.

8. The height the real image formed by a concave mirror is four times larger than the object height when the object is 30.0 cm in front of the mirror. (a) What is the radius of curvature of the mirror? (b) Use a ray diagram to locate this image.

9. A dedicated sports car enthusiast polishes the inside and outside surfaces of a hubcap that is a section of a sphere. When she looks into one side of the hubcap, she sees an image of her face 30.0 cm in back of the hubcap. She then turns the hubcap over and sees another image of her face 10.0 cm in back of the hubcap. (a) How far is her face from the hubcap? (b) What is the radius of curvature of the hubcap?

9A. A dedicated sports car enthusiast polishes the inside and outside surfaces of a hubcap that is a section of a sphere. When she looks into one side of the hubcap, she sees an image of her face a distance d in back of the hubcap. She then turns the hubcap over and sees another image of her face a distance $d/3$ in back of the hubcap. (a) How far is her face from the hubcap? (b) What is the radius of curvature of the hubcap?

10. A candle is 49 cm in front of a convex spherical mirror that has a radius of curvature of 70 cm. (a) Where is the image? (b) What is the magnification?

11. An object 2.0 cm high is placed 10 cm in front of a mirror. What type of mirror and what radius of curvature are needed for an upright image that is 4.0 cm high?

12. A ball is dropped from rest 3.00 m directly above the vertex of a concave mirror having a radius of 1.00 m and lying in a horizontal plane. (a) Describe the motion of the ball's image in the mirror. (b) At what time do the ball and its image coincide?

13. A spherical mirror is to be used to form, on a screen located 5.0 m from the object, an image five times the size of the object. (a) Describe the type of mirror required. (b) Where should the mirror be positioned relative to the object?

14. (a) A concave mirror forms an inverted image four times larger than the object. Find the focal length of the mirror if the distance between object and image is 0.60 m. (b) A convex mirror forms a virtual image half the size of the object. If the distance between image and object is 20.0 cm, determine the radius of curvature of the mirror.

15. A spherical convex mirror has a radius of 40.0 cm. Determine the position of the virtual image and magnification for object distances of (a) 30.0 cm and (b) 60.0 cm. (c) Are the images upright or inverted?

16. An object is 15 cm from the surface of a reflective spherical Christmas-tree ornament 6.0 cm in diameter. What are the magnification and position of the image?

Section 36.3 Images Formed by Refraction

17. A smooth block of ice ($n = 1.309$) rests on the floor with one face parallel to the floor. The block has a vertical thickness of 50.0 cm. Find the location of the image of a pattern in the floor covering as formed by rays that are nearly perpendicular to the block.

18. One end of a long glass rod ($n = 1.50$) is formed into a convex surface of radius 6.0 cm. An object is located in air along the axis of the rod. Find the image positions corresponding to object distances of (a) 20.0 cm, (b) 10.0 cm, and (c) 3.0 cm from the end of the rod.

19. A goldfish is swimming at 2.00 cm/s toward the right side of a rectangular tank. What is the apparent speed of the fish measured by an observer looking in from outside the right side of the tank? The index of refraction for water is 1.33.

19A. A goldfish is swimming at a speed v toward the right side of a rectangular tank. What is the apparent speed of the goldfish as measured by an observer looking in from outside the right side of the tank? The index of refraction for water is n.

20. A lens made with a material of refractive index n has a focal length f in air. When immersed in a liquid having a refractive index n_1, the lens has a focal

length f'. Derive an expression for f' in terms of f, n, and n_1.

21. A glass sphere ($n = 1.50$) of radius 15 cm has a tiny air bubble located 5.0 cm from the center. The sphere is viewed along a direction parallel to the radius containing the bubble. What is the apparent depth of the bubble below the surface of the sphere?

22. A flint glass plate ($n = 1.66$) rests on the bottom of an aquarium tank. The plate is 8.0 cm thick (vertical dimension) and covered with water ($n = 1.33$) to a depth of 12 cm. Calculate the apparent thickness of the plate as viewed from above the water. (Assume nearly normal incidence.)

23. A glass hemisphere is used as a paperweight with its flat face resting on a stack of papers. The radius of the circular cross-section is 4.0 cm, and the index of refraction of the glass is 1.55. The center of the hemisphere is directly over a letter "O" that is 2.5 mm in diameter. What is the diameter of the image of the letter as seen looking along a vertical radius?

24. A goldfish is swimming in water inside a spherical plastic bowl of index of refraction 1.33. If the goldfish is 10 cm from the wall of the 15-cm-radius bowl, where does the goldfish appear to an observer outside the bowl?

25. A transparent sphere of unknown composition is observed to form an image of the Sun on the surface of the sphere opposite the Sun. What is the refractive index of the sphere material?

Section 36.4 Thin Lenses

26. An object located 32 cm in front of a lens forms an image on a screen 8.0 cm behind the lens. (a) Find the focal length of the lens. (b) Determine the magnification. (c) Is the lens converging or diverging?

27. The left face of a biconvex lens has a radius of curvature of 12 cm, and the right face has a radius of curvature of 18 cm. The index of refraction of the glass is 1.44. (a) Calculate the focal length of the lens. (b) Calculate the focal length if the radii of curvature of the two faces are interchanged.

28. A microscope slide is placed in front of a converging lens that has a focal length of 2.44 cm. The lens forms an image of the slide 12.9 cm from the slide. How far is the lens from the slide if the image is (a) real and (b) virtual?

29. What is the image distance of an object 1.0 m in front of a converging lens of focal length 20 cm? What is the magnification of the object? (See Figure 36.23a for a ray diagram.)

30. A person looks at a gem with a jeweler's microscope—a converging lens that has a focal length of 12.5 cm. The microscope forms a virtual image 30.0 cm from the lens. (a) Determine the magnification. Is the image upright or inverted? (b) Construct a ray diagram for this arrangement.

31. Figure P36.31 shows a thin glass ($n = 1.50$) converging lens for which the radii of curvature are $R_1 = 15.0$ cm and $R_2 = -12.0$ cm. To the left of the lens is a cube having a face area of 100.0 cm². The base of the cube is on the axis of the lens, and the right face is 20.0 cm to the left of the lens. (a) Determine the focal length of the lens. (b) Draw the image of the square face formed by the lens. What type of geometric figure is this? (c) Determine the area of the image.

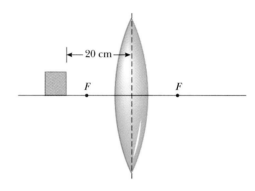

FIGURE P36.31

32. A converging lens has a focal length of 40 cm. Calculate the size of the real image of an object 4.0 cm in height for the following object distances: (a) 50 cm, (b) 60 cm, (c) 80 cm, (d) 100 cm, (e) 200 cm, (f) ∞.

33. An object is located 20 cm to the left of a diverging lens having a focal length $f = -32$ cm. Determine (a) the location and (b) the magnification of the image. (c) Construct a ray diagram for this arrangement.

34. An object is 5.00 m to the left of a flat screen. A converging lens for which the focal length is $f = 0.800$ m is placed between object and screen. (a) Show that there are two lens positions that form an image on the screen, and determine how far these positions are from the object. (b) How do the two images differ from each other?

34A. An object is at a distance d to the left of a flat screen. A converging lens having focal length $f < d$ is placed between object and screen. (a) Show that there are two lens positions that form an image on the screen, and determine how far these positions are from the object. (b) How do the two images differ from each other?

35. A slide 24.0 mm high is to be projected so that its image fills a screen 1.80 m high. The slide-screen distance is 3.00 m, as shown in Figure P36.35. (a) Determine the focal length of the lens in the projector. (b) How far from the slide should the lens of the projector be in order to form the image on the screen?

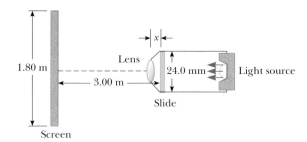

FIGURE P36.35

36. A diverging lens is used to form a virtual image of a real object. The object is positioned 80.0 cm to the left of the lens, and the image is located 40.0 cm to the left of the lens. (a) Determine the focal length of the lens. (b) If the surfaces of the lens have radii of curvature of $R_1 = -40.0$ cm and $R_2 = -50.0$ cm, what is its index of refraction?

37. The nickel's image in Figure P36.37 has twice the diameter of the nickel and is 2.84 cm from the lens. Determine the focal length of the lens.

FIGURE P36.37

*Section 36.6 The Camera and *Section 36.7 The Eye

38. A camera gives proper film exposure when set at $f/16$ and the shutter is open for $(1/32)$ s. Determine the correct exposure time if a setting of $f/8$ is used. (Assume the lighting conditions are unchanged.)

39. A camera is being used with correct exposure at $f/4$ and a shutter speed of $(1/16)$ s. In order to photograph a fast-moving subject, the shutter speed is changed to $(1/128)$ s. Find the new f-number setting needed to maintain satisfactory exposure.

40. A 1.7-m-tall woman stands 5.0 m in front of a camera equipped with a 50.0 mm focal length lens. What is the size of the image formed on the film?

41. A nearsighted woman cannot see objects clearly beyond 25 cm (the far point). If she has no astigmatism and contact lenses are prescribed, what are the power and type of lens required to correct her vision?

42. A camera that works properly with a 50.0-mm lens is to be reconfigured for underwater use. If the present double convex lens ($n = 1.50$, $R_1 = 10.0$ cm, $R_2 = -3.33$ cm) is to be retained, (a) what is its focal length under water and (b) at what distance from the lens should the film be placed when the camera is used under water?

43. If the aqueous humor of the eye has an index of refraction of 1.34 and the distance from the front of the cornea to the retina is 2.2 cm, what is the radius of curvature of the cornea for which distant objects are focused on the retina? (Assume all refraction occurs in the aqueous humor.)

44. A person sees clearly wearing eyeglasses that have a power of -4.0 diopters and sit 2.0 cm in front of the eyes. If the person wants to switch to contact lenses, which are placed directly on the eyes, what lens power should be prescribed?

45. A runner passes a spectator at 10.0 m/s. The spectator photographs the runner using a camera having a focal length of 5.00 cm and an exposure time of 20 ms. The maximum image blurring allowed in a good photograph is 0.500 mm. Find the minimum distance between runner and camera in order to obtain a good photograph.

45A. A runner passes a spectator at speed v. The spectator photographs the runner using a camera having a focal length f and an exposure time of Δt. The maximum image blurring allowed for a good photograph is Δh. Find the minimum distance between the runner and camera in order to obtain a good photograph.

46. The accommodation limits for nearsighted Nick's eyes are 18.0 cm and 80.0 cm. When he wears his glasses, he is able to see faraway objects clearly. At what minimum distance is he able to see objects clearly?

*Section 36.8 The Simple Magnifier, *Section 36.9 The Compound Microscope, and *Section 36.10 The Telescope

47. A philatelist examines the printing detail on a stamp using a convex lens of focal length 10.0 cm as a simple magnifier. The lens is held close to the eye, and the lens-to-object distance is adjusted so that the virtual image is formed at the normal near point (25 cm). Calculate the magnification.

48. A lens that has a focal length of 5.0 cm is used as a magnifying glass. (a) To obtain maximum magnification, where should the object be placed? (b) What is the magnification?

49. The Yerkes refracting telescope has a 1.0-m diameter objective lens of focal length 20 m and an eyepiece of focal length 2.5 cm. (a) Determine the magnification of the planet Mars as seen through this tele-

scope. (b) Are the martian polar caps right side up or upside down?

50. The Palomar reflecting telescope has a parabolic mirror that has an 80-m focal length. Determine the magnification achieved when an eyepiece of 2.5-cm focal length is used.

51. An astronomical telescope has an objective of focal length 75 cm and an eyepiece of focal length 4.0 cm. What is the magnifying power?

52. The distance between eyepiece and objective lens in a certain compound microscope is 23 cm. The focal length of the eyepiece is 2.5 cm, and that of the objective is 0.40 cm. What is the overall magnification of the microscope?

53. The desired overall magnification of a compound microscope is 140×. The objective alone produces a lateral magnification of 12×. Determine the required focal length of the eyepiece.

ADDITIONAL PROBLEMS

54. An object placed 10.0 cm from a concave spherical mirror produces a real image 8.0 cm from the mirror. If the object is moved to a new position 20.0 cm from the mirror, what is the position of the image? Is the latter image real or virtual?

55. Figure P36.55 shows a thin converging lens for which the radii are $R_1 = 9.00$ cm and $R_2 = -11.0$ cm. The lens is in front of a concave spherical mirror of radius $R = 8.00$ cm. (a) If its focal points F_1 and F_2 are 5.00 cm from the vertex of the lens, determine its index of refraction. (b) If the lens and mirror are 20.0 cm apart, and an object is placed 8.00 cm to the left of the lens, determine the position of the final image and its magnification as seen by the eye in the figure. (c) Is the final image inverted or upright? Explain.

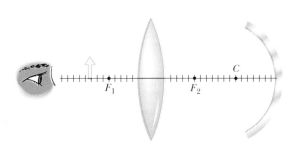

FIGURE P36.55

56. A compound microscope has an objective of focal length 0.300 cm and an eyepiece of focal length 2.50 cm. If an object is 3.40 mm from the objective, what is the magnification? (*Hint*: Use the lens equation for the objective.)

57. An object 2.00 cm tall is placed 40.0 cm to the left of a converging lens having a focal length of 30.0 cm. A diverging lens having a focal length of -20.0 cm is placed 110 cm to the right of the converging lens. (a) Determine the final position and magnification of the final image. (b) Is the image upright or inverted? (c) Repeat (a) and (b) for the case where the second lens is a converging lens having a focal length of $+20.0$ cm.

58. An object is placed 15.0 cm to the left of a converging lens ($f_1 = 10.0$ cm). A diverging lens ($f_2 = -20.0$ cm) is placed 15.0 cm to the right of the converging lens. Locate and describe the final image formed by the two lenses.

59. A parallel beam of light enters a glass hemisphere perpendicular to the flat face, as shown in Figure P36.59. The radius is $R = 6.00$ cm, and the index of refraction is $n = 1.560$. Determine the point at which the beam is focused. (Assume paraxial rays.)

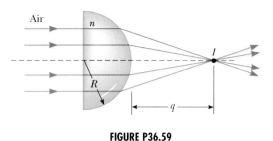

FIGURE P36.59

60. A thin lens having refractive index n is immersed in a liquid having index n'. Show that the focal length f of the lens is

$$\frac{1}{f} = \left(\frac{n}{n'} - 1\right)\left(\frac{1}{R_1} - \frac{1}{R_2}\right)$$

61. A thin lens of focal length 20.0 cm lies on a horizontal front-surfaced mirror. How far above the lens should an object be held if its image is to coincide with the object?

61A. A thin lens of focal length f lies on a horizontal front-surfaced mirror. How far above the lens should an object be held if its image is to coincide with the object?

62. Figure P36.62 shows a triangular enclosure whose inner walls are mirrors. A ray of light enters a small hole at the center of the short side. For each of the following, make a sketch showing the light path and find the angle θ for a ray that meets the stated conditions. (a) A light ray that is reflected once by each of the side mirrors and then exits through the hole. (b) A light ray that reflects only once and then exits. (c) Is there a path that reflects three times and then exits? If so, sketch the path and find θ. (d) A ray that reflects four times and then exits.

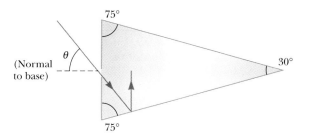

FIGURE P36.62

63. A cataract-impaired lens in an eye may be surgically removed and replaced by a manufactured lens. The focal length of the new lens is determined by the lens-to-retina distance, which is measured by a sonar-like device, and by the requirement that the implant provide for correct distant vision. (a) If the distance from lens to retina is 22.4 mm, calculate the power of the implanted lens in diopters. (b) Since there is no accommodation and the implant allows for correct distant vision, a corrective lens for close work or reading must be used. Assume a reading distance of 33 cm and calculate the power of the lens in the reading glasses.

64. The distance between an object and its upright image is 20.0 cm. If the magnification is 0.500, what is the focal length of the lens being used to form the image?

64A. The distance between an object and its upright image is d. If the magnification is M, what is the focal length of the lens being used to form the image?

65. An object is placed 12.0 cm to the left of a diverging lens of focal length -6.0 cm. A converging lens of focal length 12.0 cm is placed a distance d to the right of the diverging lens. Find the distance d so that the final image is at infinity. Draw a ray diagram for this case.

66. A converging lens has a focal length of 20.0 cm. Find the position of the image for object distances of (a) 50.0 cm, (b) 30.0 cm, and (c) 10.0 cm. (d) Determine the magnification of the lens for these object distances and whether the image is upright or inverted. (e) Draw ray diagrams for these object distances.

67. Two rays traveling parallel to the principal axis strike a plano-convex lens having a refractive index of 1.60 (Fig. P36.67). If the convex face is spherical, a ray near the edge does not strike the focal point (spherical aberration). If this face has a radius of curvature of 20.0 cm and the two rays are $h_1 = 0.500$ cm and $h_2 = 12.0$ cm from the principal axis, find the difference in the positions where each strikes the principal axis.

67A. Two rays traveling parallel to the principal axis strike a plano-convex lens having a refractive index n. If the convex face is spherical, a ray near the edge does not strike the focal point (spherical aberration). If this convex face has a radius of curvature R and the two rays are h_1 and h_2 from the principal axis, find the difference in the positions where each strikes the principal axis.

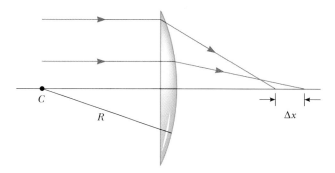

FIGURE P36.67

68. Consider light rays parallel to the principal axis approaching a concave spherical mirror. According to the mirror equation (Eq. 36.4), rays close to the principal axis focus at F, a distance $R/2$ from the mirror. What about a ray farther from the principal axis? Will it be reflected to the axis at a point closer or farther than F from the mirror? Illustrate with an accurately drawn light ray that obeys the law of reflection.

69. A colored marble is dropped into a large tank filled with benzene ($n = 1.50$). (a) What is the depth of the tank if the apparent depth of the marble when viewed from directly above the tank is 35 cm? (b) If the marble has a diameter of 1.5 cm, what is its apparent diameter when viewed from directly above the tank?

70. An object 1.0 cm in height is placed 4.0 cm to the left of a converging lens of focal length 8.0 cm. A diverging lens of focal length -16 cm is located 6.0 cm to the right of the converging lens. Find the position and size of the final image. Is it inverted or upright? Real or virtual?

71. The disk of the Sun subtends an angle of $0.50°$ at the Earth. What are the position and diameter of the solar image formed by a concave spherical mirror of radius 3.0 m?

72. A reflecting telescope with a 2.0-m focal length objective and a 10-cm focal length eyepiece is used to view the Moon. Calculate the size of the image formed at the viewer's near point, 25 cm from the eye. (The diameter of the Moon $= 3.5 \times 10^3$ km; the Earth-Moon distance $= 3.84 \times 10^5$ km.)

73. The cornea of an eye has a radius of curvature of 0.80 cm. (a) What is the focal length of the reflecting surface of the eye? (b) If a $20 gold piece 3.4 cm in diameter is held 25 cm from the cornea, what are the size and location of the reflected image?

74. Two converging lenses having focal lengths of 10.0 cm and 20.0 cm are located 50.0 cm apart as in Figure P36.74. The final image is to be located between the lenses at the position indicated. (a) How far to the left of the first lens should the object be? (b) What is the overall magnification? (c) Is the final image upright or inverted?

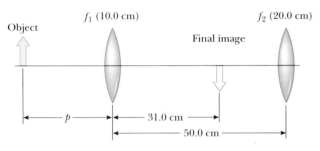

FIGURE P36.74

75. In a darkened room, a burning candle is placed 1.5 m from a white wall. A lens is placed between candle and wall at a location that causes a larger, inverted image to form on the wall. When the lens is moved 90 cm toward the wall, another image of the candle is formed. Find (a) the two object distances that produce the images stated above and (b) the focal length of the lens. (c) Characterize the second image.

76. A lens and mirror have focal lengths of +80 cm and −50 cm, respectively, and the lens is located 1.0 m to the left of the mirror. An object is placed 1.0 m to the left of the lens. Locate the final image. State whether the image is upright or inverted and determine the overall magnification.

77. A floating coin illusion consists of two parabolic mirrors, each having a focal length 7.5 cm, facing each other so that their centers are 7.5 cm apart (Fig. P36.77). If a few coins are placed on the lower mirror, an image of the coins is formed at the small opening at the center of the top mirror. Show that the final image is formed at that location and describe its characteristics. (*Note:* A very startling effect is to shine a flashlight beam on these images. Even at a glancing angle, the incoming light beam is seemingly reflected off the images! Do you understand why?)

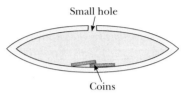

FIGURE P36.77

SPREADSHEET PROBLEM

S1. The equation for refraction by a spherical surface separating two media (Fig. 36.11) having indices of refraction n_1 and n_2 is

$$\frac{n_1}{p} + \frac{n_2}{q} = \frac{n_2 - n_1}{R}$$

This equation was derived using small-angle approximations. Spreadsheet 36.1 tests the range of validity of these approximations. The spreadsheet calculates the image distance for a series of incident angles α without making small-angle approximations. Input parameters are the indices of refraction, the radius of curvature of the surface, and the object distance. (a) Take $n_1 = 1.00$, $n_2 = 1.52$, $R = 10.0$ cm, and $p = 30.0$ cm. If the maximum acceptable error in the image distance is 3.00 percent, for what maximum angle α can the refraction equation be used? (b) Repeat part (a) using $p = 50.0$ cm. (c) Choose other values for the input parameters and repeat (a). (d) Is the error linearly proportional to α?

CHAPTER 37

Interference of Light Waves

A layer of bubbles on water, produced by soap film. The colors, produced just before the bubbles burst, are due to interference between light rays reflected from the front and back of the thin film of water making the bubble. The colors depend on the thickness of the film, ranging from black where the film is at its thinnest to red as the film gets thicker. *(Dr. Jeremy Burgess/ Science Photo Library)*

In the previous chapter on geometric optics, we used light rays to examine what happens when light passes through a lens or reflects from a mirror. The next two chapters are concerned with wave optics, which deals with the interference, diffraction, and polarization of light. These phenomena cannot be adequately explained with the ray optics of Chapter 36, but we describe how treating light as waves rather than as rays leads to a satisfying description of such phenomena.

37.1 CONDITIONS FOR INTERFERENCE

In Chapter 18, we found that two waves can add together constructively or destructively. In constructive interference, the amplitude of the resultant wave is greater than that of either of the individual waves. Light waves also interfere with each other. Fundamentally, all interference associated with light waves arises when the electromagnetic fields that constitute the individual waves combine.

1092 CHAPTER 37 *Interference of Light Waves*

To observe sustained interference in light waves, the following conditions must be met:

Conditions for interference

- The sources must be **coherent**; that is, they must maintain a constant phase with respect to each other.
- The sources must be **monochromatic,** that is, of a single wavelength.
- The superposition principle must apply.[1]

We now describe the characteristics of coherent sources. As we saw when we studied mechanical waves, two sources (producing two traveling waves) are needed to create interference. In order to produce a stable interference pattern, *the individual waves must maintain a constant phase relationship with one another.* When this is the case, the sources are said to be coherent. As an example, the sound waves emitted by two side-by-side loudspeakers driven by a single amplifier can interfere with each other because the two speakers are coherent—they respond to the amplifier in the same way at the same time.

If two light sources are placed side by side, no interference effects are observed because the light waves from one source are emitted independently of the other source; hence, the emissions from the two sources do not maintain a constant phase relationship with each other over the time of observation. Light from an ordinary light source undergoes such random changes about once every 10^{-8} s. Therefore, the conditions for constructive interference, destructive interference, or some intermediate state last for times of the order of 10^{-8} s. The result is that no interference effects are observed since the eye cannot follow such short-term changes. Such light sources are said to be **incoherent.**

A common method for producing two coherent light sources is to use one monochromatic source to illuminate a screen containing two small openings (usually in the shape of slits). The light emerging from the two slits is coherent because a single source produces the original light beam and the two slits serve only to separate the original beam into two parts (which, after all, is what was done to the sound signal from the side-by-side loudspeakers). Any random change in the light emitted by the source will occur in both beams at the same time, and as a result interference effects can be observed.

37.2 YOUNG'S DOUBLE-SLIT EXPERIMENT

Interference in light waves from two sources was first demonstrated by Thomas Young in 1801. A schematic diagram of the apparatus used in this experiment is shown in Figure 37.1a. Light is incident on a screen, in which there is a narrow slit S_0. The waves emerging from this slit arrive at a second screen, which contains two narrow, parallel slits, S_1 and S_2. These two slits serve as a pair of coherent light sources because waves emerging from them originate from the same wave front and therefore maintain a constant phase relationship. The light from the two slits produces on screen C a visible pattern of bright and dark parallel bands called **fringes** (Fig. 37.1b). When the light from S_1 and that from S_2 both arrive at a point on screen C such that constructive interference occurs at that location, a bright line appears. When the light from the two slits combines destructively at any loca-

---

[1] Later we will find that light consists of photons, and it would violate conservation of energy for one photon to annihilate another. The conditions specified above reflect the more fundamental condition that a photon can only interfere with itself.

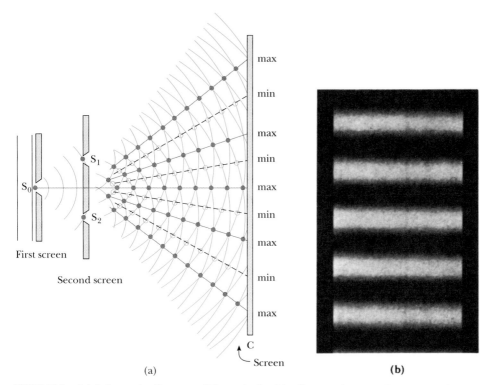

FIGURE 37.1 (a) Schematic diagram of Young's double-slit experiment. The narrow slits act as wave sources. Slits S_1 and S_2 behave as coherent sources that produce an interference pattern on screen C. (Note that this drawing is not to scale.) (b) The fringe pattern formed on screen C could look like this.

tion on the screen, a dark line results. Figure 37.2 is a photograph of an interference pattern produced by two coherent vibrating sources in a water tank.

Figure 37.3 is a schematic diagram of some of the ways the two waves can combine at the screen. In Figure 37.3a, the two waves, which leave the two slits in phase, strike the screen at the central point P. Since these waves travel an equal distance, they arrive at P in phase, and as a result constructive interference occurs at this location and a bright area is observed. In Figure 37.3b, the two light waves again start in phase, but now the upper wave has to travel one wavelength farther than the lower wave to reach point Q on the screen. Since the upper wave falls

FIGURE 37.2 An interference pattern involving water waves is produced by two vibrating sources at the water's surface. The pattern is analogous to that observed in Young's double-slit experiment. Note the regions of constructive and destructive interference. *(Richard Megna, Fundamental Photographs)*

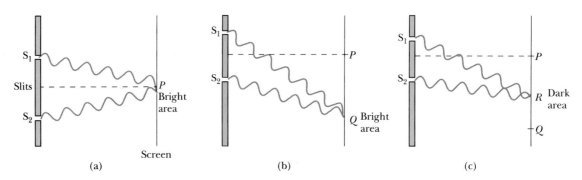

FIGURE 37.3 (a) Constructive interference occurs at *P* when the waves combine. (b) Constructive interference also occurs at *Q*. (c) Destructive interference occurs at *R* because the wave from the upper slit falls half a wavelength behind the wave from the lower slit. (Note that these figures are not drawn to scale.)

behind the lower one by exactly one wavelength, they still arrive in phase at *Q*, and so a second bright light appears at this location. Now consider point *R*, midway between *P* and *Q* in Figure 37.3c. At this location, the upper wave has fallen half a wavelength behind the lower wave. This means that the trough from the bottom wave overlaps the crest from the upper wave, giving rise to destructive interference at *R*. For this reason, a dark region is observed at this location.

We can describe Young's experiment quantitatively with the help of Figure 37.4. The screen is located a perpendicular distance *L* from the screen containing slits S_1 and S_2, which are separated by a distance *d* and the source is monochromatic. Under these conditions, the waves emerging from S_1 and S_2 have the same

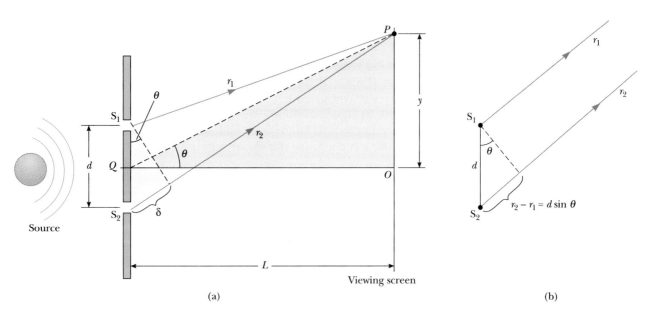

FIGURE 37.4 (a) Geometric construction for describing Young's double-slit experiment. (Note that this figure is not drawn to scale.) (b) When we use the approximation r_1 parallel to r_2, the path difference between the two rays is $r_2 - r_1 = d \sin \theta$. For approximation to be valid, it is essential that $L \gg d$.

frequency and amplitude and are in phase. The light intensity on the screen at any arbitrary point P is the resultant of the light coming from both slits. Note that, in order to reach P, a wave from the lower slit travels farther than a wave from the upper slit by a distance equal to $d \sin \theta$. This distance is called the **path difference**, δ, where

$$\delta = r_2 - r_1 = d \sin \theta \tag{37.1}$$

Path difference

This equation assumes that r_1 and r_2 are parallel, which is approximately true because L is much greater than d. The value of this path difference determines whether or not the two waves are in phase when they arrive at P. If the path difference is either zero or some integral multiple of the wavelength, the two waves are in phase at P and constructive interference results. Therefore, the condition for bright fringes, or **constructive interference**, at P is

$$\delta = d \sin \theta = m\lambda \qquad (m = 0, \pm 1, \pm 2, \ldots) \tag{37.2}$$

Conditions for constructive interference

The number m is called the **order number**. The central bright fringe at $\theta = 0$ ($m = 0$) is called the zeroth-order maximum. The first maximum on either side, when $m = \pm 1$, is called the first-order maximum, and so forth.

When the path difference is an odd multiple of $\lambda/2$, the two waves arriving at P are $180°$ out of phase and will give rise to destructive interference. Therefore, the condition for dark fringes, or **destructive interference**, at P is

$$\delta = d \sin \theta = (m + \tfrac{1}{2})\lambda \qquad (m = 0, \pm 1, \pm 2, \ldots) \tag{37.3}$$

Conditions for destructive interference

It is useful to obtain expressions for the positions of the bright and dark fringes measured vertically from O to P. In addition to our assumption that $L \gg d$, we assume $d \gg \lambda$. This situation prevails in practice because L is often of the order of 1 m while d is a fraction of a millimeter and λ is a fraction of a micrometer for visible light. Under these conditions, θ is small, and so we can use the approximation $\sin \theta \approx \tan \theta$. From the triangle OPQ in Figure 37.4, we see that

$$\sin \theta \approx \tan \theta = \frac{y}{L} \tag{37.4}$$

Using this result together with Equation 37.2, we see that the positions of the bright fringes measured from O are given by

$$y_{\text{bright}} = \frac{\lambda L}{d} m \tag{37.5}$$

Similarly, using Equations 37.3 and 37.4, we find that the dark fringes are located at

$$y_{\text{dark}} = \frac{\lambda L}{d}(m + \tfrac{1}{2}) \tag{37.6}$$

As we demonstrate in Example 37.1, Young's double-slit experiment provides a method for measuring the wavelength of light. In fact, Young used this technique to do just that. Additionally, the experiment gave the wave model of light a great deal of credibility.

EXAMPLE 37.1 Measuring the Wavelength of a Light Source

A viewing screen is separated from a double-slit source by 1.2 m. The distance between the two slits is 0.030 mm. The second-order bright fringe ($m = 2$) is 4.5 cm from the center line. (a) Determine the wavelength of the light.

Solution We can use Equation 37.5, with $m = 2$, $y_2 = 4.5 \times 10^{-2}$ m, $L = 1.2$ m, and $d = 3.0 \times 10^{-5}$ m:

$$\lambda = \frac{dy_2}{mL} = \frac{(3.0 \times 10^{-5} \text{ m})(4.5 \times 10^{-2} \text{ m})}{2 \times 1.2 \text{ m}}$$

$$= 5.6 \times 10^{-7} \text{ m} = \boxed{560 \text{ nm}}$$

(b) Calculate the distance between adjacent bright fringes.

Solution From Equation 37.5 and the results to part (a), we get

$$y_{m+1} - y_m = \frac{\lambda L(m+1)}{d} - \frac{\lambda L m}{d}$$

$$= \frac{\lambda L}{d} = \frac{(5.6 \times 10^{-7} \text{ m})(1.2 \text{ m})}{3.0 \times 10^{-5} \text{ m}}$$

$$= 2.2 \times 10^{-2} \text{ m} = \boxed{2.2 \text{ cm}}$$

EXAMPLE 37.2 The Distance Between Bright Fringes

A light source emits visible light of two wavelengths: $\lambda = 430$ nm and $\lambda' = 510$ nm. The source is used in a double-slit interference experiment in which $L = 1.5$ m and $d = 0.025$ mm. Find the separation between the third-order bright fringes.

Solution Using Equation 37.5 with $m = 3$, we find that the values of the fringe positions corresponding to these two wavelengths are

$$y_3 = \frac{\lambda L}{d} m = 3 \frac{\lambda L}{d} = 7.74 \times 10^{-2} \text{ m}$$

$$y_3' = \frac{\lambda' L}{d} m = 3 \frac{\lambda' L}{d} = 9.18 \times 10^{-2} \text{ m}$$

Hence, the separation between the two fringes is

$$\Delta y = y_3' - y_3 = \frac{3(\lambda' - \lambda)}{d} L$$

$$= 1.4 \times 10^{-2} \text{ m} = \boxed{1.4 \text{ cm}}$$

37.3 INTENSITY DISTRIBUTION OF THE DOUBLE-SLIT INTERFERENCE PATTERN

We now calculate the distribution of light intensity associated with the double-slit interference pattern. Again, suppose the two slits represent coherent sources of sinusoidal waves. Hence, the waves have the same angular frequency ω and a constant phase difference ϕ. The total electric field intensity at point P on the screen in Figure 37.5 is the vector superposition of the two waves. Assuming the two waves have the same amplitude E_0, we can write the electric field intensities at P due to each wave separately as

$$E_1 = E_0 \sin \omega t \quad \text{and} \quad E_2 = E_0 \sin(\omega t + \phi) \quad (37.7)$$

Although the waves are in phase at the slits, *their phase difference ϕ at P depends on the path difference* $\delta = r_2 - r_1 = d \sin \theta$. Because a path difference of λ (constructive interference) corresponds to a phase difference of 2π rad, while a path difference of $\lambda/2$ (destructive interference) corresponds to a phase difference of π rad, we obtain the ratio

$$\frac{\delta}{\phi} = \frac{\lambda}{2\pi}$$

$$\phi = \frac{2\pi}{\lambda} \delta = \frac{2\pi}{\lambda} d \sin \theta \quad (37.8)$$

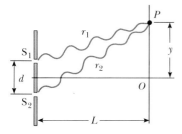

FIGURE 37.5 Construction for analyzing the double-slit interference pattern. A bright region, or intensity maximum, is observed at O.

This equation tells us precisely how the phase difference ϕ depends on the angle θ in Figure 37.4.

Using the superposition principle and Equation 37.7, we can obtain the resultant electric field at P:

$$E_P = E_1 + E_2 = E_0[\sin \omega t + \sin(\omega t + \phi)] \qquad (37.9)$$

To simplify this expression, we use the trigonometric identity

$$\sin A + \sin B = 2 \sin\left(\frac{A+B}{2}\right) \cos\left(\frac{A-B}{2}\right)$$

Taking $A = \omega t + \phi$ and $B = \omega t$, we can write Equation 37.9 in the form

$$E_P = 2E_0 \cos\left(\frac{\phi}{2}\right) \sin\left(\omega t + \frac{\phi}{2}\right) \qquad (37.10)$$

Hence, the electric field at P has the same frequency ω as the light at the slits, but its amplitude is multiplied by the factor $2 \cos(\phi/2)$. To check the consistency of this result, note that if $\phi = 0, 2\pi, 4\pi, \ldots$, the amplitude at P is $2E_0$, corresponding to the condition for constructive interference. Referring to Equation 37.8, we find that our result is consistent with Equation 37.2. Likewise, if $\phi = \pi, 3\pi, 5\pi, \ldots$, the amplitude at P is zero, which is consistent with Equation 37.3 for destructive interference.

Finally, to obtain an expression for the light intensity at P, recall that *the intensity of a wave is proportional to the square of the resultant electric field at that point* (Section 34.3). Using Equation 37.10, we can therefore express the intensity at P as

$$I \propto E_P^2 = 4E_0^2 \cos^2(\phi/2) \sin^2\left(\omega t + \frac{\phi}{2}\right)$$

Since most light-detecting instruments measure the time average light intensity and the time average value of $\sin^2(\omega t + \phi/2)$ over one cycle is $1/2$, we can write the average intensity at P as

$$I_{av} = I_0 \cos^2(\phi/2) \qquad (37.11)$$

where I_0 is the maximum possible time average light intensity. [You should note that $I_0 \propto (E_0 + E_0)^2 = (2E_0)^2 = 4E_0^2$.] Substituting Equation 37.8 into Equation 37.11, we find that

$$I_{av} = I_0 \cos^2\left(\frac{\pi d \sin \theta}{\lambda}\right) \qquad (37.12)$$

Alternatively, since $\sin \theta \approx y/L$ for small values of θ, we can write Equation 37.12 in the form

$$I_{av} = I_0 \cos^2\left(\frac{\pi d}{\lambda L} y\right) \qquad (37.13)$$

Constructive interference, which produces intensity maxima, occurs when the quantity $\pi y d/\lambda L$ is an integral multiple of π, corresponding to $y = (\lambda L/d)m$. This is consistent with Equation 37.5.

A plot of intensity distribution versus $d \sin \theta$ is given in Figure 37.6. Note that the interference pattern consists of equally spaced fringes of equal intensity. However, the result is valid only if the slit-to-screen distance L is large relative to the slit separation, and only for small values of θ.

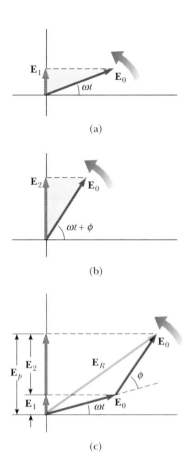

FIGURE 37.7 (a) Phasor diagram for the wave disturbance $E_1 = E_0 \sin \omega t$. The phasor is a vector of length E_0 rotating counterclockwise. (b) Phasor diagram for the wave $E_2 = E_0 \sin(\omega t + \phi)$. (c) E_R is the resultant phasor formed from the individual phasors shown in parts (a) and (b).

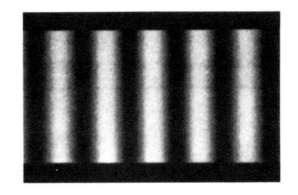

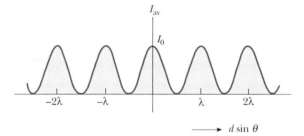

FIGURE 37.6 Intensity distribution versus $d \sin \theta$ or the double-slit pattern when the screen is far from the two slits ($L \gg d$). (Photo from M. Cagnet, M. Francon, and J. C. Thierr, Atlas of Optical Phenomena, Berlin, Springer-Verlag, 1962)

We have seen that the interference phenomena arising from two sources depend on the relative phase of the waves at a given point. Furthermore, the phase difference at a given point depends on the path difference between the two waves. The *resultant intensity at a point is proportional to the square of the resultant amplitude*. That is, the intensity is proportional to $(E_1 + E_2)^2$. It would be incorrect to calculate the resultant intensity by adding the intensities of the individual waves. This procedure would give $E_1^2 + E_2^2$, which of course is not the same thing as $(E_1 + E_2)^2$. Finally, $(E_1 + E_2)^2$ has the same average value as $E_1^2 + E_2^2$ when the time average is taken over all values of the phase difference between E_1 and E_2. Hence, there is no violation of energy conservation.

37.4 PHASOR ADDITION OF WAVES

In the previous section we combined two waves algebraically to obtain the resultant wave amplitude at some point on a screen. Unfortunately, this analytical procedure becomes cumbersome when several wave amplitudes have to be added. Since we shall eventually be interested in combining a large number of waves, we now describe a graphical procedure for this purpose.

Again, consider a sinusoidal wave whose electric field component is given by

$$E_1 = E_0 \sin \omega t$$

where E_0 is the wave amplitude and ω is the angular frequency. This wave can be represented graphically by a vector of magnitude E_0, rotating about the origin counterclockwise with an angular frequency ω, as in Figure 37.7a. Note that the phasor makes an angle ωt with the horizontal axis. The projection of the phasor on the vertical axis represents E_1, the magnitude of the wave disturbance at some time t. Hence, as the phasor rotates in a circle, the projection E_1 oscillates along the vertical axis about the origin.

Now consider a second sinusoidal wave whose electric field is given by

$$E_2 = E_0 \sin(\omega t + \phi)$$

That is, this wave has the same amplitude and frequency as E_1, but its phase is ϕ with respect to E_1. The phasor representing the wave E_2 is shown in Figure 37.7b. The resultant wave, which is the sum of E_1 and E_2, can be obtained graphically by redrawing the phasors end to end, as in Figure 37.7c, where the tail of the second phasor is placed at the tip of the first. As with vector addition, the resultant phasor $\mathbf{E}_R$ runs from the tail of the first phasor to the tip of the second. Furthermore, $\mathbf{E}_R$ rotates along with the two individual phasors at the same angular frequency ω. The projection of $\mathbf{E}_R$ along the vertical axis equals the sum of the projections of the two phasors: $E_P = E_1 + E_2$.

It is convenient to construct the phasors at $t = 0$ as in Figure 37.8. From the geometry of the triangle, we see that

$$E_R = E_0 \cos \alpha + E_0 \cos \alpha = 2E_0 \cos \alpha$$

Because the sum of the two opposite interior angles equals the exterior angle ϕ, we see that $\alpha = \phi/2$, so that

$$E_R = 2E_0 \cos(\phi/2)$$

Hence, the projection of the phasor $\mathbf{E}_R$ along the vertical axis at any time t is

$$E_P = E_R \sin\left(\omega t + \frac{\phi}{2}\right) = 2E_0 \cos\left(\frac{\phi}{2}\right) \sin\left(\omega t + \frac{\phi}{2}\right)$$

This is consistent with the result obtained algebraically, Equation 37.10. The resultant phasor has an amplitude $2E_0 \cos(\phi/2)$ and makes an angle $\phi/2$ with the first phasor. Furthermore, the average intensity at P, which varies as E_P^2, is proportional to $\cos^2(\phi/2)$, as described in Equation 37.11.

We can now describe how to obtain the resultant of several waves that have the same frequency:

- Draw the phasors representing each wave end to end, as in Figure 37.9, remembering to maintain the proper phase relationship between waves.
- The resultant represented by the phasor $\mathbf{E}_R$ is the vector sum of the individual phasors. At each instant, the projection of $\mathbf{E}_R$ along the vertical axis represents the time variation of the resultant wave. The phase angle α of the resultant wave is the angle between $\mathbf{E}_R$ and the first phasor. From the construction in Figure 37.9, drawn for four phasors, we see that the phasor of the resultant wave is given by $E_P = E_R \sin(\omega t + \alpha)$.

Phasor Diagrams for Two Coherent Sources

As an example of the phasor method, consider the interference pattern produced by two coherent sources. Figure 37.10 represents the phasor diagrams for various values of the phase difference ϕ and the corresponding values of the path difference δ, which are obtained using Equation 37.8. The intensity at a point is a maximum when $\mathbf{E}_R$ is a maximum. This occurs at $\phi = 0, 2\pi, 4\pi, \ldots$. Likewise, the intensity at some observation point is zero when $\mathbf{E}_R$ is zero. The first zero-intensity point occurs at $\phi = 180°$, corresponding to $\delta = \lambda/2$, while the other zero points (not shown) occur at $\delta = 3\lambda/2, 5\lambda/2, \ldots$. These results are in complete agreement with the analytical procedure described in the previous section.

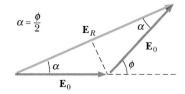

FIGURE 37.8 A reconstruction of the resultant phasor $\mathbf{E}_R$. From the geometry, note that $\alpha = \phi/2$.

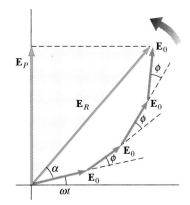

FIGURE 37.9 The phasor $\mathbf{E}_R$ is the resultant of four phasors of equal amplitude E_0. The phase of $\mathbf{E}_R$ with respect to the first phasor is α.

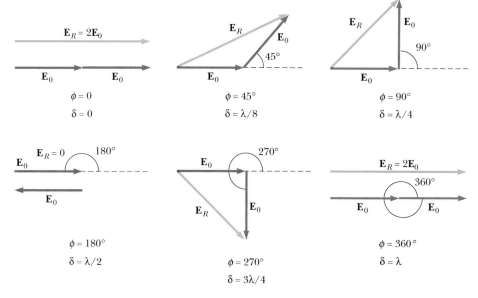

FIGURE 37.10 Phasor diagrams for the double-slit interference pattern. The resultant phasor E_R is a maximum when $\phi = 0, 2\pi, 4\pi, \ldots$ and is zero when $\phi = \pi, 3\pi, 5\pi, \ldots$.

Three-Slit Interference Pattern

Using phasor diagrams, let us analyze the interference pattern caused by three equally spaced slits. The electric fields at a point P on the screen caused by waves from the individual slits can be expressed as

$$E_1 = E_0 \sin \omega t$$
$$E_2 = E_0 \sin(\omega t + \phi)$$
$$E_3 = E_0 \sin(\omega t + 2\phi)$$

where ϕ is the phase difference between waves from adjacent slits. Hence, the resultant field at P can be obtained by using the phasor diagram shown in Figure 37.11.

The phasor diagrams for various values of ϕ are shown in Figure 37.12. Note that the resultant amplitude at P has a maximum value of $3E_0$ (called the primary

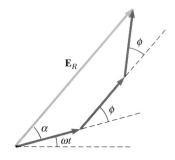

FIGURE 37.11 Phasor diagram for three equally spaced slits.

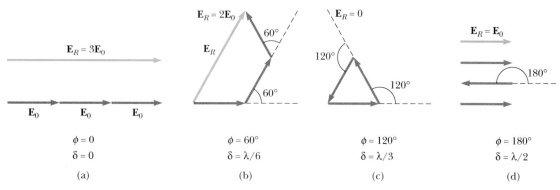

FIGURE 37.12 Phasor diagrams for three equally spaced slits at various values of ϕ. Note that there are primary maxima of amplitude $3E_0$ and secondary maxima of amplitude E_0.

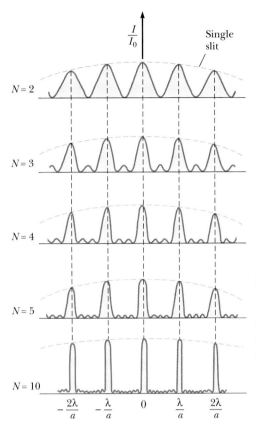

FIGURE 37.13 Multiple-slit interference patterns. As the number of slits is increased, the primary maxima (the most intense bands) become narrower but remain fixed in position and the number of secondary maxima increases. For any value of N, the decrease in intensity in maxima to the left and right of the central maximum, indicated by the dashed lines, is due to diffraction, which is discussed in Chapter 38.

maximum) when $\phi = 0, \pm 2\pi, \pm 4\pi, \ldots$. This corresponds to the case where the three phasors are aligned as in Figure 37.12a. However, we also find that secondary maxima of amplitude E_0 occur between the primary maxima when $\phi = \pm \pi, \pm 3\pi,$ $\ldots$. For these points, the wave from one slit exactly cancels that from another slit (Fig. 37.12d), which results in a total amplitude of E_0. Total destructive interference occurs whenever the three phasors form a closed triangle as in Figure 37.12c. These points where $E_0 = 0$ correspond to $\phi = \pm 2\pi/3, \pm 4\pi/3, \ldots$. You should be able to construct other phasor diagrams for values of ϕ greater than π.

Figure 37.13 shows multiple-slit interference patterns for a number of configurations. These patterns represent plots of the intensity for the various primary and secondary maxima. For three slits, note that the primary maxima are nine times more intense than the secondary maxima. This is because the intensity varies as $E_R{}^2$. Figure 37.13 also shows that as the number of slits increases, the number of secondary maxima also increases. In fact, the number of secondary maxima is always equal to $N - 2$, where N is the number of slits. Finally, as the number of slits increases, the primary maxima increase in intensity and become narrower, while the secondary maxima decrease in intensity relative to the primary maxima.

37.5 CHANGE OF PHASE DUE TO REFLECTION

Young's method for producing two coherent light sources involves illuminating a pair of slits with a single source. Another simple, yet ingenious, arrangement for producing an interference pattern with a single light source is known as Lloyd's mirror (Fig. 37.14). A light source is placed at S close to a mirror and a viewing

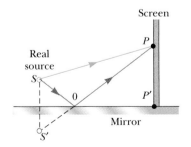

FIGURE 37.14 Lloyd's mirror. An interference pattern is produced on a screen at P as a result of the combination of the direct ray (blue) and the reflected ray (red). The reflected ray undergoes a phase change of 180°.

screen is placed at right angles to the mirror. Waves can reach P on the screen either by the direct path SP or by the path involving reflection from the mirror. The reflected ray can be treated as a ray originating from a virtual source at S'. Hence, at observation points far from the source, we would expect an interference pattern due to waves from S and S' just as is observed for two real coherent sources. An interference pattern is indeed observed. However, the positions of the dark and bright fringes are reversed relative to the pattern of two real coherent sources (Young's experiment). This is because the coherent sources at S and S' differ in phase by 180°, a phase change produced by reflection.

To illustrate this further, consider the point P', where mirror meets screen. This point is equidistant from S and S'. If path difference alone were responsible for the phase difference, we would expect to see a bright fringe at P' (since the path difference is zero for this point), corresponding to the central fringe of the two-slit interference pattern. Instead, we observe a dark fringe at P' because of the 180° phase change produced by reflection. In general,

> an electromagnetic wave undergoes a phase change of 180° upon reflection from a medium of higher index of refraction than the one in which the wave is traveling.

It is useful to draw an analogy between reflected light waves and the reflections of a transverse wave on a stretched string when the wave meets a boundary (Section

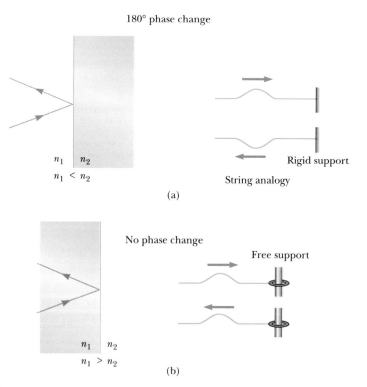

FIGURE 37.15 (a) A ray traveling in medium 1 reflecting from the surface of medium 2 undergoes a 180° phase change. The right side shows the analogy with a reflected pulse on a string fixed at one end. (b) A ray traveling in medium 1 reflecting from the surface of medium 2 with $n_1 > n_2$ undergoes no phase change. The right side shows the analogy with a reflected pulse on a string whose end is free.

16.6). The reflected pulse on a string undergoes a phase change of 180° when reflected from the boundary of a medium of higher index of refraction, but no phase change when reflected from the boundary of a medium of lower index of refraction. Similarly, an electromagnetic wave undergoes a 180° phase change when reflected from a boundary leading to a medium of higher index of refraction, but no phase change when reflected from a boundary leading to a medium of lower index of refraction. In either case, the part of the wave that crosses the boundary undergoes no phase change. These rules, summarized in Figure 37.15, can be deduced from Maxwell's equations, but the treatment is beyond the scope of this text.

37.6 INTERFERENCE IN THIN FILMS

Interference effects are commonly observed in thin films, such as thin layers of oil on water and soap bubbles. The varied colors observed when white light is incident on such films result from the interference of waves reflected from the two surfaces of the film.

Consider a film of uniform thickness t and index of refraction n, as in Figure 37.16. Let us assume that the light rays traveling in air are nearly normal to the two surfaces of the film. To determine whether the reflected rays interfere constructively or destructively, first note the following facts:

- A wave traveling from a medium of index of refraction n_1 toward a medium of index of refraction n_2 undergoes a 180° phase change upon reflection when $n_2 > n_1$. There is no phase change in the reflected wave if $n_2 < n_1$.
- The wavelength of light λ_n in a medium whose refraction index is n (Section 35.4) is

$$\lambda_n = \frac{\lambda}{n} \tag{37.14}$$

where λ is the wavelength of light in free space.

Let us apply these rules to the film of Figure 37.16. Ray 1, which is reflected from the upper surface (A), undergoes a phase change of 180° with respect to the incident wave, and ray 2, which is reflected from the lower surface (B), undergoes no phase change. Therefore, ray 1 is 180° out of phase with ray 2, which is equivalent to a path difference of $\lambda_n/2$. However, we must also consider that ray 2 travels an extra distance $2t$ before the waves recombine. For example, if $2t = \lambda_n/2$, rays 1 and 2 recombine in phase and the result is constructive interference. In general, the condition for constructive interference is

$$2t = (m + \tfrac{1}{2})\lambda_n \quad (m = 0, 1, 2, \ldots) \tag{37.15}$$

This condition takes into account two factors: (a) the difference in optical path length for the two rays (the term $m\lambda_n$) and (b) the 180° phase change upon reflection (the term $\lambda_n/2$). Because $\lambda_n = \lambda/n$, we can write Equation 37.15 as

$$2nt = (m + \tfrac{1}{2})\lambda \quad (m = 0, 1, 2, \ldots) \tag{37.16}$$

If the extra distance $2t$ traveled by ray 2 corresponds to a multiple of λ_n, the two waves combine out of phase and the result is destructive interference. The general

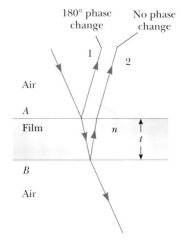

FIGURE 37.16 Interference in light reflected from a thin film is due to a combination of rays reflected from the upper and lower surfaces of the film.

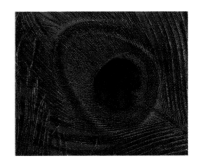

The brilliant colors in a peacock's feathers are due to interference, rather than absorption and reflection. The multilayer structure of the feathers causes constructive interference for certain colors such as blues and greens. The colors change as you view a peacock's feather from different angles. Other creatures such as butterflies exhibit similar interference effects. *(Werner H. Muller/Peter Arnold, Inc.)*

(Left) A thin film of oil on water displays interference, as shown by the pattern of colors when white light is incident on the film. The film thickness varies, thereby producing the interesting color pattern. *(Peter Aprahamian/Science Photo Library).* (Right) Interference in a vertical soap film of variable thickness. The top of the film appears darkest where the film is thinnest. *(© 1983 Larry Mulvehill, Photo Researchers)*

equation for destructive interference is

$$2nt = m\lambda \qquad (m = 0, 1, 2, \ldots) \qquad (37.17)$$

Conditions for destructive interference in thin films

The foregoing conditions for constructive and destructive interference are valid only when the medium above the top surface of the film is the same as the medium below the bottom surface. The surrounding medium may have a refractive index less than or greater than that of the film. In either case, the rays reflected from the two surfaces are out of phase by 180°. If the film is placed between two different media, one with $n < n_{\text{film}}$ and the other with $n > n_{\text{film}}$, the conditions for constructive and destructive interference are reversed. In this case, either there is a phase change of 180° for both ray 1 reflecting from surface A and ray 2 reflecting from surface B or there is no phase change for either ray; hence, the net change in relative phase due to the reflections is zero.

Newton's Rings

Another method for observing interference of light waves is to place a plano-convex lens (one having one flat side and one convex side) on top of a flat glass surface as in Figure 37.17a. With this arrangement, the air film between the glass surfaces varies in thickness from zero at the point of contact to some value t at P. If the radius of curvature of the lens R is very large compared with the distance r, and if the system is viewed from above using light of wavelength λ, a pattern of light and dark rings is observed. A photograph of such a pattern is shown in Figure 37.17b. These circular fringes, discovered by Newton, are called **Newton's rings.** Newton's particle model of light explained the origin of the rings by assuming intermittent behavior of the particle-medium interaction.

The interference effect is due to the combination of ray 1, reflected from the flat plate, with ray 2, reflected from the lower part of the lens. Ray 1 undergoes a phase change of 180° upon reflection, because it is reflected from a medium of higher refractive index, whereas ray 2 undergoes no phase change. Hence, the conditions for constructive and destructive interference are given by Equations 37.16 and 37.17, respectively, with $n = 1$ because the film is air. Point O is dark, as seen in Figure 37.17b, because ray 1, reflected from the flat surface, undergoes a 180° phase change with respect to ray 2. Using the geometry shown in Figure

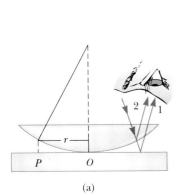 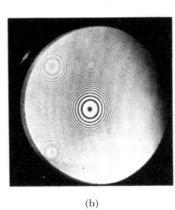 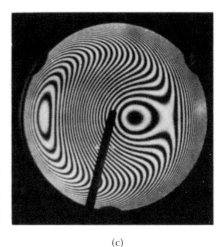

(a) (b) (c)

FIGURE 37.17 (a) The combination of rays reflected from the flat plate and the curved lens surface gives rise to an interference pattern known as Newton's rings. (b) Photograph of Newton's rings. *(Courtesy of Bausch and Lomb Optical Co.)* (c) This asymmetrical interference pattern indicates imperfections in the lens. *(From Physical Science Study Committee,* College Physics, *Lexington, Mass., Heath, 1968)*

37.17a, expressions for the radii of the bright and dark bands can be obtained in terms of the radius of curvature R and wavelength λ. For example, the dark rings have radii given by $r \approx \sqrt{m\lambda R/n}$. The details are left as a problem for the reader (Problem 67). By measuring the radii of the rings, the wavelength can be obtained, provided R is known. Conversely, if the wavelength is accurately known, this effect can be used to obtain R.

One of the important uses of Newton's rings is in the testing of optical lenses. A circular pattern like that pictured in Figure 37.17b is obtained only when the lens is ground to a perfectly spherical curvature. Variations from such symmetry produce a pattern like that in Figure 37.17c. These variations indicate how the lens must be ground and polished to remove the imperfections.

Problem-Solving Strategy
Thin-Film Interference

The following ideas should be kept in mind when you work thin-film interference problems:

- Identify the thin film causing the interference.
- The type of interference that occurs is determined by the phase relationship between the portion of the wave reflected at the upper surface of the film and the portion reflected at the lower surface.
- Phase differences between the two portions of the wave have two causes: (1) differences in the distances traveled by the two portions and (2) phase changes that may occur upon reflection.
- When distance traveled and phase changes upon reflection are both taken into account, the interference is constructive if the path difference between the two waves is an integral multiple of λ, and destructive if the path difference is $\lambda/2$, $3\lambda/2$, $5\lambda/2$, and so forth.

EXAMPLE 37.3 Interference in a Soap Film

Calculate the minimum thickness of a soap bubble film ($n = 1.33$) that results in constructive interference in the reflected light if the film is illuminated with light whose wavelength in free space is 600 nm.

Solution The minimum film thickness for constructive interference in the reflected light corresponds to $m = 0$ in Equation 37.16. This gives $2nt = \lambda/2$, or

$$t = \frac{\lambda}{4n} = \frac{600 \text{ nm}}{4(1.33)} = \boxed{113 \text{ nm}}$$

Exercise What other film thicknesses produce constructive interference?

Answer 338 nm, 564 nm, 789 nm, and so on.

EXAMPLE 37.4 Nonreflecting Coatings for Solar Cells

Semiconductors such as silicon are used to fabricate solar cells—devices that generate electricity when exposed to sunlight. Solar cells are often coated with a transparent thin film, such as silicon monoxide (SiO, $n = 1.45$), in order to minimize reflective losses from the surface. A silicon solar cell ($n = 3.5$) is coated with a thin film of silicon monoxide for this purpose (Fig. 37.18). Determine the minimum film thickness that produces the least reflection at a wavelength of 552 nm, which is the center of the visible spectrum.

Reasoning The reflected light is a minimum when rays 1 and 2 in Figure 37.18 meet the condition of destructive interference. Note that both rays undergo a 180° phase change upon reflection in this case, one from the upper and one from the lower surface. Hence, the net change in phase is zero due to reflection, and the condition for a reflection minimum requires a path difference of $\lambda_n/2$; hence, $2t = \lambda/2n$.

Solution Since $2t = \lambda/2n$, the required thickness is

$$t = \frac{\lambda}{4n} = \frac{550 \text{ nm}}{4(1.45)} = \boxed{94.8 \text{ nm}}$$

Typically, such antireflecting coatings reduce the reflective loss from 30% (with no coating) to 10% (with coating), thereby increasing the cell's efficiency, since more light is available to create charge carriers in the cell. In reality, the coating is never perfectly nonreflecting because the required thickness is wavelength-dependent and the incident light covers a wide range of wavelengths.

Glass lenses used in cameras and other optical instruments are usually coated with a transparent thin film, such as magnesium fluoride (MgF_2), to reduce or eliminate unwanted reflection. More important, such coatings enhance the transmission of light through the lenses.

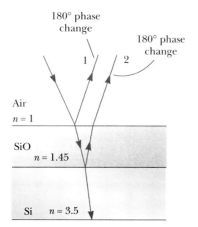

FIGURE 37.18 (Example 37.4) Reflective losses from a silicon solar cell are minimized by coating it with a thin film of silicon monoxide.

EXAMPLE 37.5 Interference in a Wedge-Shaped Film

A thin, wedge-shaped film of refractive index n is illuminated with monochromatic light of wavelength λ, as illustrated in Figure 37.19. Describe the interference pattern observed for this case.

Reasoning and Solution The interference pattern is that of a thin film of variable thickness surrounded by air. Hence, the pattern is a series of alternating bright and dark parallel bands. A dark band corresponding to destructive interference appears at point O, the apex, because the upper reflected ray undergoes a 180° phase change while the lower one does not. According to Equation 37.17, other dark bands appear when $2nt = m\lambda$, so that $t_1 = \lambda/2n$, $t_2 = \lambda/n$, $t_3 = 3\lambda/2n$, and so on. Similarly, bright bands are observed when the thickness satisfies the condition $2nt = (m + \frac{1}{2})\lambda$, corresponding to thicknesses of $\lambda/4n$, $3\lambda/4n$, $5\lambda/4n$, and so on. If white light is used, bands of different colors are observed at different points, corresponding to the different wavelengths of light.

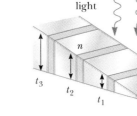

FIGURE 37.19 (Example 37.5) Interference bands in reflected light can be observed by illuminating a wedge-shaped film with monochromatic light. The dark blue areas correspond to positions of destructive interference.

*37.7 THE MICHELSON INTERFEROMETER

The **interferometer,** invented by the American physicist A. A. Michelson (1852–1931), splits a light beam into two parts and then recombines them to form an interference pattern. The device can be used to measure wavelengths or other lengths accurately.

A schematic diagram of the interferometer is shown in Figure 37.20. A beam of

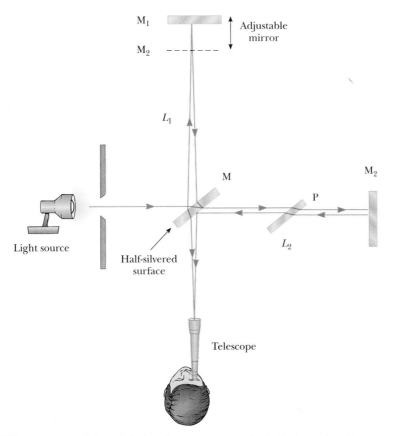

FIGURE 37.20 Diagram of the Michelson interferometer. A single beam is split into two rays by the partially silvered mirror M. The path difference between the two rays is varied with the adjustable mirror M_1.

light provided by a monochromatic source is split into two rays by a partially silvered mirror M inclined at 45° to the incident light beam. One ray is reflected vertically upward toward mirror M_1, while the second ray is transmitted horizontally through M toward mirror M_2. Hence, the two rays travel separate paths L_1 and L_2. After reflecting from M_1 and M_2, the two rays eventually recombine to produce an interference pattern, which can be viewed through a telescope. The glass plate P, equal in thickness to mirror M, is placed in the path of the horizontal ray in order to ensure that the two rays travel the same distance through glass.

The interference condition for the two rays is determined by the difference in their optical path lengths. When the two rays are viewed as shown, the image of M_2 is at M_2' parallel to M_1. Hence, M_2' and M_1 form the equivalent of an air film. The effective thickness of the air film is varied by moving mirror M_1 along the direction of the light beam with a finely threaded screw. Under these conditions, the interference pattern is a series of bright and dark circular rings. If a dark circle appears at the center of the pattern, the two rays interfere destructively. If M_1 is then moved a distance $\lambda/4$, the path difference changes by $\lambda/2$ (twice the separation between M_1 and M_2'). The two rays now interfere constructively, giving a bright circle in the middle. As M_1 is moved an additional distance $\lambda/4$, another dark circle appears. Thus, we see that successive dark or bright circles are formed each time M_1 is moved a distance of $\lambda/4$. The wavelength of light is then measured by counting the number of fringe shifts for a given displacement of M_1. Conversely, if the wavelength is accurately known (as with a laser beam), mirror displacements can be measured to within a fraction of the wavelength.

Since the interferometer can accurately measure displacement, it is often used to make highly precise measurements of the length of mechanical components.

SUMMARY

Interference in light waves occurs whenever two or more waves overlap at a given point. A sustained interference pattern is observed if (1) the sources are coherent (that is, they maintain a constant phase relationship with one another), (2) the sources have identical wavelengths, and (3) the linear superposition principle is applicable.

In Young's double-slit experiment, two slits separated by a distance d are illuminated by a single-wavelength light source. An interference pattern consisting of bright and dark fringes is observed on a viewing screen. The condition for bright fringes (**constructive interference**) is

$$d \sin \theta = m\lambda \qquad (m = 0, \pm 1, \pm 2, \ldots) \tag{37.2}$$

The condition for dark fringes (**destructive interference**) is

$$d \sin \theta = (m + \tfrac{1}{2})\lambda \qquad (m = 0, \pm 1, \pm 2, \ldots) \tag{37.3}$$

The number m is called the **order number** of the fringe.

The **average intensity** of the double-slit interference pattern is

$$I_{av} = I_0 \cos^2\left(\frac{\pi d \sin \theta}{\lambda}\right) \tag{37.12}$$

where I_0 is the maximum intensity on the screen.

When a series of N slits is illuminated, the diffraction pattern produced can be viewed as interference arising from the superposition of a large number of waves.

It is convenient to use phasor diagrams to simplify the analysis of interference from three or more equally spaced slits.

A wave traveling from a medium of index of refraction n_1 toward a medium of index of refraction n_2 undergoes a 180° phase change upon reflection when $n_2 > n_1$. There is no phase change in the reflected wave if $n_2 < n_1$.

The condition for constructive interference in a film of thickness t and refractive index n with a common medium on both sides of the film is

$$2nt = (m + \tfrac{1}{2})\lambda \qquad (m = 0, 1, 2, \ldots) \qquad (37.16)$$

Similarly, the condition for destructive interference in thin films is

$$2nt = m\lambda \qquad (m = 0, 1, 2, \ldots) \qquad (37.17)$$

QUESTIONS

1. What is the necessary condition on the path-length difference between two waves that interfere (a) constructively and (b) destructively?
2. Explain why two flashlights held close together do not produce an interference pattern on a distant screen.
3. If Young's double-slit experiment were performed underwater, how would the observed interference pattern be affected?
4. What is the difference between interference and diffraction?
5. In Young's double-slit experiment, why do we use monochromatic light? If white light is used, how would the pattern change?
6. As a soap bubble evaporates, it appears black just before it breaks. Explain this phenomenon in terms of the phase changes that occur upon reflection from the two surfaces of the soap film.
7. An oil film on water appears brightest at the outer regions, where it is thinnest. From this information, what can you say about the index of refraction of oil relative to that of water?
8. A soap film on a wire loop and held in air appears black in the thinnest regions when observed by reflected light and shows a variety of colors in thicker regions, as in Figure 37.21. Explain.
9. A simple way of observing an interference pattern is to look at a distant light source through a stretched handkerchief or an opened umbrella. Explain how this works.

FIGURE 37.21 (Question 8).

10. In order to observe interference in a thin film, why must the film not be very thick (on the order of a few wavelengths)?
11. A lens with outer radius of curvature R and index of refraction n rests on a flat glass plate and the combination is illuminated with white light from above. Is there a dark spot or a light spot at the center of the lens? What does it mean if the observed rings are noncircular?
12. Why is the lens on a good-quality camera coated with a thin film?
13. Would it be possible to place a nonreflective coating on an airplane to cancel radar waves of wavelength 3 cm?
14. Why is it so much easier to perform interference experiments with a laser than with an ordinary light source?

PROBLEMS

Section 37.2 Young's Double-Slit Experiment

1. A pair of narrow, parallel slits separated by 0.25 mm is illuminated by green light ($\lambda = 546.1$ nm). The interference pattern is observed on a screen 1.2 m away from the plane of the slits. Calculate the distance (a) from the central maximum to the first bright region on either side of the central maximum and (b) between the first and second dark bands.

□ indicates problems that have full solutions available in the Student Solutions Manual and Study Guide.

2. A laser beam ($\lambda = 632.8$ nm) is incident on two slits 0.20 mm apart. Approximately how far apart are the bright interference lines on a screen 5.0 m away from the double slits?

3. A Young's interference experiment is performed with monochromatic light. The separation between the slits is 0.50 mm, and the interference pattern on a screen 3.3 m away shows the first maximum 3.4 mm from the center of the pattern. What is the wavelength?

4. Light ($\lambda = 442$ nm) passes through a double-slit system that has a slit separation $d = 0.40$ mm. Determine how far away a screen must be placed so that dark fringes appear directly opposite both slits.

5. On a day when the speed of sound is 354 m/s, a 2000-Hz sound wave impinges on two slits 30.0 cm apart. (a) At what angle is the first maximum located? (b) If the sound wave is replaced by 3.00-cm microwaves, what slit separation gives the same angle for the first maximum? (c) If the slit separation is 1.00 μm, light of what frequency gives the same first-maximum angle?

6. A double slit with a spacing of 0.083 mm between the slits is 2.5 m from a screen. (a) If yellow light of wavelength 570 nm strikes the double slit, what is the separation between the zeroth- and first-order maxima on the screen? (b) If blue light of wavelength 410 nm strikes the double slit, what is the separation between the second- and fourth-order maxima? (c) Repeat parts (a) and (b) for the minima.

7. A riverside warehouse has two open doors as in Figure P37.7. A boat on the river sounds its horn. To person A the sound is loud and clear. To person B, the sound is barely audible. The principal wavelength of the sound waves is 3.00 m. Assuming B is at the position of the first minimum, determine the distance between the doors, center to center.

8. Monochromatic light illuminates a double-slit system with a slit separation $d = 0.30$ mm. The second-order maximum occurs at $y = 4.0$ mm on a screen 1.0 m from the slits. Determine (a) the wavelength, (b) the position (y) of the third-order maximum, and (c) the angular position (θ) of the $m = 1$ minimum.

9. A radio transmitter A, operating at 60.0 MHz is 10.0 m from a similar transmitter B that is 180° out of phase with A. How far must an observer move from A toward B along the line connecting the two in order to move between two points where the two beams are in phase?

9A. A radio transmitter A, operating at a frequency f, is a distance d from a similar transmitter B that is 180° out of phase with A. How far must an observer move from A toward B along the line connecting the two in order to move between two points where the two beams are in phase?

10. Two radio antennas separated by 300 m as in Figure P37.10 simultaneously transmit identical signals at the same wavelength. A radio in a car traveling due north receives the signals. (a) If the car is at the position of the second maximum, what is the wavelength of the signals? (b) How much farther must the car travel to encounter the next minimum in reception? (*Caution:* Do not use small-angle approximations in this problem.)

10A. Two radio antennas separated by a distance d as in Figure P37.10 simultaneously transmit identical signals at the same wavelength. A radio in a car traveling due north receives the signals. (a) If the car is at the position of the second maximum, what is the wavelength of the signals? (b) How much farther must the car travel to encounter the next minimum in reception? (*Caution:* Do not use small-angle approximations in this problem.)

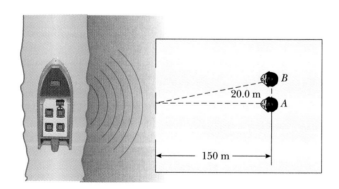

FIGURE P37.7

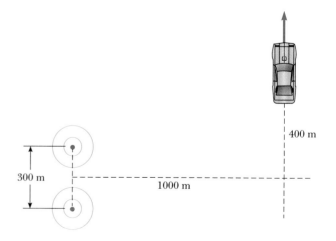

FIGURE P37.10

11. Two slits are separated by a distance d. Coherent light rays of wavelength λ strike the slits at an angle θ_1 as in Figure P37.11. If an interference maximum is

formed at an angle θ_2 far from the slits, show that $d(\sin\theta_2 - \sin\theta_1) = m\lambda$, where m is an integer.

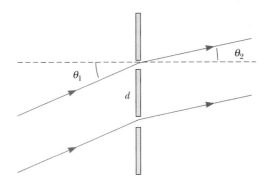

FIGURE P37.11

12. In a double-slit arrangement of Figure 37.4, $d = 0.15$ mm, $L = 140$ cm, $\lambda = 643$ nm, and $y = 1.8$ cm. (a) What is the path difference δ for the two rays from the two slits arriving at P? (b) Express this path difference in terms of λ. (c) Does P correspond to a maximum, a minimum, or an intermediate condition?

Section 37.3 Intensity Distribution of the Double-Slit Interference Pattern

13. In Figure 37.4, let $L = 120$ cm and $d = 0.25$ cm. The slits are illuminated with coherent 600-nm light. Calculate the distance y above the central maximum for which the average intensity on the screen is 75% of the maximum.
14. Two slits are separated by 0.18 mm. An interference pattern is formed on a screen 80 cm away by 656.3-nm light. Calculate the fraction of the maximum intensity 0.60 cm above the central maximum.
15. In Figure 37.4, $d = 0.20$ mm, $L = 160$ cm, and $y = 1.0$ mm. What wavelength results in an average intensity at P that is 36% of the maximum?
16. Monochromatic coherent light of amplitude E_0 and angular frequency ω passes through three parallel slits each separated by a distance d from its neighbor. (a) Show that the time-averaged intensity as a function of the angle θ is

$$I(\theta) = I_0\left[1 + 2\cos\left(\frac{2\pi d \sin\theta}{\lambda}\right)\right]^2$$

(b) Determine the ratio of the intensities of the two maxima.

17. In a Young's double-slit experiment using 350 nm light, a thin piece of Plexiglas ($n = 1.51$) covers one of the slits. If the center point on the screen is a dark spot, what is the minimum thickness of the Plexiglas?

17A. In a Young's double-slit experiment using light of wavelength λ, a thin piece of Plexiglas having index of refraction n covers one of the slits. If the center point on the screen is a dark spot, what is the minimum thickness of the Plexiglas?

18. In Figure 37.4 let $L = 1.2$ m and $d = 0.12$ mm and assume that the slit system is illuminated with monochromatic 500-nm light. Calculate the phase difference between the two wave fronts arriving at P when (a) $\theta = 0.50°$ and (b) $y = 5.0$ mm. (c) What is the value of θ for which (1) the phase difference is 0.333 rad and (2) the path difference is $\lambda/4$?
19. Two slits are separated by 0.32 mm. A beam of 500-nm light strikes them, producing an interference pattern. Determine the number of maxima observed in the angular range $-30° < \theta < 30°$.
20. The intensity on the screen at a certain point in a double-slit interference pattern is 64% of the maximum value. (a) What minimum phase difference (in radians) between sources produces this result? (b) Express this phase difference as a path difference for 486.1-nm light.
21. At a particular location in a Young's interference pattern, the intensity on the screen is 6.4% of maximum. (a) What minimum phase difference (in radians) between sources produces this result? (b) Determine the path difference for 587.5-nm light.
22. Monochromatic light ($\lambda = 632.8$ nm) is incident on two fine parallel slits separated by 0.20 mm. What is the distance to the first maximum and its intensity (relative to the central maximum) on a screen 2.0 m beyond the slits?
23. Two narrow parallel slits separated by 0.85 mm are illuminated by 600-nm light, and the viewing screen is 2.80 m away from the slits. (a) What is the phase difference between the two interfering waves on a screen at a point 2.50 mm from the central bright fringe? (b) What is the ratio of the intensity at this point to the intensity at the center of a bright fringe?

Section 37.4 Phasor Addition of Waves

24. The electric fields from three coherent sources are described by $E_1 = E_0 \sin\omega t$, $E_2 = E_0 \sin(\omega t + \phi)$, and $E_3 = E_0 \sin(\omega t + 2\phi)$. Let the resultant field be represented by $E_P = E_R \sin(\omega t + \alpha)$. Use phasors to find E_R and α when (a) $\phi = 20°$, (b) $\phi = 60°$, (c) $\phi = 120°$. (d) Repeat when $\phi = (3\pi/2)$ rad.
25. Suppose the slit openings in a Young's double-slit experiment have different sizes so that the electric field and intensity from one slit are different from those of the other slit. If $E_1 = E_{10}\sin(\omega t)$ and $E_2 = E_{20}\sin(\omega t + \phi)$, show that the resultant electric field is $E = E_0 \sin(\omega t + \theta)$, where

$$E_0 = \sqrt{E_{10}^2 + E_{20}^2 + 2E_{10}E_{20}\cos\phi}$$

and

$$\sin\theta = \frac{E_{20}\sin\phi}{E_0}$$

26. Use phasors to find the resultant (magnitude and phase angle) of two fields represented by $E_1 = 12\sin\omega t$ and $E_2 = 18\sin(\omega t + 60°)$. (Note that in this case the amplitudes of the two fields are unequal.)

27. Determine the resultant of the two waves $E_1 = 6.0\sin(100\pi t)$ and $E_2 = 8.0\sin(100\pi t + \pi/2)$.

28. Two coherent waves are described by

$$E_1 = E_0 \sin\left(\frac{2\pi x_1}{\lambda} - 2\pi ft + \frac{\pi}{6}\right)$$

$$E_2 = E_0 \sin\left(\frac{2\pi x_2}{\lambda} - 2\pi ft + \frac{\pi}{8}\right)$$

Determine the relationship between x_1 and x_2 that produces constructive interference when the two waves are superposed.

29. When illuminated, four equally spaced parallel slits act as multiple coherent sources, each differing in phase from the adjacent one by an angle ϕ. Use a phasor diagram to determine the smallest value of ϕ for which the resultant of the four waves (assumed to be of equal amplitude) is zero.

30. Sketch a phasor diagram to illustrate the resultant of $E_1 = E_{01} \sin\omega t$ and $E_2 = E_{02}\sin(\omega t + \phi)$, where $E_{02} = 1.5 E_{01}$ and $\pi/6 \le \phi \le \pi/3$. Use the sketch and the law of cosines to show that, for two coherent waves, the resultant intensity can be written in the form $I_R = I_1 + I_2 + 2\sqrt{I_1 I_2}\cos\phi$.

31. Consider N coherent sources described by $E_1 = E_0 \sin(\omega t + \phi)$, $E_2 = E_0 \sin(\omega t + 2\phi)$, $E_3 = E_0 \sin(\omega t + 3\phi)$, ..., $E_N = E_0 \sin(\omega t + N\phi)$. Find the minimum value of ϕ for which $E_R = E_1 + E_2 + E_3 + \cdots + E_N$ is zero.

Section 37.6 Interference in Thin Films

32. A material having an index of refraction of 1.30 is used to coat a piece of glass ($n = 1.50$). What should be the minimum thickness of this film to minimize reflection of 500-nm light?

33. A film of MgF_2 ($n = 1.38$) having thickness 1.00×10^{-5} cm is used to coat a camera lens. Are any wavelengths in the visible spectrum intensified in the reflected light?

34. A soap bubble of index of refraction 1.33 strongly reflects both the red and the green components of white light. What film thickness allows this to happen? (In air, $\lambda_{red} = 700$ nm, $\lambda_{green} = 500$ nm.)

35. A thin film of oil ($n = 1.25$) covers a smooth wet pavement. When viewed perpendicular to the pavement, the film appears to be predominantly red (640 nm) and has no blue (512 nm). How thick is it?

36. A thin layer of oil ($n = 1.25$) is floating on water. How thick is the oil in the region that reflects green light ($\lambda = 525$ nm)?

37. A thin layer of liquid methylene iodide ($n = 1.756$) is sandwiched between two flat parallel plates of glass. What must be the thickness of the liquid layer if normally incident 600-nm light is to be strongly reflected?

38. A beam of 580-nm light passes through two closely spaced glass plates, as shown in Figure P37.38. For what minimum nonzero value of the plate separation, d, is the transmitted light bright?

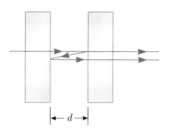

FIGURE P37.38

39. An oil film ($n = 1.45$) floating on water is illuminated by white light at normal incidence. The film is 280 nm thick. Find (a) the dominant observed color in the reflected light and (b) the dominant color in the transmitted light. Explain your reasoning.

40. A possible means for making an airplane invisible to radar is to coat the plane with an antireflective polymer. If radar waves have a wavelength of 3.00 cm and the index of refraction of the polymer is $n = 1.50$, how thick would you make the coating?

41. A thin film of cryolite ($n = 1.35$) is applied to a camera lens ($n = 1.50$). The coating is designed to reflect wavelengths at the blue end of the spectrum and transmit wavelengths in the near infrared. What minimum thickness gives high reflectivity at 450 nm and high transmission at 900 nm?

42. Two rectangular flat glass plates ($n = 1.52$) are in contact along one end and separated along the other end by a sheet of paper 4.0×10^{-3} cm thick (Fig. P37.42). The top plate is illuminated by monochromatic light ($\lambda = 546.1$ nm). Calculate the number of dark parallel bands crossing the top plate (include the dark band at zero thickness along the edge of contact between the two plates).

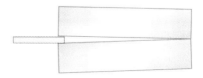

FIGURE P37.42

43. An air wedge is formed between two glass plates separated at one edge by a very fine wire as in Figure P37.42. When the wedge is illuminated from above by 600-nm light, 30 dark fringes are observed. Calculate the radius of the wire.

44. When a liquid is introduced into the air space between the lens and the plate in a Newton's-rings apparatus, the diameter of the tenth ring changes from 1.50 to 1.31 cm. Find the index of refraction of the liquid.

44A. When a liquid is introduced into the air space between the lens and the plate in a Newton's-rings apparatus, the diameter of the tenth ring changes from d_1 to d_2. Find the index of refraction of the liquid.

*Section 37.7 The Michelson Interferometer

45. Mirror M_1 in Figure 37.20 is displaced a distance ΔL. During this displacement, 250 fringes (formation of successive dark or bright bands) are counted. The light being used has a wavelength of 632.8 nm. Calculate the displacement ΔL.

46. Monochromatic light is beamed into a Michelson interferometer. The movable mirror is displaced 0.382 mm, causing the interferometer pattern to reproduce itself 1700 times. Determine the wavelength of the light. What color is it?

47. Light of wavelength 550.5 nm is used to calibrate a Michelson interferometer and mirror M_1 is moved 0.18 mm. How many dark fringes are counted?

48. One leg of a Michelson interferometer contains an evacuated cylinder that is 3.00 cm long and has a glass plate on each end. A gas is slowly leaked into the cylinder until a pressure of 1.00 atm is reached. If 35 bright fringes pass on the screen when 633-nm light is used, what is the index of refraction of the gas?

48A. One leg of a Michelson interferometer contains an evacuated cylinder that is of length L and has a glass plate on each end. A gas is slowly leaked into the cylinder until a pressure of P is reached. If N bright fringes pass on the screen when light of wavelength λ is used, what is the index of refraction of the gas?

ADDITIONAL PROBLEMS

49. Figure P37.49 shows a radio-wave transmitter and a receiver separated by $d = 600$ m and both $h = 30.0$ m high. The receiver can receive both direct signals from the transmitter and indirect ones reflected off the ground. Assuming that the ground is level between the two towers and that a $\lambda/2$ phase shift occurs upon reflection, determine the longest wavelengths that interfere (a) constructively and (b) destructively.

49A. Figure P37.49 shows a radio-wave transmitter and a receiver separated by a distance d and both a height h. The receiver can receive both direct signals from the transmitter and indirect ones reflected off the ground. Assuming that the ground is level between the towers and that a $\lambda/2$ phase shift occurs upon reflection, determine the longest wavelengths that interfere (a) constructively and (b) destructively.

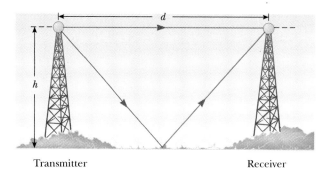

FIGURE P37.49

50. Interference effects are produced at point P on a screen as a result of direct rays from a 500-nm source and reflected rays off the mirror, as in Figure 37.14. If the source is 100 m to the left of the screen, and 1.00 cm above the mirror, find the distance y (in millimeters) to the first dark band above the mirror.

51. Astronomers observed a 60-MHz radio source both directly and by reflection from the sea. If the receiving dish is 20 m above sea level, what is the angle of the radio source above the horizon at first maximum?

52. The waves from a radio station can reach a home receiver by two paths. One is a straight-line path from transmitter to home, a distance of 30 km. The second path is by reflection from the ionosphere (a layer of ionized air molecules near the top of the atmosphere). Assume this reflection takes place at a point midway between receiver and transmitter. If the wavelength broadcast by the radio station is 350 m, find the minimum height of the ionospheric layer that produces destructive interference between the direct and reflected beams. (Assume no phase changes on reflection.)

53. Measurements are made of the intensity distribution in a Young's interference pattern (Fig. 37.6). At a particular value of y, it is found that $I/I_0 = 0.81$ when 600-nm light is used. What wavelength of light should be used to reduce the relative intensity at the same location to 64%?

54. Waves broadcast by a 1500-kHz radio station arrive at a home receiver by two paths. One is a direct path,

and the second is from reflection off an airplane directly above the receiver. The airplane is approximately 100 m above the receiver, and the direct distance from station to home is 20.0 km. What is the exact height of the airplane if destructive interference is occurring? (Assume no phase change on reflection.)

54A. Waves broadcast by a radio station at frequency f arrive at a home receiver by two paths. One is a direct path, and the second is from reflection off an airplane directly above the receiver. The airplane is at an approximate height h above the receiver, and the direct distance from station to home is d. What is the exact height of the airplane if destructive interference is occurring? (Assume no phase change on reflection.)

55. In a Young's interference experiment, the two slits are separated by 0.15 mm, and the incident light includes light of wavelengths $\lambda_1 = 540$ nm and $\lambda_2 = 450$ nm. The overlapping interference patterns are formed on a screen 1.40 m from the slits. Calculate the minimum distance from the center of the screen to the point where a bright line of the λ_1 light coincides with a bright line of the λ_2 light.

56. In a Newton's-ring experiment, a plano-convex glass ($n = 1.52$) lens of diameter 10.0 cm is placed on a flat plate as in Figure 37.17a. When 650-nm light is incident normally, 55 bright rings are observed, with the last one right on the edge of the lens. (a) What is the radius of curvature of the convex surface of the lens? (b) What is the focal length of the lens?

57. Young's double-slit experiment is performed with 589-nm light and a slits-to-screen distance of 2.00 m. The tenth interference minimum is observed 7.26 mm from the central maximum. Determine the spacing of the slits.

58. A soap film 500 nm thick has an index of refraction of 1.35 and is illuminated with white light. (a) If the film is part of a bubble in air, what color is the reflected light? (b) If the film is on a flat glass plate, what color is the reflected light?

59. A hair is placed at one edge between two flat glass plates 8.00 cm long. When this arrangement is illuminated with 600-nm light, 121 dark bands are counted, starting at the point of contact of the two plates. How thick is the hair?

60. A glass plate ($n = 1.61$) is covered with a thin uniform layer of oil ($n = 1.20$). A nonmonochromatic light beam in air is incident normally on the oil surface. Observation of the reflected beam shows destructive interference at 500 nm and constructive interference at 750 nm with no intervening maxima or minima. Calculate the thickness of the oil layer.

61. A piece of transparent material having an index of refraction n is cut into the shape of a wedge, as shown in Figure P37.61. The angle of the wedge is small, and monochromatic light of wavelength λ is normally incident from above. If the height of the wedge is h and the width is ℓ, show that bright fringes occur at the positions $x = \lambda\ell(m + \frac{1}{2})/2hn$ and dark fringes occur at the positions $x = \lambda\ell m/2hn$, where $m = 0, 1, 2, \ldots$ and x is measured as shown.

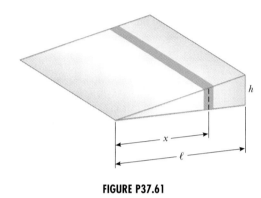

FIGURE P37.61

62. An air wedge is formed between two glass plates in contact along one edge and slightly separated at the opposite edge. When the plates are illuminated with monochromatic light from above, the reflected light has 85 dark fringes. Calculate the number of dark fringes that would appear if water ($n = 1.33$) were to replace the air between the plates.

63. The condition for constructive interference by reflection from a thin film in air as developed in Section 37.6 assumes nearly normal incidence. (a) Show that if the light is incident on the film at an angle $\phi_1 \gg 0$ (relative to the normal), then the condition for constructive interference is $2nt\cos\theta_2 = (m + \frac{1}{2})\lambda$, where θ_2 is the angle of refraction. (b) Calculate the minimum thickness for constructive interference if 590-nm light is incident at an angle of 30° on a film that has an index of refraction of 1.38.

64. Use phasor addition to find the resultant amplitude and phase constant when the following three harmonic functions are combined: $E_1 = \sin(\omega t + \pi/6)$, $E_2 = 3.0\sin(\omega t + 7\pi/2)$, $E_3 = 6.0\sin(\omega t + 4\pi/3)$.

65. A planoconvex lens having a radius of curvature $r = 4.00$ m is placed on a concave reflecting surface having a radius of curvature $R = 12.0$ m, as in Figure P37.65. Determine the radius of the 100th bright ring if 500-nm light is incident normal to the flat surface of the lens.

66. A soap film ($n = 1.33$) is contained within a rectangular wire frame. The frame is held vertically so that the film drains downward and becomes thicker at the bottom than at the top, where the thickness is essentially zero. The film is viewed in white light with near-normal incidence, and the first violet ($\lambda = 420$ nm) interference band is observed 3.0 cm from the top

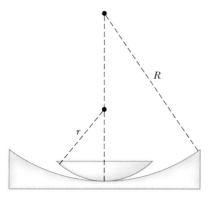

FIGURE P37.65

edge of the film. (a) Locate the first red ($\lambda = 680$ nm) interference band. (b) Determine the film thickness at the positions of the violet and red bands. (c) What is the wedge angle of the film?

67. (a) A lens is made with its lower surface described by the function $y = f(x)$ rotated about the y axis (Fig. P37.67). The lens is placed on a flat glass plate. Light of wavelength λ is incident straight down. Show that dark rings form with radii.

$$r = f^{-1}\left(\frac{m\lambda}{2n}\right)$$

where n is the index of refraction of the medium surrounding the lens, m is a non-negative integer, and f^{-1} is the inverse of the function f. (b) Show that the dark rings correspond to the Newton's rings described in the text, where

$$f(x) = R - \sqrt{R^2 - x^2}$$

for $0 < x < R$, R is the radius of curvature of the convex lens, and y is small compared to R.

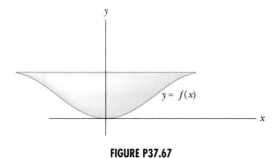

FIGURE P37.67

68. Interference fringes are produced using Lloyd's mirror and a 606-nm source as in Figure 37.14. If fringes 1.2 mm apart are formed on a screen 2.0 m from the real source S, find the vertical distance h of the source above the reflecting surface.

69. (a) Both sides of a uniform film that has index of refraction n and thickness d are in contact with air. For normal incidence of light, an intensity minimum is observed in the reflected light at λ_2 and an intensity maximum is observed at λ_1, where $\lambda_1 > \lambda_2$. If there are no intensity minima observed between λ_1 and λ_2, show that the integer m in Equations 37.16 and 37.17 is given by $m = \lambda_1/2(\lambda_1 - \lambda_2)$. (b) Determine the thickness of the film if $n = 1.40$, $\lambda_1 = 500$ nm, and $\lambda_2 = 370$ nm.

70. Consider the double-slit arrangement shown in Figure P37.70, where the separation d is 0.30 mm and the distance L is 1.00 m. A sheet of transparent plastic ($n = 1.50$) 0.050 mm thick (about the thickness of this page) is placed over the upper slit. As a result, the central maximum of the interference pattern moves upward a distance y'. Find this distance.

70A. Consider the double-slit arrangement shown in Figure P37.70, where the slit separation is d and the slit-to-screen distance is L. A sheet of transparent plastic having an index of refraction n and thickness t is placed over the upper slit. As a result, the central maximum of the interference pattern moves upward a distance y'. Find y'.

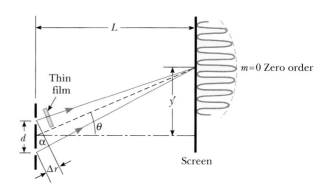

FIGURE P37.70

71. Slit 1 of a double slit is wider than slit 2 so that the light from slit 1 has an amplitude three times that of the light from slit 2. Show that for this situation, Equation 37.11 has the form $I = (4I_0/9)(1 + 3\cos^2 \phi/2)$.

SPREADSHEET PROBLEMS

S1. To calculate the intensity distribution of the interference pattern for equally spaced sources, it is necessary to add a series of terms such as: $E_1 = A_0 \sin \alpha$, $E_2 = A_0 \sin(\alpha + \phi)$, $E_3 = A_0 \sin(\alpha + 2\phi)$, where ϕ is the phase difference caused by different path lengths. For N sources the time-averaged relative intensity as a function of the phase angle ϕ is

$$I_{rel} = (f_N(\phi))^2 + (g_N(\phi))^2$$

where

$$f_N(\phi) = \sum_{n=0}^{N-1} \cos(n\phi) \qquad g_N(\phi) = \sum_{n=0}^{N-1} \sin(n\phi)$$

and I_{rel} is the ratio of the average intensity for N sources to that of one source. Spreadsheet 37.1 calculates and plots I_{rel} versus phase angle ϕ for up to a maximum of six sources. The coefficients a, b, c, d, e, and f used in the spreadsheet determine the number of sources. For two sources, for example, set $a = 1$, $b = 1$, and $c = d = e = f = 0$. For three sources set $a = b = c = 1$, $d = e = f = 0$, and so on. (a) For three sources, what is the ratio of the intensity of the principal maxima to that of a single source? (b) What is the ratio of the intensity of the secondary maxima to that of the principal maximum?

S2. Use Spreadsheet 37.1 to calculate the intensity pattern for four equally spaced sources. (a) What is the ratio of the intensity of the principal maxima to that of a single source? (b) What is the ratio of the intensity of the secondary maxima to that of the principal maxima? (c) Repeat parts (a) and (b) for five and six sources. (d) Are all the secondary maxima of equal intensity?

C H A P T E R   3 8

Diffraction and Polarization

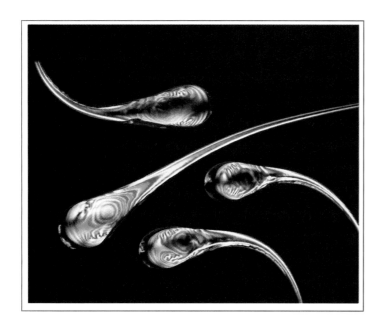

These glass objects, called Prince Rupert drops, are made by dripping molten glass into water. The photograph was made by placing the objects between two crossed polarizers. The patterns that are observed represent the strain distribution in the glass. Studies of such patterns led to the development of tempered glass. *(James L. Amos/Peter Arnold, Inc.)*

When light waves pass through a small aperture, an interference pattern is observed rather than a sharp spot of light, showing that light spreads beyond the aperture into regions where a shadow would be expected if light traveled in straight lines. Other waves, such as sound waves and water waves, also have this property of being able to bend around corners. This phenomenon, known as diffraction, can be regarded as interference from a great number of coherent wave sources. In other words, diffraction and interference are basically equivalent.

In Chapter 34, we learned that electromagnetic waves are transverse. That is, the electric and magnetic field vectors are perpendicular to the direction of propagation. In this chapter, we see that under certain conditions, light waves can be polarized in various ways, such as by passing light through polarizing sheets.

38.1 INTRODUCTION TO DIFFRACTION

In Section 37.2 we learned that when two slits are illuminated by a single-wavelength light source, an interference pattern is formed on a viewing screen. If the light truly traveled in straight-line paths after passing through the slits, as in Figure

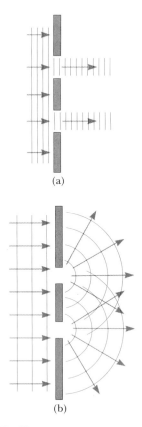

FIGURE 38.1 (a) If light waves did not spread out after passing through the slits, no interference would occur. (b) The light waves from the two slits overlap as they spread out, filling the expected shadowed regions with light and producing interference fringes.

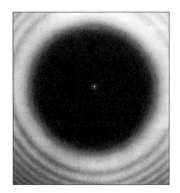

FIGURE 38.3 Diffraction pattern of a penny, taken with the penny midway between screen and source. *(Courtesy of P. M. Rinard, from* Am. J. Phys. *44:70, 1976)*

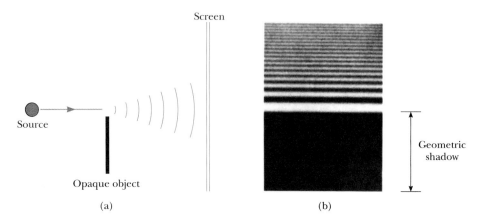

FIGURE 38.2 (a) We expect the shaded region to be completely shielded from the light by the opaque object. Instead, the light bends around the object and enters what "should be" a shadowed region. (b) Diffraction pattern of a straight edge.

38.1a, the waves would not overlap and no interference pattern would be seen. Instead, Huygens' principle requires that the waves spread out from the slits as shown in Figure 38.1b. In other words, the light deviates from a straight-line path and enters the region that would otherwise be shadowed. This divergence of light from its initial line of travel is called **diffraction**.

In general, diffraction occurs when waves pass through small openings, around obstacles, or past sharp edges. As an example of diffraction, consider the following. When an opaque object is placed between a point source of light and a screen, as in Figure 38.2a, the boundary between the shadowed and illuminated regions on the screen is not sharp. A careful inspection of the boundary shows that a small amount of light bends into the shadowed region. The region outside the shadow contains alternating light and dark bands, as in Figure 38.2b. The intensity in the first bright band is greater than the intensity in the region of uniform illumination.

Figure 38.3 shows the diffraction pattern and shadow of a penny. There are a bright spot at the center, circular fringes near the shadow's edge, and another set of fringes outside the shadow. The center bright spot can be explained only through the use of the wave theory of light, which predicts constructive interference at this point. From the viewpoint of geometric optics (light as a collection of particles), we expect the center of the shadow to be dark because that part of the viewing screen is completely shielded by the penny.

It is interesting to point out an historical incident that occurred shortly before the central bright spot was first observed. One of the supporters of geometric optics, Simeon Poisson, argued that if Augustin Fresnel's wave theory of light were valid, then a central bright spot should be observed in any shadow. Because the spot was observed shortly thereafter, Poisson's prediction reinforced the wave theory rather than disproved it.

Diffraction phenomena are usually classified into two types. **Fraunhofer diffraction** occurs when the rays reaching a viewing screen are approximately parallel. This can be achieved experimentally either by placing the screen far from the opening as in Figure 38.4a or by using a converging lens to focus the parallel rays on the screen. A bright fringe is observed along the axis at $\theta = 0$, with alternating dark and bright fringes on either side of the central bright one. Figure 38.4b is a photograph of a single-slit Fraunhofer diffraction pattern.

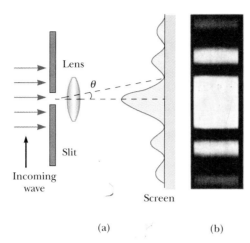

FIGURE 38.4 (a) Fraunhofer diffraction pattern of a single slit. The pattern consists of a central bright region flanked by much weaker maxima alternating with dark bands. (Note that this is not to scale.) (b) Photograph of a single-slit Fraunhofer diffraction pattern. *(From M. Cagnet, M. Francon, and J. C. Thierr,* Atlas of Optical Phenomena, *Berlin, Springer-Verlag, 1962, plate 18)*

When the observing screen is placed a finite distance from the slit and no lens is used to focus parallel rays, the observed pattern is called a **Fresnel diffraction** pattern (Fig. 38.5). The diffraction patterns shown in Figures 38.2b and 38.3 are examples of Fresnel diffraction. Because Fresnel diffraction is difficult to treat quantitatively, the following discussion is restricted to Fraunhofer diffraction.

38.2 SINGLE-SLIT DIFFRACTION

Up until now, we have assumed that slits are point sources of light. In this section, we abandon this assumption and determine how the finite width of slits is the basis for understanding Fraunhofer diffraction.

We can deduce some important features of this problem by examining waves coming from various portions of the slit, as shown in Figure 38.6. According to Huygens' principle, *each portion of the slit acts as a source of waves.* Hence, light from one portion of the slit can interfere with light from another portion, and the resultant intensity on the screen depends on the direction θ.

To analyze the diffraction pattern, it is convenient to divide the slit in two halves, as in Figure 38.6. All the waves that originate from the slit are in phase. Consider waves 1 and 3, which originate from a segment just above the bottom and just above the center of the slit, respectively. Wave 1 travels farther than wave 3 by an amount equal to the path difference $(a/2) \sin \theta$, where a is the width of the slit. Similarly, the path difference between waves 2 and 4 is also $(a/2) \sin \theta$. If this path difference is exactly one-half wavelength (corresponding to a phase difference of 180°), the two waves cancel each other and destructive interference results. This is true, in fact, for any two waves that originate at points separated by half the slit width because the phase difference between two such points is 180°. Therefore, waves from the upper half of the slit interfere destructively with waves from the lower half of the slit when

$$\frac{a}{2} \sin \theta = \frac{\lambda}{2}$$

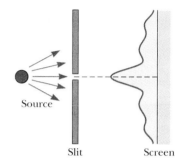

FIGURE 38.5 A Fresnel diffraction pattern of a single slit is observed when the rays are not parallel and the observing screen is a finite distance from the slit. (Note that this is not to scale.)

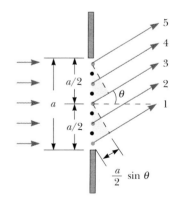

FIGURE 38.6 Diffraction of light by a narrow slit of width a. Each portion of the slit acts as a point source of waves. The path difference between rays 1 and 3 or between rays 2 and 4 is $(a/2) \sin \theta$. (Note that this is not to scale.)

1120 CHAPTER 38 Diffraction and Polarization

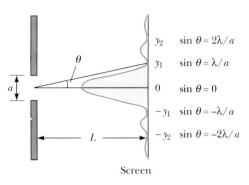

FIGURE 38.7 Positions of the minima for the Fraunhofer diffraction pattern of a single slit of width a. The pattern is obtained only if $L \gg a$. (Note that this is not to scale.)

or when

$$\sin\theta = \frac{\lambda}{a}$$

If we divide the slit into four parts rather than two and use similar reasoning, we find that the screen is also dark when

$$\sin\theta = \frac{2\lambda}{a}$$

Likewise, we can divide the slits into six parts and show that darkness occurs on the screen when

$$\sin\theta = \frac{3\lambda}{a}$$

Therefore, the general condition for destructive interference is

Condition for destructive interference

$$\sin\theta = m\frac{\lambda}{a} \qquad (m = \pm 1, \pm 2, \pm 3, \ldots) \qquad (38.1)$$

Equation 38.1 gives the values of θ for which the diffraction pattern has zero intensity—that is, a dark fringe is formed. However, Equation 38.1 tells us nothing about the variation in intensity along the screen. The general features of the intensity distribution are shown in Figure 38.7. A broad central bright fringe is observed, flanked by much weaker, bright fringes alternating with dark fringes. The various dark fringes (points of zero intensity) occur at the values of θ that satisfy Equation 38.1. The position of the points of constructive interference lies approximately halfway between the dark fringes. Note that the central bright fringe is twice as wide as the weaker maxima.

EXAMPLE 38.1 Where Are the Dark Fringes?

Light of wavelength 580 nm is incident on a slit of width 0.300 mm. The observing screen is 2.00 m from the slit. Find the positions of the first dark fringes and the width of the central bright fringe.

Solution The first dark fringes that flank the central bright fringe correspond to $m = \pm 1$ in Equation 38.1. Hence, we find that

$$\sin\theta = \pm\frac{\lambda}{a} = \pm\frac{5.80 \times 10^{-7}\ \text{m}}{0.300 \times 10^{-3}\ \text{m}} = \pm 1.93 \times 10^{-3}$$

From the triangle in Figure 38.7, note that $\tan\theta = y_1/L$. Since θ is very small, we can use the approximation $\sin\theta \approx \tan\theta$, so that $\sin\theta \approx y_1/L$. Therefore, the positions of the first minima measured from the central axis are given by

$$y_1 \approx L \sin \theta = \pm L \frac{\lambda}{a} = \pm 3.87 \times 10^{-3} \text{ m}$$

The positive and negative signs correspond to the dark fringes on either side of the central bright fringe. Hence, the width of the central bright fringe is equal to $2|y_1| = 7.73 \times 10^{-3}$ m = 7.73 mm. Note that this value is much larger than the width of the slit. However, as the width of the slit is increased, the diffraction pattern narrows, corresponding to smaller values of θ. In fact, for large values of a, the various maxima and minima are so closely spaced that the only thing observed is a large central bright area, which resembles the geometric image of the slit. This matter is of great importance in the design of lenses used in telescopes, microscopes, and other optical instruments.

Exercise Determine the width of the first-order bright fringe.

Answer 3.87 mm.

Intensity of the Single-Slit Diffraction Pattern

We can use phasors to determine the intensity distribution for a single-slit diffraction pattern. Imagine a slit divided into a large number of small zones, each of width Δy as in Figure 38.8. Each zone acts as a source of coherent radiation, and each contributes an incremental electric field amplitude ΔE at some point P on the screen. The total electric field amplitude E at P is obtained by summing the contributions from all the zones. The light intensity at P is the square of the amplitude.

The incremental electric field amplitudes between adjacent zones are out of phase with one another by an amount $\Delta \beta$. The phase difference $\Delta \beta$ is related to the path difference $\Delta y \sin \theta$ between adjacent zones by

$$\text{Phase difference} = \left(\frac{2\pi}{\lambda}\right) \text{path difference}$$

$$\Delta \beta = \frac{2\pi}{\lambda} \Delta y \sin \theta \tag{38.2}$$

To find the total electric field amplitude on the screen at any angle θ, we sum the incremental amplitudes ΔE due to each zone. For small values of θ, we can assume that all the ΔE values are the same. It is convenient to use phasor diagrams for various angles as in Figure 38.9. When $\theta = 0$, all phasors are aligned as in Figure 38.9a, because all the waves from the various zones are in phase. In this case, the total amplitude of the center of the screen is $E_0 = N \Delta E$, where N is the number of zones. The amplitude E_θ at some small angle θ is shown in Figure 38.9b, where each phasor differs in phase from an adjacent one by an amount $\Delta \beta$. In this case, E_θ is the vector sum of the incremental amplitudes, and hence is given by the length of the chord. Therefore, $E_\theta < E_0$. The total phase difference β between

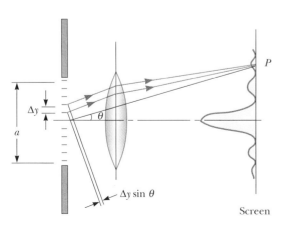

FIGURE 38.8 Fraunhofer diffraction by a single slit. The intensity at the point P is the resultant of all the incremental fields from zones of width Δy leaving the slit.

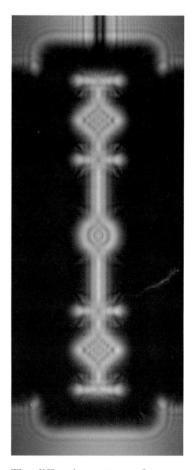

The diffraction pattern of a razor blade observed under monochromatic light. (© Ken Kay, 1987/Fundamental Photographs)

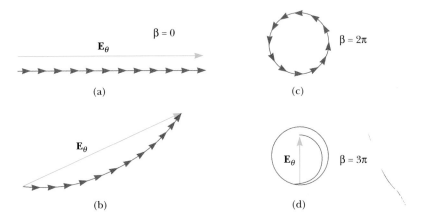

FIGURE 38.9 Phasor diagrams for obtaining the various maxima and minima of a single-slit diffraction pattern.

waves from the top and bottom portions of the slit is

$$\beta = N\Delta\beta = \frac{2\pi}{\lambda} N\Delta y \sin\theta = \frac{2\pi}{\lambda} a \sin\theta \qquad (38.3)$$

where $a = N\Delta y$ is the width of the slit.

As θ increases, the chain of phasors eventually forms a closed path as in Figure 38.9c. At this point, the vector sum is zero, and so $E_\theta = 0$, corresponding to the first minimum on the screen. Noting that $\beta = N\Delta\beta = 2\pi$ in this situation, we see from Equation 38.3 that

$$2\pi = \frac{2\pi}{\lambda} a \sin\theta$$

$$\sin\theta = \frac{\lambda}{a}$$

That is, the first minimum in the diffraction pattern occurs when $\sin\theta = \lambda/a$, which agrees with Equation 38.1.

At larger values of θ, the spiral chain of phasors continues. For example, Figure 38.9d represents the situation corresponding to the second maximum, which occurs when $\beta \cong 360° + 180° = 540°$ (3π rad). The second minimum (two complete spirals, not shown) corresponds to $\beta = 720°$ (4π rad), which satisfies the condition $\sin\theta = 2\lambda/a$.

The total amplitude and intensity at any point on the screen can be obtained by considering the limiting case where Δy becomes infinitesimal (dy) and $N \to \infty$. In this limit, the phasor chains in Figure 38.9 become smooth curves, as in Figure 38.10. From this figure, we see that at some angle θ, the wave amplitude on the screen, E_θ, is equal to the chord length, while E_0 is the arc length. From the triangle whose angle is $\beta/2$, we see that

$$\sin\frac{\beta}{2} = \frac{E_\theta/2}{R}$$

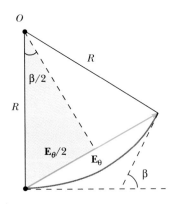

FIGURE 38.10 Phasor diagram for a large number of coherent sources. Note that all the ends of the phasors lie on a circular arc of radius R. The resultant amplitude E_θ equals the length of the chord.

where R is the radius of curvature. But the arc length E_0 is equal to the product $R\beta$, where β is in radians. Combining this with the expression above gives

$$E_\theta = 2R\sin\frac{\beta}{2} = 2\left(\frac{E_0}{\beta}\right)\sin\frac{\beta}{2} = E_0\left[\frac{\sin(\beta/2)}{\beta/2}\right]$$

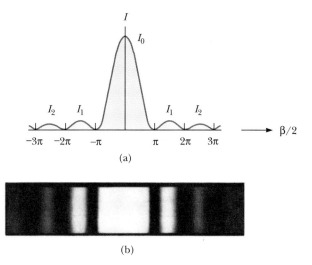

FIGURE 38.11 (a) A plot of the intensity I versus $\beta/2$ for the single-slit Fraunhofer diffraction pattern. (b) Photograph of a single-slit Fraunhofer diffraction pattern. *(From M. Cagnet, M. Francon, and J. C. Thierr,* Atlas of Optical Phenomena, *Berlin, Springer-Verlag, 1962, plate 18)*

Because the resultant intensity I_θ at P is proportional to the square of the amplitude E_θ, we find

$$I_\theta = I_0 \left[\frac{\sin(\beta/2)}{\beta/2} \right]^2 \quad (38.4)$$

Intensity of a single-slit Fraunhofer diffraction pattern

where I_0 is the intensity at $\theta = 0$ (the central maximum) and $\beta = 2\pi a \sin\theta/\lambda$. Substitution of this expression for β into Equation 38.4 gives

$$I_\theta = I_0 \left[\frac{\sin(\pi a \sin\theta/\lambda)}{\pi a \sin\theta/\lambda} \right]^2 \quad (38.5)$$

From this result, we see that minima occur when

$$\frac{\pi a \sin\theta}{\lambda} = m\pi$$

$$\sin\theta = m\frac{\lambda}{a}$$

Condition for intensity minima

where $m = \pm 1, \pm 2, \pm 3, \ldots$. This is in agreement with our earlier result, given by Equation 38.1.

Figure 38.11a represents a plot of Equation 38.4, and a photograph of a single-slit Fraunhofer diffraction pattern is shown in Figure 38.11b. Most of the light intensity is concentrated in the central bright fringe.

EXAMPLE 38.2 Relative Intensities of the Maxima

Find the ratio of intensities of the secondary maxima to the intensity of the central maximum for the single-slit Fraunhofer diffraction pattern.

Solution To a good approximation, the secondary maxima lie midway between the zero points. From Figure 38.11a, we see that this corresponds to $\beta/2$ values of $3\pi/2$, $5\pi/2$,

$7\pi/2, \ldots$. Substituting these into Equation 38.4 gives for the first two ratios

$$\frac{I_1}{I_0} = \left[\frac{\sin(3\pi/2)}{(3\pi/2)}\right]^2 = \frac{1}{9\pi^2/4} = 0.045$$

$$\frac{I_2}{I_0} = \left[\frac{\sin(5\pi/2)}{5\pi/2}\right]^2 = \frac{1}{25\pi^2/4} = 0.016$$

That is, the secondary maximum (the one adjacent to the central maximum) has an intensity of 4.5% that of the central bright fringe, and the next second-order maximum has an intensity of 1.6% that of the central bright fringe.

Exercise Determine the intensity of the secondary maximum corresponding to $m = 3$ relative to the central maximum.

Answer 0.0083.

38.3 RESOLUTION OF SINGLE-SLIT AND CIRCULAR APERTURES

The ability of optical systems to distinguish between closely spaced objects is limited because of the wave nature of light. To understand this difficulty, consider Figure 38.12, which shows two light sources far from a narrow slit of width a. The sources can be considered as two noncoherent point sources, S_1 and S_2. For example, they could be two distant stars. If no diffraction occurred, two distinct bright spots (or images) would be observed on the viewing screen. However, because of diffraction, each source is imaged as a bright central region flanked by weaker bright and dark bands. What is observed on the screen is the sum of two diffraction patterns, one from S_1 and the other from S_2.

If the two sources are separated enough to keep their central maxima from overlapping, as in Figure 38.12a, their images can be distinguished and are said to be resolved. If the sources are close together, however, as in Figure 38.12b, the two central maxima overlap and the images are not resolved. To decide when two images are resolved, the following criterion is used:

When the central maximum of one image falls on the first minimum of another image, the images are said to be just resolved. This limiting condition of resolution is known as **Rayleigh's criterion.**

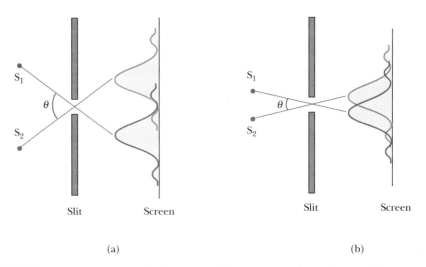

FIGURE 38.12 Two point sources far from a small aperture each produce a diffraction pattern. (a) The angle subtended by the sources at the aperture is large enough so that the diffraction patterns are distinguishable. (b) The angle subtended by the sources is so small that their diffraction patterns overlap and the images are not well resolved. (Note that the angles are greatly exaggerated.)

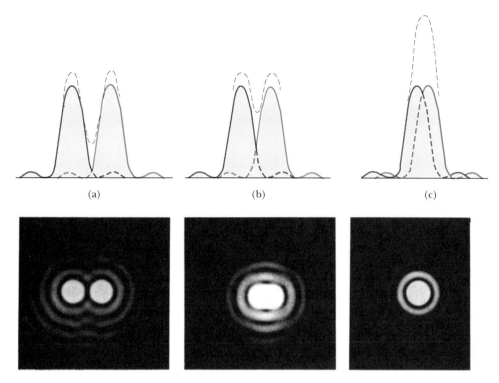

FIGURE 38.13 The diffraction patterns of two point sources (solid curves) and the resultant pattern (dashed curves) for various angular separations of the sources. In each case, the dashed curve is the sum of the two solid curves. (a) The sources are far apart, and the patterns are well resolved. (b) The sources are closer together, and the patterns are just resolved. (c) The sources are so close together that the patterns are not resolved. *(From M. Cagnet, M. Francon, and J. C. Thierr,* Atlas of Optical Phenomena, *Berlin, Springer-Verlag, 1962, plate 16)*

Figure 38.13 shows diffraction patterns for three situations. When the objects are far apart, their images are well resolved (Fig. 38.13a). The images are just resolved when the angular separation of the objects satisfies Rayleigh's criterion (Fig. 38.13b). Finally, the images are not resolved in Figure 38.13c.

From Rayleigh's criterion, we can determine the minimum angular separation, θ_{min}, subtended by the sources at the slit so that their images are just resolved. In Section 38.2, we found that the first minimum in a single-slit diffraction pattern occurs at the angle for which

$$\sin \theta = \frac{\lambda}{a}$$

where a is the width of the slit. According to Rayleigh's criterion, this expression gives the smallest angular separation for which the two images are resolved. Because $\lambda \ll a$ in most situations, $\sin \theta$ is small and we can use the approximation $\sin \theta \approx \theta$. Therefore, the limiting angle of resolution for a slit of width a is

$$\theta_{min} = \frac{\lambda}{a} \qquad (38.6)$$

where θ_{min} is expressed in radians. Hence, the angle subtended by the two sources at the slit must be greater than λ/a if the images are to be resolved.

Many optical systems use circular apertures rather than rectangular slits. The diffraction pattern of a circular aperture, illustrated in Figure 38.14, consists of a

FIGURE 38.14 The diffraction pattern of a circular aperture consists of a central bright disk surrounded by concentric bright and dark rings. *(From M. Cagnet, M. Francon, and J. C. Thierr,* Atlas of Optical Phenomena, *Berlin, Springer-Verlag, 1962, plate 34)*

Limiting angle of resolution for a circular aperture

central circular bright disk surrounded by progressively fainter rings. The limiting angle of resolution of the circular aperture is

$$\theta_{min} = 1.22 \frac{\lambda}{D} \qquad (38.7)$$

where D is the diameter of the aperture. Note that Equation 38.7 is similar to Equation 38.6 except for the factor of 1.22, which arises from a complex mathematical analysis of diffraction from the circular aperture.

CONCEPTUAL EXAMPLE 38.3 Can You See an Atom?

Explain why it is theoretically impossible to see an object as small as an atom regardless of the quality of the light microscope being used.

Reasoning In order to "see" an object, the wavelength of the light in the microscope must be comparable to the size of the object. An atom is much smaller than the wavelength of light in the visible region of the spectrum, so an atom can never be seen using visible light.

EXAMPLE 38.4 Limiting Resolution of a Microscope

Light of wavelength 589 nm is used to view an object under a microscope. If the aperture of the objective has a diameter of 0.900 cm, (a) find the limiting angle of resolution.

Solution (a) From Equation 38.7, we find the limiting angle of resolution to be

$$\theta_{min} = 1.22 \left(\frac{589 \times 10^{-9} \text{ m}}{0.900 \times 10^{-2} \text{ m}} \right) = 7.98 \times 10^{-5} \text{ rad}$$

This means that any two points on the object subtending an angle less than approximately 8×10^{-5} rad at the objective cannot be distinguished in the image.

(b) Using visible light of any wavelength you desire, what is the maximum limit of resolution for this microscope?

Solution To obtain the smallest angle corresponding to the maximum limit of resolution, we have to use the shortest wavelength available in the visible spectrum. Violet light (400 nm) gives us a limiting angle of resolution of

$$\theta_{min} = 1.22 \left(\frac{400 \times 10^{-9} \text{ m}}{0.900 \times 10^{-2} \text{ m}} \right) = 5.42 \times 10^{-5} \text{ rad}$$

(c) Suppose water ($n = 1.33$) fills the space between object and objective. What effect would this have on resolving power?

Solution In this case, the wavelength of the sodium light in the water is found by $\lambda_w = \lambda_a/n$ (Eq. 35.7). Thus, we have

$$\lambda_w = \frac{\lambda_a}{n} = \frac{589 \text{ nm}}{1.33} = 443 \text{ nm}$$

The limiting angle of resolution at this wavelength is

$$\theta_m = 1.22 \left(\frac{443 \times 10^{-9} \text{ m}}{0.900 \times 10^{-2} \text{ m}} \right) = 6.00 \times 10^{-5} \text{ rad}$$

EXAMPLE 38.5 Resolution of a Telescope

The Hale telescope at Mount Palomar has a diameter of 200 in. What is its limiting angle of resolution for 600-nm light?

Solution Because $D = 200$ in. $= 5.08$ m and $\lambda = 6.00 \times 10^{-7}$ m, Equation 38.7 gives

$$\theta_{min} = 1.22 \frac{\lambda}{D} = 1.22 \left(\frac{6.00 \times 10^{-7} \text{ m}}{5.08 \text{ m}} \right)$$

$$= 1.44 \times 10^{-7} \text{ rad} \cong 0.03 \text{ s of arc}$$

Therefore, any two stars that subtend an angle greater than or equal to this value are resolved (assuming ideal atmospheric conditions).

The Hale telescope can never reach its diffraction limit. Instead, its limiting angle of resolution is always set by atmospheric blurring. This seeing limit is usually approximately 1 s of arc and is never smaller than approximately 0.1 s of arc. (This is one of the reasons for the current interest in a large space telescope.)

Exercise The large radio telescope at Arecibo, Puerto Rico, has a diameter of 305 m and is designed to detect 0.75-m radio waves. Calculate the minimum angle of resolution for this telescope and compare your answer with that of the Hale telescope.

Answer 3.0×10^{-3} rad (10 min of arc), more than 10 000 times larger than the Hale minimum.

EXAMPLE 38.6 Resolution of the Eye

Calculate the limiting angle of resolution for the eye assuming its resolution is limited only by diffraction. Take the wavelength to be 500 nm and assume a pupil diameter of 2.00 mm.

Solution We use Equation 38.7, taking $\lambda = 500$ nm and $D = 2.00$ mm. This gives

$$\theta_{min} = 1.22 \frac{\lambda}{D} = 1.22 \left(\frac{5.00 \times 10^{-7} \text{ m}}{2.00 \times 10^{-3} \text{ m}} \right)$$

$$= 3.05 \times 10^{-4} \text{ rad} = 0.0175°$$

We can use this result to calculate the minimum separation d between two point sources that the eye can distinguish if they are a distance L from the observer (Fig. 38.15). Since θ_{min} is small, we see that

$$\sin \theta_{min} \approx \theta_{min} \approx \frac{d}{L}$$

$$d = L \theta_{min}$$

For example, if the objects are 25 cm from the eye (the near

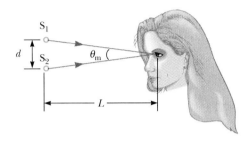

FIGURE 38.15 (Example 38.6) Two point sources separated by a distance d as observed by the eye.

point), then

$$d = (25 \text{ cm})(3.05 \times 10^{-4} \text{ rad}) = 7.6 \times 10^{-3} \text{ cm}$$

This is approximately equal to the thickness of a human hair.

Exercise Suppose the eye is dilated to a diameter of 5.0 mm and is looking at two point sources 3.0 cm away. How far apart must the sources be if this eye is to resolve them?

Answer 0.037 cm.

38.4 THE DIFFRACTION GRATING

The diffraction grating, a useful device for analyzing light sources, consists of a large number of equally spaced parallel slits. A grating can be made by cutting parallel lines on a glass plate with a precision ruling machine. In a transmission grating, the space between any two lines is transparent to the light and hence acts as a separate slit. Gratings with many lines very close to each other can have very small slit spacings. For example, a grating ruled with 5000 lines/cm has a slit spacing $d = (1/5000)$ cm $= 2.00 \times 10^{-4}$ cm.

A section of a diffraction grating is illustrated in Figure 38.16. A plane wave is incident from the left, normal to the plane of the grating. A converging lens brings the rays together at the point P. The pattern observed on the screen is the result of the combined effects of interference and diffraction. Each slit produces diffraction, and the diffracted beams interfere with each other to produce the final pattern. Moreover, each slit acts as a source of waves, where all waves start at the slits in phase. However, for some arbitrary direction θ measured from the horizontal, the waves must travel different path lengths before reaching P. From Figure 38.16, note that the path difference between waves from any two adjacent slits is

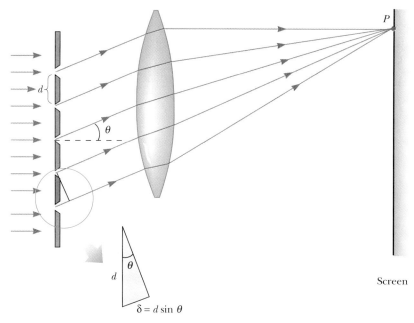

FIGURE 38.16 Side view of a diffraction grating. The slit separation is d, and the path difference between adjacent slits is $d \sin \theta$.

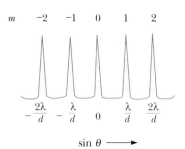

FIGURE 38.17 Intensity versus $\sin \theta$ for a diffraction grating. The zeroth-, first-, and second-order maxima are shown.

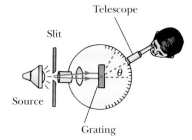

FIGURE 38.18 Diagram of a diffraction grating spectrometer. The collimated beam incident on the grating is diffracted into the various orders at the angles θ that satisfy the equation $d \sin \theta = m\lambda$, where $m = 0, 1, 2,\ldots$. (In most spectrometers, the angle of incidence is approximately equal to the angle of diffraction, so $m\lambda = d(\sin \theta + \sin \theta')$, with $\theta \approx \pm \theta'$.)

equal to $d \sin \theta$. If this path difference equals one wavelength or some integral multiple of a wavelength, waves from all slits are in phase at P and a bright line is observed. Therefore, the condition for maxima in the interference pattern at the angle θ is

$$d \sin \theta = m\lambda \qquad (m = 0, 1, 2, 3, \ldots) \qquad (38.8)$$

This expression can be used to calculate the wavelength from a knowledge of the grating spacing and the angle of deviation θ. If the incident radiation contains several wavelengths, the mth order maximum for each wavelength occurs at a specific angle. All wavelengths are seen at $\theta = 0$, corresponding to $m = 0$, the zeroth-order maximum. The first-order maximum ($m = 1$) is observed at an angle that satisfies the relationship $\sin \theta = \lambda/d$; the second-order maximum ($m = 2$) is observed at a larger angle θ, and so on.

The intensity distribution for a diffraction grating using a monochromatic source is shown in Figure 38.17. Note the sharpness of the principal maxima and the broadness of dark areas. This is in contrast to the broad bright fringes characteristic of the two-slit interference pattern (Fig. 37.1).

A simple arrangement used to measure orders of a diffraction pattern is shown in Figure 38.18. This is a form of a diffraction grating spectrometer. The light to be analyzed passes through a slit, and a parallel beam of light exits from the collimator, which is perpendicular to the grating. The diffracted light leaves the grating at angles that satisfy Equation 38.8, and a telescope is used to view the image of the slit. The wavelength can be determined by measuring the precise angles at which the images of the slit appear for the various orders.

CONCEPTUAL EXAMPLE 38.7 The Compact Disc Is a Diffraction Grating

Light reflected from the surface of a compact disc has a multicolored appearance as shown in Figure 38.19. Furthermore, the observation depends on the orientation of the disc relative to the eye and the position of the light source. Explain how this works.

Reasoning The surface of a compact disc has a spiral grooved track (with a spacing of approximately 1 μm) that acts as a reflection grating. The light scattered by these closely spaced grooves interferes constructively only in certain directions that depend on the wavelength and on the direction of the incident light. Any one section of the disc serves as a diffraction grating for white light, sending different colors in different directions. The different colors you see when viewing one section of the disc change as the light source, the disc, or you move to change the angles of incidence or diffraction.

FIGURE 38.19 (Conceptual Example 38.7) A compact disc observed under white light. The colors that are observed in the reflected light and their intensities depend on the orientation of the disc relative to the eye and to the light source. Can you explain how this works? *(Kristen Brochmann/Fundamental Photographs)*

EXAMPLE 38.8 The Orders of a Diffraction Grating

Monochromatic light from a helium–neon laser ($\lambda = 632.8$ nm) is incident normally on a diffraction grating containing 6000 lines/cm. Find the angles at which the first-order, second-order, and third-order maximum can be observed.

Solution First, we must calculate the slit separation, which is equal to the inverse of the number of lines per cm:

$$d = (1/6000) \text{ cm} = 1.667 \times 10^{-4} \text{ cm} = 1667 \text{ nm}$$

For the first-order maximum ($m = 1$), we get

$$\sin \theta_1 = \frac{\lambda}{d} = \frac{632.8 \text{ nm}}{1667 \text{ nm}} = 0.3797$$

$$\theta_1 = 22.31°$$

For $m = 2$, we find

$$\sin \theta_2 = \frac{2\lambda}{d} = \frac{2(632.8 \text{ nm})}{1667 \text{ nm}} = 0.7592$$

$$\theta_2 = 49.41°$$

For $m = 3$, we find $\sin \theta_3 = 1.139$. Since $\sin \theta$ cannot exceed unity, this does not represent a realistic solution. Hence, only zeroth-, first-, and second-order maxima are observed for this situation.

Resolving Power of the Diffraction Grating

The diffraction grating is useful for measuring wavelengths accurately. Like the prism, the diffraction grating can be used to disperse a spectrum into its components. Of the two devices, the grating may be more precise if one wants to distinguish between two closely spaced wavelengths.

If λ_1 and λ_2 are the two nearly equal wavelengths between which the spectrometer can just barely distinguish, the **resolving power** R is defined as

Resolving power

$$R \equiv \frac{\lambda}{\lambda_2 - \lambda_1} = \frac{\lambda}{\Delta\lambda} \qquad (38.9)$$

where $\lambda = (\lambda_1 + \lambda_2)/2$ and $\Delta\lambda = \lambda_2 - \lambda_1$. Thus, a grating with a high resolving power can distinguish small differences in wavelength. Furthermore, if N lines of the grating are illuminated, it can be shown that the resolving power in the mth-order diffraction equals the product Nm:

Resolving power of a grating

$$R = Nm \qquad (38.10)$$

Thus, resolving power increases with increasing order number. Furthermore, R is large for a grating that has a large number of illuminated slits. Note that for $m = 0$, $R = 0$, which signifies that *all wavelengths are indistinguishable* for the zeroth-order maximum. However, consider the second-order diffraction pattern ($m = 2$) of a grating that has 5000 rulings illuminated by the light source. The resolving power of such a grating in second order is $R = 5000 \times 2 = 10\,000$. Therefore, the minimum wavelength separation between two spectral lines that can be just resolved, assuming a mean wavelength of 600 nm, is $\Delta\lambda = \lambda/R = 6.00 \times 10^{-2}$ nm. For the third-order principal maximum, $R = 15\,000$ and $\Delta\lambda = 4.00 \times 10^{-2}$ nm, and so on.

EXAMPLE 38.9 Resolving the Sodium Spectral Lines

Two strong lines in the spectrum of sodium have wavelengths of 589.00 nm and 589.59 nm. (a) What must the resolving power of a grating be in order to distinguish these wavelengths?

Solution

$$R = \frac{\lambda}{\Delta\lambda} = \frac{589.30 \text{ nm}}{589.59 \text{ nm} - 589.00 \text{ nm}} = \frac{589.30}{0.59} = \boxed{999}$$

(b) In order to resolve these lines in the second-order spectrum ($m = 2$), how many lines of the grating must be illuminated?

Solution From Equation 38.10 and the results to part (a), we find

$$N = \frac{R}{m} = \frac{999}{2} = \boxed{500 \text{ lines}}$$

---

*38.5 DIFFRACTION OF X-RAYS BY CRYSTALS

In principle, the wavelength of any electromagnetic wave can be determined if a grating of the proper spacing (of the order of λ) is available. X-rays, discovered by W. Roentgen (1845–1923) in 1895, are electromagnetic waves of very short wavelength (of the order of $= 0.1$ nm). Obviously, it would be impossible to construct a grating having such a small spacing. However, the atomic spacing in a solid is known to be about 10^{-10} m. In 1913, Max von Laue (1879–1960) suggested that the regular array of atoms in a crystal could act as a three-dimensional diffraction grating for x-rays. Subsequent experiments confirmed this prediction. The diffraction patterns that are observed are complicated because of the three-dimensional nature of the crystal. Nevertheless, x-ray diffraction has proved to be an invaluable technique for elucidating crystalline structures and for understanding the structure of matter.[1]

---

[1] For more details on this subject, see Sir Lawrence Bragg, "X-Ray Crystallography," *Scientific American*, July 1968.

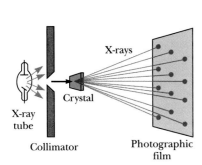

FIGURE 38.20 Schematic diagram of the technique used to observe the diffraction of x-rays by a crystal. The array of spots formed on the film is called a Laue pattern.

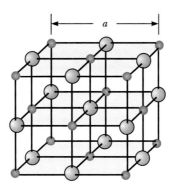

FIGURE 38.21 A model of the crystalline structure of sodium chloride. The blue spheres represent the Cl^- ions, and the red spheres represent the Na^+ ions. The length of the cube edge is $a = 0.562737$ nm.

Figure 38.20 is one experimental arrangement for observing x-ray diffraction from a crystal. A collimated beam of x-rays containing a continuous range of wavelengths is incident on a crystal. The diffracted beams are very intense in certain directions, corresponding to constructive interference from waves reflected from layers of atoms in the crystal. The diffracted beams can be detected by a counter or photographic film, forming an array of spots known as a *Laue pattern*. The crystalline structure is deduced by analyzing the positions and intensities of the various spots in the pattern.

The arrangement of atoms in a crystal of NaCl is shown in Figure 38.21. The smaller, dark spheres represent Na^+ ions and the larger spheres represent Cl^- ions. Each unit cell (the set of atoms that repeats through the crystal) contains four Na^+ and four Cl^- ions. The unit cell is a cube whose edge length is a.

A careful examination of the NaCl structure shows that the ions lie in discrete planes, the shaded areas in Figure 38.21. Now suppose an incident x-ray beam makes an angle θ with one of the planes, as in Figure 38.22. The beam can be reflected from both the upper plane and the lower one. However, the geometric construction in Figure 38.22 shows that the beam reflected from the lower plane travels farther than the beam reflected from the upper one. The effective path difference between the two beams is $2d \sin \theta$. The two beams reinforce each other

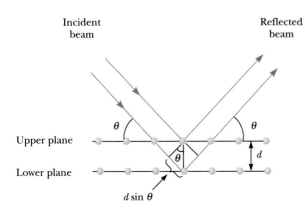

FIGURE 38.22 A two-dimensional description of the reflection of an x-ray beam from two parallel crystalline planes separated by a distance d. The beam reflected from the lower plane travels farther than the one reflected from the upper plane by a distance equal to $2d \sin \theta$.

Bragg's law

(constructive interference) when this path difference equals some integral multiple of the wavelength λ. The same is true for reflection from the entire family of parallel planes. Hence, the condition for constructive interference (maxima in the reflected wave) is

$$2d \sin \theta = m\lambda \quad (m = 1, 2, 3, \ldots) \quad (38.11)$$

This condition is known as **Bragg's law** after W. L. Bragg (1890–1971), who first derived the relationship. If the wavelength and diffraction angle are measured, Equation 38.11 can be used to calculate the spacing between atomic planes.

38.6 POLARIZATION OF LIGHT WAVES

In Chapter 34 we described the transverse nature of light and all other electromagnetic waves. Figure 38.23 shows that the electric and magnetic vectors associated with an electromagnetic wave are at right angles to each other and also to the direction of wave propagation. Polarization is firm evidence of the transverse nature of electromagnetic waves.

An ordinary beam of light consists of a large number of waves emitted by the atoms or molecules of the light source. Each atom produces a wave with its own orientation of **E**, as in Figure 38.23, corresponding to the direction of atomic vibration. The direction of polarization of the electromagnetic wave is defined to be the direction in which **E** is vibrating. However, because all directions of vibration are possible, the resultant electromagnetic wave is a superposition of waves produced by the individual atomic sources. The result is an **unpolarized** light wave, described in Figure 38.24a. The direction of wave propagation in this figure is perpendicular to the page. Note that *all* directions of the electric field vector, lying in a plane perpendicular to the direction of propagation, are equally probable. At any given point and at some instant of time, there is only one resultant electric field, hence you should not be misled by the meaning of Figure 38.24a.

A wave is said to be **linearly polarized** if **E** vibrates in the same direction *at all times* at a particular point, as in Figure 38.24b. (Sometimes such a wave is described as *plane-polarized*, or simply *polarized*.) The wave described in Figure 38.23 is an example of a wave linearly polarized in the *y* direction. As the wave propagates in the *x* direction, **E** is always in the *y* direction. The plane formed by **E** and the direction of propagation is called the *plane of polarization* of the wave. In Figure 38.23, the plane of polarization is the *xy* plane. It is possible to obtain a linearly polarized beam from an unpolarized beam by removing all waves from the beam except those whose electric field vectors oscillate in a single plane. We now discuss four processes for producing polarized light from unpolarized light.

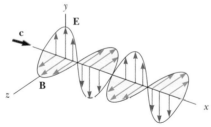

FIGURE 38.23 Schematic diagram of an electromagnetic wave propagating in the *x* direction. The electric field vector **E** vibrates in the *xy* plane, and the magnetic field vector **B** vibrates in the *xz* plane.

Polarization by Selective Absorption

The most common technique for polarizing light is to use a material that transmits waves whose electric field vectors vibrate in a given plane and absorbs waves whose electric field vectors vibrate in other directions.

In 1938, E. H. Land discovered a material, which he called **Polaroid,** that polarizes light through selective absorption by oriented molecules. This material is fabricated in thin sheets of long-chain hydrocarbons. The sheets are manufactured such that the molecules are aligned in long chains. After a sheet is dipped into a solution containing iodine, the molecules become good electrical conductors. However, the conduction takes place primarily along the hydrocarbon chains since the electrons of the molecules can move easily only along the chains. As a result, the molecules readily absorb light whose electric field vector is parallel to their length and transmit light whose electric field vector is perpendicular to their length. It is common to refer to the direction perpendicular to the molecular chains as the **transmission axis.** In an ideal polarizer, all light with **E** parallel to the transmission axis is transmitted, and all light with **E** perpendicular to the transmission axis is absorbed.

Figure 38.25 represents an unpolarized light beam incident on a first polarizing sheet, called the **polarizer,** where the transmission axis is indicated by the straight lines on the polarizer. The light passing through this sheet is polarized vertically as shown, where the transmitted electric field vector is $\mathbf{E}_0$. A second polarizing sheet, called the **analyzer,** intercepts this beam because the analyzer transmission axis is set at an angle θ to the polarizer axis. The component of $\mathbf{E}_0$ perpendicular to the axis of the analyzer is completely absorbed, and the component of $\mathbf{E}_0$ parallel to the axis of the analyzer is $E_0 \cos \theta$. Note that when light emerges from a polarizer, it loses all information on its original plane of polarization and has the polarization of the most recent polarizer. Because the transmitted intensity varies as the square of the transmitted amplitude, we conclude that the intensity of the transmitted (polarized) light varies as

$$I = I_0 \cos^2 \theta \tag{38.12}$$

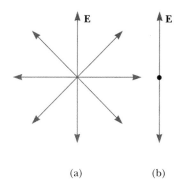

FIGURE 38.24 (a) An unpolarized light beam viewed along the direction of propagation (perpendicular to the page). The transverse electric field vector can vibrate in any direction with equal probability. (b) A linearly polarized light beam with the electric field vector vibrating in the vertical direction.

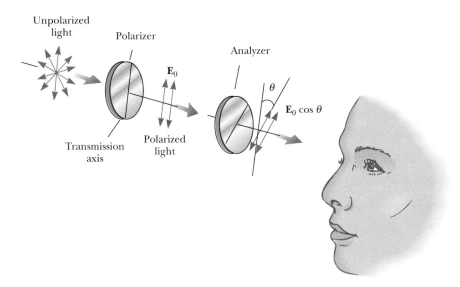

FIGURE 38.25 Two polarizing sheets whose transmission axes make an angle θ with each other. Only a fraction of the polarized light incident on the analyzer is transmitted.

where I_0 is the intensity of the polarized wave incident on the analyzer. This expression, known as **Malus's law**,[2] applies to any two polarizing materials whose transmission axes are at an angle θ to each other.

From this expression, note that the transmitted intensity is maximum when the transmission axes are parallel ($\theta = 0$ or $180°$) and zero (complete absorption by the analyzer) when the transmission axes are perpendicular to each other. This variation in transmitted intensity through a pair of polarizing sheets is illustrated in Figure 38.26.

Polarization by Reflection

When an unpolarized light beam is reflected from a surface, the reflected light is completely polarized, partially polarized, or unpolarized, depending on the angle of incidence. If the angle of incidence is either 0 or $90°$, the reflected beam is unpolarized. For intermediate angles of incidence, however, the reflected light is polarized to some extent, and for one particular angle of incidence, the reflected light is completely polarized. Let us now investigate reflection at that special angle.

Suppose an unpolarized light beam is incident on a surface as in Figure 38.27a. The beam can be described by two electric field components, one parallel to the surface (represented by the dots) and the other perpendicular both to the first component (represented by the red arrows) and to the direction of propagation. It is found that the parallel component reflects more strongly than the perpendicular component, and this results in a partially polarized reflected beam. Furthermore, the refracted beam is also partially polarized.

Now suppose the angle of incidence, θ_1, is varied until the angle between the reflected and refracted beams is $90°$ (Fig. 38.27b). At this particular angle of incidence, the reflected beam is completely polarized, with its electric field vector parallel to the surface, while the refracted beam is partially polarized. The angle of incidence at which this occurs is called the **polarizing angle**, θ_p.

The polarizing angle

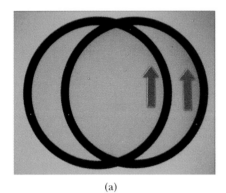
(a)
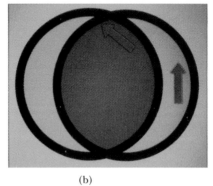
(b)
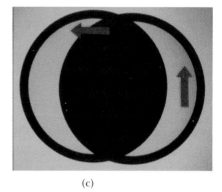
(c)

FIGURE 38.26 The intensity of light transmitted through two polarizers depends on the relative orientation of their transmission axes. (a) The transmitted light has maximum intensity when the transmission axes are aligned with each other. (b) The transmitted light intensity diminishes when the transmission axes are at an angle of $45°$ with each other. (c) The transmitted light intensity is a minimum when the transmission axes are at right angles to each other. *(Henry Leap and Jim Lehman)*

---

[2] Named after its discoverer, E. L. Malus (1775–1812). Actually, Malus first discovered that reflected light was polarized by viewing it through a calcite ($CaCO_3$) crystal.

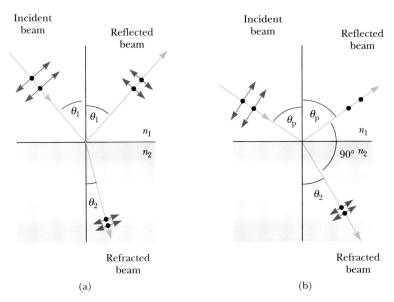

FIGURE 38.27 (a) When unpolarized light is incident on a reflecting surface, the reflected and refracted beams are partially polarized. (b) The reflected beam is completely polarized when the angle of incidence equals the polarizing angle, θ_p, which satisfies the equation $n = \tan \theta_p$.

An expression relating the polarizing angle to the index of refraction of the reflecting substance can be obtained from Figure 38.27b. From this figure, we see that $\theta_p + 90° + \theta_2 = 180°$, so that $\theta_2 = 90° - \theta_p$. Using Snell's law and taking $n_1 = 1.00$ and $n_2 = n$, we have

$$n = \frac{\sin \theta_1}{\sin \theta_2} = \frac{\sin \theta_p}{\sin \theta_2}$$

Because $\sin \theta_2 = \sin(90° - \theta_p) = \cos \theta_p$, the expression for n can be written as $n = \sin \theta_p / \cos \theta_p$, or

$$n = \tan \theta_p \tag{38.13}$$

Brewster's law

This expression is called **Brewster's law**, and the polarizing angle θ_p is sometimes called **Brewster's angle**, after its discoverer, Sir David Brewster (1781–1868). For example, the Brewster's angle for crown glass ($n = 1.52$) is $\theta_p = \tan^{-1}(1.52) = 56.7°$. Because n varies with wavelength for a given substance, the Brewster's angle is also a function of wavelength.

Polarization by reflection is a common phenomenon. Sunlight reflected from water, glass, and snow is partially polarized. If the surface is horizontal, the electric field vector of the reflected light has a strong horizontal component. Sunglasses made of polarizing material reduce the glare of reflected light. The transmission axes of the lenses are oriented vertically so as to absorb the strong horizontal component of the reflected light.

Polarization by Double Refraction

Solids can be classified on the basis of internal structure. Those in which the atoms are arranged in a specific order are called *crystalline;* the sodium chloride structure

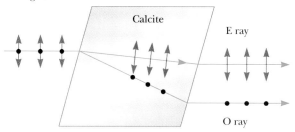

FIGURE 38.28 Unpolarized light incident on a calcite crystal splits into an ordinary (O) ray and an extraordinary (E) ray. These two rays are polarized in mutually perpendicular directions. (Note that this is not to scale.)

of Figure 38.21 is just one example of crystalline solids. Those in which the atoms are distributed randomly are called *amorphous*. When light travels through an amorphous material, such as glass, it travels with a speed that is the same in all directions. That is, glass has a single index of refraction. In certain crystalline materials, however, such as calcite and quartz, the speed of light is not the same in all directions. Such materials are characterized by two indices of refraction. Hence, they are often referred to as **double-refracting** or **birefringent** materials.

When unpolarized light enters a calcite crystal, it splits into two plane-polarized rays that travel with different speeds, corresponding to two angles of refraction, as in Figure 38.28. The two rays are polarized in two mutually perpendicular directions, as indicated by the dots and arrows. One ray, called the **ordinary** (O) **ray**, is characterized by an index of refraction, n_O, that is the same in all directions. This means that if one could place a point source of light inside the crystal, as in Figure 38.29, the ordinary waves would spread out from the source as spheres.

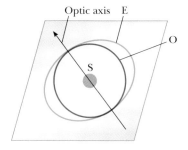

FIGURE 38.29 A point source, S, inside a double-refracting crystal produces a spherical wave front corresponding to the ordinary ray and an elliptical wavefront corresponding to the extraordinary ray. The two waves propagate with the same velocity along the optic axis.

The second plane-polarized ray, called the **extraordinary** (E) **ray**, travels with different speeds in different directions and hence is characterized by an index of refraction, n_E, that varies with the direction of propagation. A point source of light inside such a crystal would send out an extraordinary wave having wave fronts that are elliptical in cross-section (Fig. 38.29). Note from Figure 38.29 that there is one direction, called the **optic axis,** along which the ordinary and extraordinary rays have the same speed, corresponding to the direction for which $n_O = n_E$. The difference in speed for the two rays is a maximum in the direction perpendicular to the optic axis. For example, in calcite, $n_O = 1.658$ at a wavelength of 589.3 nm and n_E varies from 1.658 along the optic axis to 1.486 perpendicular to the optic axis. Values for n_O and n_E for various double-refracting crystals are given in Table 38.1.

If a piece of calcite is placed on a sheet of paper and then we look through the crystal at any writing on the paper, two images are seen, as shown in Figure 38.30.

FIGURE 38.30 A calcite crystal produces a double image because it is a birefringent (double-refracting) material. *(Henry Leap and Jim Lehman)*

TABLE 38.1 Indices of Refraction for Some Double-Refracting Crystals at a Wavelength of 589.3 nm

Crystal	n_O	n_E	n_O/n_E
Calcite ($CaCO_3$)	1.658	1.486	1.116
Quartz (SiO_2)	1.544	1.553	0.994
Sodium nitrate ($NaNO_3$)	1.587	1.336	1.188
Sodium sulfite ($NaSO_3$)	1.565	1.515	1.033
Zinc chloride ($ZnCl_2$)	1.687	1.713	0.985
Zinc sulfide (ZnS)	2.356	2.378	0.991

As can be seen from Figure 38.28, these two images correspond to one formed by the ordinary ray and the second formed by the extraordinary ray. If the two images are viewed through a sheet of rotating polarizing glass, they alternately appear and disappear because the ordinary and extraordinary rays are plane-polarized along mutually perpendicular directions.

Polarization by Scattering

When light is incident on any material, the electrons in the material can absorb and reradiate part of the light. Such absorption and reradiation of light by electrons in the gas molecules that make up air are what causes sunlight reaching an observer on the Earth to be partially polarized. You can observe this effect—called **scattering**—by looking directly up through a pair of sunglasses whose lenses are made of polarizing material. Less light passes through at certain orientations of the lenses than at others.

Figure 38.31 illustrates how sunlight becomes partially polarized. An incident unpolarized beam of sunlight is traveling in the horizontal direction and on the verge of striking a molecule of one of the gases that make up air. This beam striking the gas molecule sets the electrons of the molecule into vibration. These vibrating charges act like the vibrating charges in an antenna, except that these charges are vibrating in a complicated pattern. The horizontal part of the electric field vector in the incident wave causes the charges to vibrate horizontally, and the vertical part of the vector simultaneously causes them to vibrate vertically. A horizontally polarized wave is emitted by the electrons as a result of their horizontal motion, and a vertically polarized wave is emitted parallel to the Earth as a result of their vertical motion.

Some phenomena involving the scattering of light in the atmosphere can be understood as follows: When light of various wavelengths is incident on gas molecules of diameter d, where $d \ll \lambda$, the relative intensity of the scattered light varies as $1/\lambda^4$. This condition is satisfied for scattering of sunlight in the Earth's atmosphere. Hence shorter wavelengths (blue light) are scattered more efficiently than longer wavelengths (red light). When sunlight is scattered by gas molecules in the air, therefore, the shorter wavelength radiation (blue part) is scattered more intensely than the longer wavelength radiation (red part), and as a result the sky appears blue.

Optical Activity

Many important applications of polarized light involve materials that display **optical activity**. A substance is said to be optically active if it rotates the plane of polarization of transmitted light. The angle through which the light is rotated by a specific material depends on the length of the sample and on the concentration if the substance is in solution. One optically active material is a solution of the common sugar dextrose. A standard method for determining the concentration of sugar solutions is to measure the rotation produced by a fixed length of the solution.

A material is optically active because of an asymmetry in the shape of its constituent molecules. For example, some proteins are optically active because of their spiral shape. Other materials, such as glass and plastic, become optically active when stressed. Suppose an unstressed piece of plastic is placed between a polarizer and an analyzer so that light passes from polarizer to plastic to analyzer. When the analyzer axis is perpendicular to the polarizer axis, none of the polarized light passes through the analyzer. In other words, the unstressed plastic has no

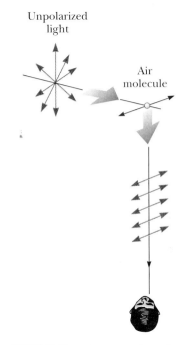

FIGURE 38.31 The scattering of unpolarized sunlight by air molecules. The light observed at right angles is plane-polarized because the vibrating molecule has a horizontal component of vibration.

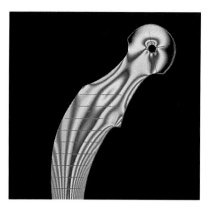 

FIGURE 38.32 (a) Photograph showing strain distribution in a plastic model of a hip replacement used in a medical research laboratory. The pattern is produced when the plastic model is viewed between a polarizer and an analyzer oriented perpendicularly to the polarizer. *(Sepp Seitz, 1981)* (b) A plastic model of an arch structure under load conditions observed between two crossed polarizers. Such patterns are useful in the optimum design of architectural components. *(Peter Aprahamian/Science Photo Library)*

effect on the light passing through it. If the plastic is stressed, however, the regions of greatest stress rotate the polarized light through the largest angles. Hence, a series of bright and dark bands is observed in the transmitted light, with the bright bands corresponding to regions of greatest stress.

Engineers often use this technique, called *optical stress analysis,* to assist in designing structures ranging from bridges to small tools. A plastic model is built and analyzed under different load conditions to determine regions of potential weakness and failure under stress. Some examples of plastic models under stress are shown in Figure 38.32.

SUMMARY

Diffraction arises from interference in a very large number of coherent sources. Diffraction accounts for the deviation of light from a straight-line path when the light passes through an aperture or around obstacles.

The **Fraunhofer diffraction pattern** produced by a single slit of width a on a distant screen consists of a central bright maximum and alternating bright and dark regions of much lower intensities. The angles θ at which the diffraction pattern has zero intensity, corresponding to destructive interference, are given by

$$\sin\theta = m\frac{\lambda}{a} \quad (m = \pm 1, \pm 2, \pm 3, \ldots) \tag{38.1}$$

How the intensity I of a single-slit diffraction pattern varies with angle θ is given by

$$I_\theta = I_0 \left[\frac{\sin(\beta/2)}{\beta/2}\right]^2 \tag{38.4}$$

where $\beta = 2\pi a \sin\theta/\lambda$ and I_0 is the intensity at $\theta = 0$.

Rayleigh's criterion, which is a limiting condition of resolution, says that two images formed by an aperture are just distinguishable if the central maximum of the diffraction pattern for one image falls on the first minimum of the diffraction pattern for the other image. The limiting angle of resolution for a slit of width a is $\theta_{min} = \lambda/a$, and the limiting angle of resolution for a circular aperture of diameter D is given by $\theta_{min} = 1.22\lambda/D$.

A **diffraction grating** consists of a large number of equally spaced, identical slits. The condition for intensity maxima in the interference pattern of a diffraction grating for normal incidence is

$$d \sin \theta = m\lambda \qquad (m = 0, 1, 2, 3, \ldots) \qquad (38.8)$$

where d is the spacing between adjacent slits and m is the order number of the diffraction pattern. The resolving power of a diffraction grating in the mth order of the diffraction pattern is

$$R = Nm \qquad (38.10)$$

where N is the number of rulings in the grating.

When polarized light of intensity I_0 is incident on a polarizing film, the light transmitted through the film has an intensity equal to $I_0 \cos^2 \theta$, where θ is the angle between the transmission axis of the polarizer and the electric field vector of the incident light.

In general, reflected light is partially polarized. However, the reflected light is completely polarized when the angle of incidence is such that the angle between the reflected and refracted beams is 90°. This angle of incidence, called the **polarizing angle** θ_p, satisfies **Brewster's law**:

$$n = \tan \theta_p \qquad (38.13)$$

where n is the index of refraction of the reflecting medium.

QUESTIONS

1. What is the difference between Fraunhofer and Fresnel diffraction?
2. Although we can hear around corners, we cannot see around corners. How can you explain this in view of the fact that sound and light are both waves?
3. Describe the change in width of the central maximum of the single-slit diffraction pattern as the width of the slit is made narrower.
4. Assuming that the headlights of a car are point sources, estimate the maximum distance from an observer to the car at which the headlights are distinguishable from each other.
5. A laser beam is incident at a shallow angle on a machinist's ruler that has a finely calibrated scale. The rulings on the scale give rise to a diffraction pattern on a screen. Discuss how you can use this technique to obtain a measure of the wavelength of the laser light.
6. Certain sunglasses use a polarizing material to reduce the intensity of light reflected from shiny surfaces. What orientation of polarization should the material have to be most effective?
7. The path of a light beam can be made visible by placing dust in the air (perhaps by shaking a blackboard eraser in the path of the light beam). Explain why the beam can be seen under these circumstances.
8. Is light from the sky polarized? Why is it that clouds seen through Polaroid glasses stand out in bold contrast to the sky?
9. If a coin is glued to a glass sheet and this arrangement is held in front of a laser beam, the projected shadow has diffraction rings around its edge and a bright spot in the center. How is this possible?
10. If a fine wire is stretched across the path of a laser beam, is it possible to produce a diffraction pattern?
11. How could the index of refraction of a flat piece of dark obsidian glass be determined?

PROBLEMS

Review Problem

A wide screen is placed 1.00 m from a single slit of width 0.0200 mm. A beam of white light ($\lambda = 400 - 750$ nm is incident on the slit. Find (a) the number of spectra formed on the screen (*Hint*: The order number is a maximum for $\theta = 90°$), (b) the location of the second-order maximum at 750 nm, (c) the width of the second-order maximum at 400 nm, (d) the fractional light intensity (I/I_0) at 5.00 mm from the center when a 500-nm transmission filter is placed between the beam and the slit, (e) the resolution of the slit for 600-nm light, (f) the minimum separation between two candles (1.00 km from the slit) in order to be resolved by the slit, (g) the maximum resolving power (at 750 nm) of a diffraction grating that is 4.00 cm wide and has a line spacing of 0.0200 mm.

Section 38.2 Single-Slit Diffraction

1. Light from a He-Ne laser ($\lambda = 632.8$ nm) is incident on a single slit. What is the minimum width for which no diffraction minima are observed?

2. A Fraunhofer diffraction pattern is produced on a screen 140 cm from a single slit. The distance from the center of the central maximum to the first-order maximum is $1.00 \times 10^4 \lambda$. Calculate the slit width.

3. The second-order bright fringe in a single-slit diffraction pattern is 1.4 mm from the center of the central maximum. The screen is 80 cm from a slit of width 0.80 mm. Assuming monochromatic incident light, calculate the wavelength.

4. Helium-neon laser light ($\lambda = 632.8$ nm) is sent through a 0.300-mm-wide single slit. What is the width of the central maximum on a screen 1.00 m from the slit?

5. The pupil of a cat's eye narrows to a slit of width 0.50 mm in daylight. What is the angular resolution? (Use 500-nm light in your calculation.)

6. Light of wavelength 587.5 nm illuminates a single slit 0.75 mm in width. (a) At what distance from the slit should a screen be located if the first minimum in the diffraction pattern is to be 0.85 mm from the center of the screen? (b) What is the width of the central maximum?

7. A screen is placed 50 cm from a single slit, which is illuminated with 690-nm light. If the distance between the first and third minima in the diffraction pattern is 3.0 mm, what is the width of the slit?

8. In Equation 38.4, let $\beta/2 \equiv \phi$ and show that $I = 0.5I_0$ when $\sin \phi = \phi/\sqrt{2}$.

9. The equation $\sin \phi = \phi/\sqrt{2}$ found in Problem 8 is known as a *transcendental equation*, which can be solved graphically. To illustrate this, let $\phi = \beta/2$, $y_1 = \sin \phi$, and $y_2 = \phi/\sqrt{2}$. Plot y_1 and y_2 on the same set of axes over a range from $\phi = 1$ rad to $\phi = \pi/2$ rad. Determine ϕ from the point of intersection of the two curves.

10. A beam of green light is diffracted by a slit of width 0.55 mm. The diffraction pattern forms on a wall 2.06 m beyond the slit. The distance between the positions of zero intensity ($m = \pm 1$) is 4.1 mm. Estimate the wavelength of the laser light.

11. A diffraction pattern is formed on a screen 120 cm away from a 0.40-mm-wide slit. Monochromatic 546.1-nm light is used. Calculate the fractional intensity I/I_0 at a point on the screen 4.1 mm from the center of the principal maximum.

12. If the light in Figure 38.6 strikes the single slit at an angle of β from the perpendicular direction, show that Equation 38.1, the condition for destructive interference, must be modified to read

$$\sin \theta = m\left(\frac{\lambda}{a}\right) - \sin \beta$$

Section 38.3 Resolution of Single-Slit and Circular Apertures

13. A helium-neon laser emits light that has a wavelength of 632.8 nm. The circular aperture through which the beam emerges has a diameter of 0.50 cm. Estimate the diameter of the beam 10.0 km from the laser.

14. The Moon is approximately 400 000 km from the Earth. Can two lunar craters 50 km apart be resolved by a telescope on the Earth if the telescope mirror has a diameter of 15 cm? Can craters 1.0 km apart be resolved? Take the wavelength to be 700 nm, and justify your answers with approximate calculations.

15. On the night of April 18, 1775, a signal was to be sent from the Old North Church steeple to Paul Revere, who was 1.8 miles away: "One if by land, two if by sea." At what minimum separation did the sexton have to set the lanterns so that Paul Revere could receive the correct message? Assume that Paul Revere's pupils had a diameter of 4.00 mm at night, and that the candlelight had a predominant wavelength of 580 nm.

16. If we were to send a ruby laser beam ($\lambda = 694.3$ nm) outward from the barrel of a 2.7-m-diameter telescope, what would be the diameter of the big red spot when the beam hit the Moon 384 000 km away? (Neglect atmospheric dispersion.)

□ indicates problems that have full solutions available in the Student Solutions Manual and Study Guide.

17. Suppose you are standing on a straight highway and watching a car moving away from you at 20.0 m/s. The air is perfectly clear, and after 2.00 min you see only one taillight. If the diameter of your pupil is 7.00 mm and the index of refraction of your eye is 1.33, estimate the width of the car.

17A. Suppose you are standing on a straight highway and watching a car moving away from you at a speed v. The air is perfectly clear, and after a time t you see only one taillight. If the diameter of your pupil is d and the index of refraction of your eye is n, estimate the width of the car.

18. Find the radius of a star image formed on the retina of the eye if the aperture diameter (the pupil) at night is 0.70 cm, and the length of the eye is 3.0 cm. Assume the wavelength of starlight in the eye is 500 nm.

19. At what distance could one theoretically distinguish two automobile headlights separated by 1.4 m? Assume a pupil diameter of 6.0 mm and yellow headlights (λ = 580 nm). The index of refraction in the eye is approximately 1.33.

20. A binary star system in the constellation Orion has an angular separation between the two stars of 1.0×10^{-5} rad. If λ = 500 nm, what is the smallest diameter the telescope can have and just resolve the two stars?

21. The Impressionist painter Georges Seurat created paintings with an enormous number of dots of pure pigment approximately 2.0 mm in diameter. The idea was to have colors such as red and green next to each other to form a scintillating canvas. Outside what distance would one be unable to discern individual dots on the canvas? (Assume λ = 500 nm within the eye and a pupil diameter of 4.0 mm.)

22. The angular resolution of a radio telescope is to be 0.10° when the incident waves have a wavelength of 3.0 mm. Approximately what minimum diameter is required for the telescope's receiving dish?

23. A circular radar antenna on a navy ship has a diameter of 2.1 m and radiates at a frequency of 15 GHz. Two small boats are located 9.0 km away from the ship. How close together could the boats be and still be detected as two objects?

24. When Mars is nearest the Earth, the distance separating the two planets is 88.6×10^6 km. Mars is viewed through a telescope whose mirror has a diameter of 30 cm. (a) If the wavelength of the light is 590 nm, what is the angular resolution of the telescope? (b) What is the smallest distance that can be resolved between two points on Mars?

Section 38.4 The Diffraction Grating

25. Show that, whenever a continuous visible spectrum is passed through a diffraction grating of any spacing size, the third-order fringe of the violet end of the visible spectrum always overlaps the second-order fringe of the red light at the other end.

26. A beam of 541-nm light is incident on a diffraction grating that has 400 lines/mm. (a) Determine the angle of the second-order ray. (b) If the entire apparatus is immersed in water, determine the new second-order angle of diffraction. (c) Show that the two diffracted rays of parts (a) and (b) are related through the law of refraction.

27. A grating with 250 lines/mm is used with an incandescent light source. Assume the visible spectrum to range in wavelength from 400 to 700 nm. In how many orders can one see (a) the entire visible spectrum and (b) the short-wavelength region?

28. Two wavelengths λ and $\lambda + \Delta\lambda$ ($\Delta\lambda \ll \lambda$) are incident on a diffraction grating. Show that the angular separation between the mth-order spectra is

$$\Delta\theta = \frac{\Delta\lambda}{\sqrt{(d/m)^2 - \lambda^2}}$$

where d is the grating constant and m is the order number.

29. The full width of a 3.00-cm-wide grating is illuminated by a sodium discharge tube. The lines in the grating are uniformly spaced at 775 nm. Calculate the angular separation in the first-order spectrum between the two wavelengths forming the sodium doublet (λ_1 = 589.0 nm and λ_2 = 589.6 nm).

30. Light from an argon laser strikes a diffraction grating that has 5310 lines per centimeter. The central and first-order principal maxima are separated by 0.488 m on a wall 1.72 m from the grating. Determine the wavelength of the laser light.

31. A source emits 531.62-nm and 531.81-nm light. (a) What minimum number of lines is required for a grating that resolves the two wavelengths in the first-order spectrum? (b) Determine the slit spacing for a grating 1.32 cm wide that has the required minimum number of lines.

32. A diffraction grating has 800 rulings per millimeter. A beam of light containing wavelengths from 500 to 700 nm hits the grating. Do the spectra of different orders overlap? Explain.

33. A diffraction grating has 4200 rulings per centimeter. On a screen 2.0 m from the grating, it is found that, for a particular order m, the maxima corresponding to two closely spaced wavelengths of sodium (589.0 and 589.6 nm) are separated by 1.5 mm. Determine the value of m.

34. A helium-neon laser (λ = 632.8 nm) is used to calibrate a diffraction grating. If the first-order maximum occurs at 20.5°, what is the line spacing, d?

35. White light is spread out into its spectral components by a diffraction grating. If the grating has 2000 lines

per centimeter, at what angle does red light ($\lambda = 640$ nm) appear in first order?

36. Two spectral lines in a mixture of hydrogen (H_2) and deuterium (D_2) gas have wavelengths of 656.30 nm and 656.48 nm, respectively. What is the minimum number of lines a diffraction grating must have to resolve these two wavelengths in first order?

37. Monochromatic 632.8-nm light is incident on a diffraction grating containing 4000 lines per centimeter. Determine the angle of the first-order maximum.

*Section 38.5 Diffraction of X-Rays by Crystals

38. Potassium iodide (KI) has the same crystalline structure as NaCl, with $d = 0.353$ nm. A monochromatic x-ray beam shows a diffraction maximum when the grazing angle is 7.6°. Calculate the x-ray wavelength. (Assume first order.)

39. Monochromatic x-rays of the K_α line of potassium from a nickel target ($\lambda = 0.166$ nm) are incident on a KCl crystal surface. The interplanar distance in KCl is 0.314 nm. At what angle (relative to the surface) should the beam be directed so that a second-order maximum is observed?

40. A monochromatic x-ray beam is incident on a NaCl crystal surface that has an interplanar spacing of 0.281 nm. The second-order maximum in the reflected beam is found when the angle between incident beam and surface is 20.5°. Determine the wavelength of the x-rays.

41. Certain cubic crystals (such as NaCl) do not diffract the odd orders ($m = 1, 3, 5, \ldots$) because of destructive interference from the unit cell planes. For such a crystal having interplanar spacing of 0.2810 nm and illuminated with 0.0956-nm x-rays, find (a) the angles at which constructive interference occurs and (b) the radii of the concentric circles such patterns make on a screen 10.0 cm from the crystal.

42. The first-order diffraction is observed at +12.6° for a crystal in which the interplanar spacing is 0.240 nm. How many other orders can be observed?

43. X-rays of wavelength 0.140 nm are reflected from a NaCl crystal, and the first-order maximum occurs at an angle of 14.4°. What value does this give for the interplanar spacing of NaCl?

44. A wavelength of 0.129 nm characterizes K_β x-rays from zinc. When a beam of these x-rays is incident on the surface of a crystal whose structure is similar to that of NaCl, a first-order maximum is observed at 8.15°. Calculate the interplanar spacing based on this information.

45. If the interplanar spacing of NaCl is 0.281 nm, what is the predicted angle at which 0.140-nm x-rays are diffracted in a first-order maximum?

46. In an x-ray diffraction experiment using x-rays of $\lambda = 0.500 \times 10^{-10}$ m, a first-order maximum occurred at 5.0°. Find the crystal plane spacing.

Section 38.6 Polarization of Light Waves

47. The angle of incidence of a light beam onto a reflecting surface is continuously variable. The reflected ray is found to be completely polarized when the angle of incidence is 48°. What is the index of refraction of the reflecting material?

48. Light is reflected from a smooth ice surface, and the reflected ray is completely polarized. Determine the angle of incidence. ($n = 1.309$ for ice)

49. A light beam is incident on heavy flint glass ($n = 1.65$) at the polarizing angle. Calculate the angle of refraction for the transmitted ray.

50. How far above the horizon is the Moon when its image reflected in calm water is completely polarized? ($n_{water} = 1.33$)

51. Consider a polarizer-analyzer arrangement as shown in Figure 38.25. At what angle is the axis of the analyzer to the axis of the polarizer if the original beam intensity is reduced by (a) 10, (b) 50, and (c) 90 percent after passing through both sheets?

52. An unpolarized light beam in water reflects off a glass plate ($n = 1.570$). At what angle should the beam be incident on the plate to be totally polarized upon reflection?

53. Plane-polarized light is incident on a single polarizing disk with the direction of E_0 parallel to the direction of the transmission axis. Through what angle should the disk be rotated so that the intensity in the transmitted beam is reduced by a factor of (a) 3, (b) 5, (c) 10?

54. Three polarizing disks whose planes are parallel are centered on a common axis. The direction of the transmission axis in each case is shown in Figure P38.54 relative to the common vertical direction. A plane-polarized beam of light with E_0 parallel to the vertical reference direction is incident from the left on the first disk with intensity $I_i = 10$ units (arbitrary). Calculate the transmitted intensity I_f when (a) $\theta_1 = 20°$, $\theta_2 = 40°$, and $\theta_3 = 60°$; (b) $\theta_1 = 0°$, $\theta_2 = 30°$, and $\theta_3 = 60°$.

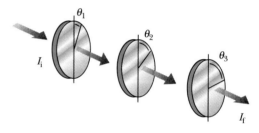

FIGURE P38.54

55. The critical angle for sapphire surrounded by air is 34.4°. Calculate the polarizing angle for sapphire.

56. For a particular transparent medium surrounded by air, show that the critical angle for internal reflection and the polarizing angle are related by $\cot \theta_p = \sin \theta_c$.

57. If the polarizing angle for cubic zirconia (ZrO_2) is 65.6°, what is the index of refraction for this material?

ADDITIONAL PROBLEMS

58. A solar eclipse is projected through a pinhole of diameter 0.50 mm and strikes a screen 2.0 meters away. (a) What is the diameter of the image? (b) What is the radius of the first diffraction minimum? Both the Sun and Moon have angular diameters of very nearly 0.50°. (Assume $\lambda = 550$ nm.)

59. The hydrogen spectrum has a red line at 656 nm and a blue line at 434 nm. What is the angular separation between two spectral lines obtained with a diffraction grating that has 4500 lines/cm?

60. What are the approximate dimensions of the smallest object on Earth that astronauts can resolve by eye when they are orbiting 250 km above the Earth? Assume $\lambda = 500$ nm light in the eye and a pupil diameter of 5.0 mm.

61. Grote Reber was a pioneer in radio astronomy. He constructed a radio telescope with a 10-m diameter receiving dish. What was the telescope's angular resolution for 2.0-m radio waves?

62. An unpolarized beam of light is reflected from a glass surface, and the reflected beam is linearly polarized when the light is incident from air at 58.6°. What is the refractive index of the glass?

63. A 750-nm light beam hits the flat surface of a certain liquid, and the beam is split into a reflected ray and a refracted ray. If the reflected rays are completely polarized at grazing angle 36°, what is the wavelength of the refracted ray?

64. Light strikes a water surface at the polarizing angle. The part of the beam refracted into the water strikes a submerged glass slab (index of refraction = 1.50) as in Figure P38.64. If the light reflected from the upper surface of the slab is completely polarized, find the angle between water surface and glass slab.

65. A diffraction grating of length 4.00 cm contains 6000 rulings over a width of 2.00 cm. (a) What is the resolving power of this grating in the first three orders? (b) If two monochromatic waves incident on this grating have a mean wavelength of 400 nm, what is their wavelength separation if they are just resolved in the third order?

66. An American standard television picture is composed of approximately 485 horizontal lines of varying light intensity. Assume that your ability to resolve the lines is limited only by the Rayleigh criterion and that the pupils of your eyes are 5.0 mm in diameter. Calculate the ratio of minimum viewing distance to the vertical dimension of the picture such that you will not be able to resolve the lines. Assume that the average wavelength of the light coming from the screen is 550 nm.

67. Light of wavelength 500 nm is incident normally on a diffraction grating. If the third-order maximum of the diffraction pattern is observed at 32°, (a) what is the number of rulings per centimeter for the grating? (b) Determine the total number of primary maxima that can be observed in this situation.

68. (a) If light traveling in a medium for which the index of refraction is n_1 is incident at an angle θ on the surface of a medium of index n_2 so that the angle between the reflected and refracted rays is β, show that

$$\tan \theta = \frac{n_2 \sin \beta}{n_1 - n_2 \cos \beta}$$

(*Hint:* Use the identity $\sin(A + B) = \sin A \cos B + \cos A \sin B$.) (b) Show that this expression for $\tan \theta$ reduces to Brewster's law when $\beta = 90°$, $n_1 = 1$, and $n_2 = n$.

69. Two polarizing sheets are placed together with their transmission axes crossed so that no light is transmitted. A third sheet is inserted between them with its transmission axis at an angle of 45° with respect to each of the other axes. Find the fraction of incident unpolarized light intensity transmitted by the three-sheet combination. (Assume each polarizing sheet is ideal.)

70. Figure P38.70a is a three-dimensional sketch of a birefringent crystal. The dotted lines illustrate how a thin parallel-faced slab of material could be cut from the larger specimen with the optic axis of the crystal parallel to the faces of the plate. A section cut from the crystal in this manner is known as a *retardation plate*. When a beam of light is incident on the plate perpendicular to the direction of the optic axis, as shown in Figure P38.70b, the O ray and the E ray travel along a straight line, but with different speeds. (a) Let the thickness of the plate be d and show that

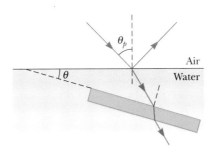

FIGURE P38.64

the phase difference between the O ray and the E ray is

$$\theta = \frac{2\pi d}{\lambda_0} |n_O - n_E|$$

where λ_0 is the wavelength in air. (Recall that the optical path length in a material is the product of geometric path and index of refraction.) (b) If in a particular case the incident light has a wavelength of 550 nm, find the minimum value of d for a quartz plate for which $\theta = \pi/2$. Such a plate is called a quarter-wave plate. (Use values of n_O and n_E from Table 38.1.)

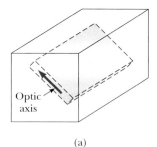

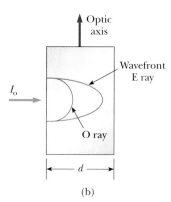

FIGURE P38.70

71. You want to rotate the plane of polarization of a polarized light beam by 45° for a maximum intensity reduction of 10 percent. (a) How many sheets of perfect polarizers do you need to use to achieve your goal? (b) What is the angle between the adjacent polarizers?

72. In Figure P38.54, suppose that the left and right polarizing disks have their transmission axes perpendicular to each other. Also, let the center disk be rotated on the common axis with an angular speed ω. Show that if unpolarized light is incident on the left disk with an intensity I_0, the intensity of the beam emerging from the right disk is

$$I = \frac{1}{16} I_0 (1 - \cos 4\omega t)$$

This means that the intensity of the emerging beam is modulated at a rate that is four times the rate of rotation of the center disk. [*Hint:* Use the trigonometric identities $\cos^2 \theta = (1 + \cos 2\theta)/2$ and $\sin^2 \theta = (1 - \cos 2\theta)/2$, and recall that $\theta = \omega t$.]

73. Suppose that the single slit in Figure 38.7 is 6.0 cm wide and in front of a microwave source operating at 7.5 GHz. (a) Calculate the angle subtended by the first minimum in the diffraction pattern. (b) What is the relative intensity I/I_0 at $\theta = 15°$? (c) Consider the case when there are two such sources, separated laterally by 20 cm, behind the slit. What must the maximum distance between the plane of the sources and the slit be if the diffraction patterns are to be resolved? (In this case, the approximation $\sin \theta \approx \tan \theta$ is not valid because of the relatively small value of a/λ.)

74. Light of wavelength 632.8 nm illuminates a single slit, and a diffraction pattern is formed on a screen 1.00 m from the slit. Using the following data, plot relative intensity versus distance. Choose an appropriate value for the slit width a and, on the same graph used for the experimental data, plot the theoretical expression for the relative intensity

$$\frac{I_\theta}{I_0} = \frac{\sin^2(\beta/2)}{(\beta/2)^2}$$

What value of a gives the best fit of theory and experiment?

Relative Intensity	Distance from Center of Central Maximum (mm)
0.95	0.8
0.80	1.6
0.60	2.4
0.39	3.2
0.21	4.0
0.079	4.8
0.014	5.6
0.003	6.5
0.015	7.3
0.036	8.1
0.047	8.9
0.043	9.7
0.029	10.5
0.013	11.3
0.002	12.1
0.0003	12.9
0.005	13.7
0.012	14.5
0.016	15.3
0.015	16.1
0.010	16.9
0.0044	17.7
0.0006	18.5
0.0003	19.3
0.003	20.2

75. From Equation 38.4 show that, in the Fraunhofer diffraction pattern of a single slit, the angular width of the central maximum at the point where $I = 0.5 I_0$ is $\Delta\theta = 0.886\lambda/a$. (*Hint:* In Equation 38.4, let $\beta/2 = \phi$ and solve the resulting transcendental equation graphically; see Problem 9.)

76. Another method to solve the equation $\phi = \sqrt{2} \sin \phi$ in Problem 9 is to use a calculator, guess a first value of ϕ, see if it fits, and continue to update your estimate until the equation balances. How many steps (iterations) did this take?

SPREADSHEET PROBLEM

S1. Figure 38.11 shows the relative intensity of a single-slit Fraunhofer diffraction pattern as a function of the parameter $\beta/2 = \pi a \sin \theta/\lambda$. Spreadsheet 38.1 plots the relative intensity I/I_0 as a function of θ, where θ is defined in Figure 38.8. Examine the cases where $\lambda = a$, $\lambda = 0.5a$, $\lambda = 0.1a$, and $\lambda = 0.05a$, and explain your results.

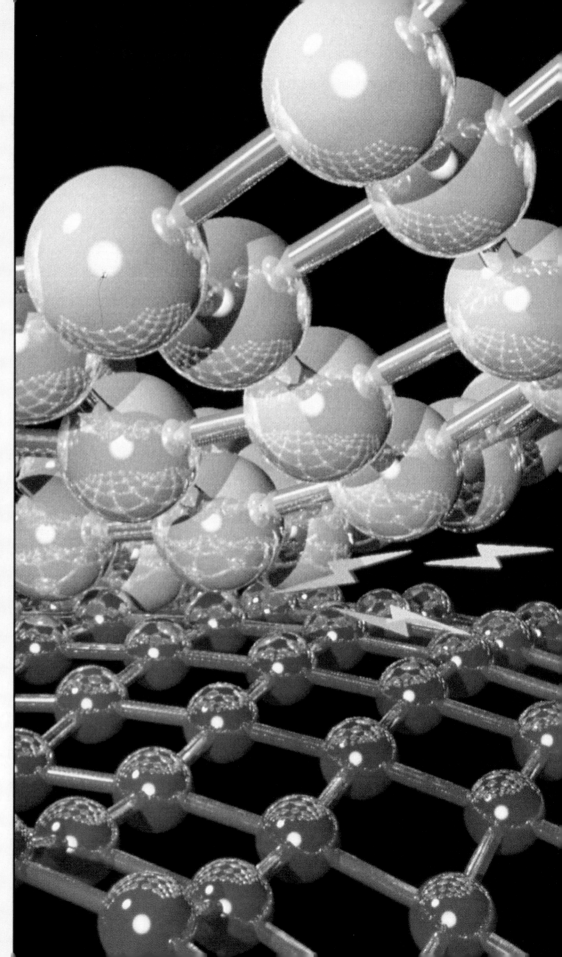

Computer simulation of a two-dimensional array of spheres as they would appear as they move past an observer moving at a relativistic speed. *(Mel Pruitt)*

PART VI

Modern Physics

"Imagination is more important than knowledge."

ALBERT EINSTEIN

At the end of the 19th century, scientists believed that they had learned most of what there was to know about physics. Newton's laws of motion and his universal theory of gravitation, Maxwell's theoretical work in unifying electricity and magnetism, and the laws of thermodynamics and kinetic theory were highly successful in explaining a wide variety of phenomena.

As the century turned, however, a major revolution shook the world of physics. In 1900 Planck provided the basic ideas that led to the formulation of the quantum theory, and in 1905 Albert Einstein formulated his brilliant special theory of relativity. The excitement of the times is captured in Einstein's own words: "It was a marvelous time to be alive." Both ideas were to have a profound effect on our understanding of nature. Within a few decades, these theories inspired new developments and theories in the fields of atomic physics, nuclear physics, and condensed-matter physics.

In Chapter 39 we introduce the special theory of relativity. Although the concepts underlying this theory seem to contradict our common sense, the theory provides us with a new and deeper view of physical laws.

You should keep in mind that, although modern physics has been developed during this century and has led to a multitude of important technological achievements, the story is still incomplete. Discoveries will continue to evolve during our lifetime, and many of these discoveries will deepen or refine our understanding of nature and the world around us. It is still a "marvelous time to be alive."

CHAPTER 39

Relativity

Albert Einstein (1879–1955), one of the greatest physicists of all times, is best known for developing the theory of relativity. He is shown here in a playful mood riding a bicycle. The photograph was taken in 1933 in Santa Barbara, California. *(From the California Institute of Technology archives)*

Most of our everyday experiences and observations have to do with objects that move at speeds much less than that of light. Newtonian mechanics and early ideas on space and time were formulated to describe the motion of such objects. This formalism is very successful in describing a wide range of phenomena that occur at low speeds. It fails, however, when applied to particles whose speeds approach that of light. Experimentally, the predictions of Newtonian theory can be tested at high speeds by accelerating electrons or other charged particles through a large electric potential difference. For example, it is possible to accelerate an electron to a speed of $0.99c$ (where c is the speed of light) by using a potential difference of several million volts. According to Newtonian mechanics, if the potential difference (as well as the corresponding energy) is increased by a factor of 4, the electron speed should jump to $1.98c$. However, experiments show that the speed of the electron — as well as the speeds of all other particles in the Universe — always remains less than the speed of light, regardless of the size of the accelerating voltage. Because it places no upper limit on speed, Newtonian mechanics is contrary to modern experimental results and is clearly a limited theory.

In 1905, at the age of only 26, Einstein published his special theory of relativity. Regarding the theory, Einstein wrote:

> The relativity theory arose from necessity, from serious and deep contradictions in the old theory from which there seemed no escape. The strength of the new theory lies in the consistency and simplicity with which it solves all these difficulties, using only a few very convincing assumptions. . . .[1]

Although Einstein made many important contributions to science, the theory of relativity alone represents one of the greatest intellectual achievements of the 20th century. With this theory, experimental observations can be correctly predicted over the range of speeds from $v = 0$ to speeds approaching the speed of light. Newtonian mechanics, which was accepted for over 200 years, is in fact a special case of Einstein's theory. This chapter gives an introduction to the special theory of relativity, with emphasis on some of its consequences.

Special relativity covers phenomena such as the slowing down of clocks and the contraction of lengths in moving reference frames as measured by a laboratory observer. We also discuss the relativistic forms of momentum and energy, and some consequences of the famous mass-energy formula, $E = mc^2$.

In addition to its well-known and essential role in theoretical physics, the theory of relativity has practical applications, including the design of accelerators and other devices that utilize high-speed particles. These devices will not work if designed according to nonrelativistic principles. We shall have occasion to use relativity in some subsequent chapters of this text, most often presenting only the outcome of relativistic effects.

39.1 THE PRINCIPLE OF NEWTONIAN RELATIVITY

To describe a physical event, it is necessary to establish a frame of reference. You should recall from Chapter 5 that Newton's laws are valid in all inertial frames of reference. Because an inertial frame is defined as one in which Newton's first law is valid, it can be said that *an inertial system is one in which a free body experiences no acceleration.* Furthermore, any system moving with constant velocity with respect to an inertial system must also be an inertial system.

Inertial frame of reference

There is no preferred frame. This means that the results of an experiment performed in a vehicle moving with uniform velocity will be identical to the results of the same experiment performed in a stationary vehicle. The formal statement of this result is called the **principle of Newtonian relativity**:

> The laws of mechanics must be the same in all inertial frames of reference.

Let us consider an observation that illustrates the equivalence of the laws of mechanics in different inertial frames. Consider a pickup truck moving with a constant velocity as in Figure 39.1a. If a passenger in the truck throws a ball straight up in the air, the passenger observes that the ball moves in a vertical path. The

---

[1] A. Einstein and L. Infeld, *The Evolution of Physics,* New York, Simon and Schuster, 1961.

(a)　　　　　　　　　　　　　　(b)

FIGURE 39.1 (a) The observer in the truck sees the ball move in a vertical path when thrown upward. (b) The Earth observer views the path of the ball to be a parabola.

motion of the ball appears to be precisely the same as if the ball were thrown by a person at rest on Earth. The law of gravity and the equations of motion under constant acceleration are obeyed whether the truck is at rest or in uniform motion. Now consider the same experiment viewed by an observer at rest on Earth. This stationary observer sees the path of the ball as a parabola, as in Figure 39.1b. Furthermore, according to this observer, the ball has a horizontal component of velocity equal to the velocity of the truck. Although the two observers disagree on certain aspects of the experiment, they agree on the validity of Newton's laws and classical principles like conservation of energy and conservation of momentum. This agreement implies that no mechanical experiment can detect any difference between the two inertial frames. The only thing that can be detected is the relative motion of one frame with respect to the other. That is, the notion of absolute motion through space is meaningless, as is the notion of a preferred reference frame.

Suppose that some physical phenomenon, which we call an event, occurs in an inertial frame. The event's location and time of occurrence can be specified by the coordinates (x, y, z, t). We would like to be able to transform these coordinates from one inertial frame to another moving with uniform relative velocity. This is accomplished by using what is called a *Galilean transformation*, which owes its origin to Galileo.

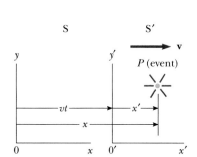

FIGURE 39.2 An event occurs at a point P. The event is observed by two observers in inertial frames S and S', where S' moves with a velocity v relative to S.

Consider two inertial frames S and S' (Fig. 39.2). The system S' moves with a constant velocity **v** along the xx' axes, where **v** is measured relative to S. We assume that an event occurs at the point P and that the origins of S and S' coincide at $t = 0$. An observer in S describes the event with space-time coordinates (x, y, z, t), while an observer in S' uses (x', y', z', t') to describe the same event. As we can see from Figure 39.2, these coordinates are related by the equations

$$\begin{aligned} x' &= x - vt \\ y' &= y \\ z' &= z \\ t' &= t \end{aligned} \quad (39.1)$$

Galilean transformation of coordinates

These equations constitute what is known as a **Galilean transformation of coordi-**

nates. Note that time is assumed to be the same in both inertial systems. That is, within the framework of classical mechanics, all clocks run at the same rate, regardless of their velocity, so that the time at which an event occurs for an observer in S is the same as the time for the same event in S'. Consequently, the time interval between two successive events should be the same for both observers. Although this assumption may seem obvious, it turns out to be incorrect when treating situations in which v is comparable to the speed of light. This point of equal time intervals represents one of the profound differences between Newtonian concepts and Einstein's theory of relativity.

Now suppose two events are separated by a distance dx and a time interval dt as measured by an observer in S. It follows from Equation 39.1 that the corresponding distance dx' measured by an observer in S' is $dx' = dx - v\,dt$, where dx is the distance between the two events measured by an observer in S. Since $dt = dt'$, we find that

$$\frac{dx'}{dt} = \frac{dx}{dt} - v$$
$$u'_x = u_x - v \qquad (39.2)$$

Galilean addition law for velocities

where u_x and u'_x are the instantaneous velocities of the object relative to S and S', respectively. This result, which is called the **Galilean addition law for velocities** (or Galilean velocity transformation), is used in everyday observations and is consistent with our intuitive notion of time and space. As we will soon see, however, it leads to serious contradictions when applied to electromagnetic waves.

The Speed of Light

It is quite natural to ask whether the principle of Newtonian relativity in mechanics also applies to electricity, magnetism, and optics, and the answer is no. Recall from Chapter 34 that Maxwell in the 1860s showed that the speed of light in free space was given by $c = 3.00 \times 10^8$ m/s. Physicists of the late 1800s thought that light waves required a definite medium to move in called the *ether*, and that the speed of light was only c in a special, absolute frame at rest with respect to the ether. The Galilean addition law of velocities was expected to hold in any other frame moving at speed v relative to the absolute ether frame.

Since the existence of a preferred, absolute ether frame would show that light was similar to other classical waves, and that Newtonian ideas of an absolute frame were true, considerable importance was attached to establishing the existence of the ether frame. Because the speed of light is enormous, experiments involving light traveling in media moving at then attainable laboratory speeds had not been capable of detecting small changes of the size of $c \pm v$ prior to the late 1800s. Scientists of the period, realizing that the Earth moved around the Sun at 30 km/s, decided to use the Earth as the moving frame in an attempt to improve their chances of detecting these small changes in light speed.

From our point of view of observers fixed on Earth, we may say that we are stationary and that the absolute ether frame containing the medium for light propagation moves past us with speed v. Determining the speed of light under these circumstances is just like determining the speed of an aircraft by measuring the speed of the air next to it, and consequently we speak of an "ether wind" blowing through our apparatus fixed to the Earth.

A direct method for detecting an ether wind would be to measure its influence

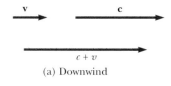

(a) Downwind

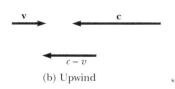

(b) Upwind

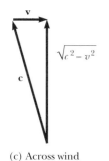

(c) Across wind

FIGURE 39.3 If the velocity of the ether wind relative to the Earth is **v**, and **c** is the velocity of light relative to the ether, the speed of light relative to the Earth is (a) $c + v$ in the downwind direction, (b) $c - v$ in the upwind direction, and (c) $(c^2 - v^2)^{1/2}$ in the direction perpendicular to the wind.

on the speed of light using an apparatus fixed in a frame of reference on Earth. If v is the speed of the ether relative to Earth, then the speed of light should have its maximum value, $c + v$, when propagating downwind as in Figure 39.3a. Likewise, the speed of light should have its minimum value, $c - v$, when propagating upwind as in Figure 39.3b, and some intermediate value, $(c^2 - v^2)^{1/2}$, in the direction perpendicular to the ether wind as in Figure 39.3c. If the Sun is assumed to be at rest in the ether, then the velocity of the ether wind would be equal to the orbital velocity of the Earth around the Sun, which has a magnitude of approximately 3×10^4 m/s. Since $c = 3 \times 10^8$ m/s, it should be possible to detect a change in speed of about 1 part in 10^4 for measurements in the upwind or downwind directions. However, as we shall see in the next section, all attempts to detect such changes and establish the existence of the ether (and hence the absolute frame) proved futile!

If it is assumed that the laws of electricity and magnetism are the same in all inertial frames, a paradox concerning the speed of light immediately arises. This can be understood by recalling that according to Maxwell's equations, the speed of light always has the fixed value $(\mu_0 \epsilon_0)^{-1/2} \approx 3.00 \times 10^8$ m/s, a result in direct contradiction to what would be expected based on the Galilean law. According to Galileo, the speed of light should not be the same in all inertial frames.

For example, suppose a light pulse is sent out by an observer S' standing in a boxcar moving with a velocity **v** (Fig. 39.4). The light pulse has a speed c relative to S'. According to Newtonian relativity, the speed of the pulse relative to the stationary observer S outside the boxcar should be $c + v$. This is in obvious contradiction to Einstein's theory, which, we shall see, postulates that the speed of the light pulse is the same for all observers.

In order to resolve this paradox, it must be concluded that either (1) the laws of electricity and magnetism are not the same for all inertial frames or (2) the Galilean law of addition for velocities is incorrect. If we assume the first alternative, we are forced to abandon the seemingly obvious notions of absolute time and absolute length that form the basis for the Galilean transformations. If we assume that the second alternative is true, then there must exist a preferred reference frame in which the speed of light has the value c and the speed must be greater or less than this value in any other reference frame, in accordance with the Galilean addition law for velocities. It is useful to draw an analogy with sound waves propagating through air. The speed of sound in air is about 330 m/s when measured in a

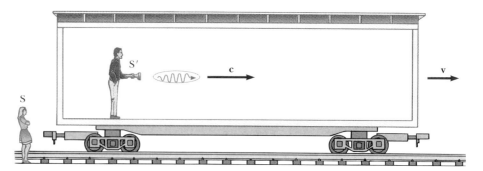

FIGURE 39.4 A pulse of light is sent out by a person in a moving boxcar. According to Newtonian relativity, the speed of the pulse should be $c + v$ relative to a stationary observer.

1. A. Piccard	11. L. Brillouin	21. I. Langmuir
2. E. Henriot	12. P. Debye	22. M. Planck
3. P. Ehrenfest	13. M. Knudsen	23. M. Curie
4. E. Herzen	14. W.L. Bragg	24. H.A. Lorentz
5. Th. de Donder	15. H.A. Kramers	25. A. Einstein
6. E. Schroedinger	16. P.A.M. Dirac	26. P. Langevin
7. E. Verschaffelt	17. A.H. Compton	27. C.E. Guye
8. W. Pauli	18. L.V. de Broglie	28. C.T.R. Wilson
9. W. Heisenberg	19. M. Born	29. O.W. Richardson
10. R.H. Fowler	20. N. Bohr	

The "architects" of modern physics. This unique photograph shows many eminent scientists who participated in the fifth international congress of physics held in 1927 by the Solvay Institute in Brussels. At this and similar conferences, held regularly from 1911 on, scientists were able to discuss and share the many dramatic developments in atomic and nuclear physics. This elite company of scientists includes fifteen Nobel Prize winners in physics and three in chemistry. *(Photograph courtesy of AIP Niels Bohr Library)*

reference frame in which the air is stationary. However, the speed is greater or less than this value when measured from a reference frame that is moving relative to the sound source.

39.2 THE MICHELSON–MORLEY EXPERIMENT

The most famous experiment designed to detect small changes in the speed of light was performed in 1881 by Albert A. Michelson (1852–1931) and repeated under various conditions in 1887 with Edward W. Morley (1838–1923). We state at the outset that the outcome of the experiment was null, thus contradicting the ether hypothesis. The experiment was designed to determine the velocity of the Earth relative to the hypothetical ether. The experimental tool used was the Michelson interferometer, shown in Figure 39.5. One of the arms of the interferometer is aligned along the direction of the motion of Earth through space. The Earth moving through the ether is equivalent to the ether flowing past the Earth in the opposite direction with speed v as in Figure 39.5. This ether wind blowing in the direction opposite the direction of Earth's motion should cause the speed of light measured in the Earth's frame of reference to be $c - v$ as the light approaches mirror M_2 in Figure 39.5 and $c + v$ after reflection. The speed v is the speed of the Earth through space, and hence the speed of the ether wind, while c is the speed of light in the ether frame. The two beams of light reflected from M_1 and M_2 recombine, and an interference pattern consisting of alternating dark and bright fringes is formed.

During the experiment, the interference pattern was observed while the interferometer was rotated through an angle of 90°. This rotation changed the speed of the ether wind along the direction of the arms of the interferometer. The effect of this rotation should have been to cause the fringe pattern to shift slightly but measurably. Measurements failed to show any change in the interference pattern! The Michelson–Morley experiment was repeated at different times of the year when the ether wind was expected to change direction and magnitude, but the results were always the same: *No fringe shift of the magnitude required was ever observed.*[2]

The null results of the Michelson–Morley experiment not only contradicted the ether hypothesis, it also meant that it was impossible to measure the absolute velocity of the Earth with respect to the ether frame. However, as we see in the next section, Einstein offered a postulate for his theory of relativity that places quite a different interpretation on these null results. In later years, when more was known about the nature of light, the idea of an ether that permeates all of space was relegated to the ash heap of worn-out concepts. Light is now understood to be *an electromagnetic wave, which requires no medium for its propagation*. As a result, the idea of having an ether in which these waves could travel became unnecessary.

Details of the Michelson–Morley Experiment

To understand the outcome of the Michelson–Morley experiment, let us assume that the two arms of the interferometer in Figure 39.5 are of equal length L. First consider the light beam traveling along arm 1 parallel to the direction of the ether wind. As the light beam moves to the right, its speed is reduced by the wind

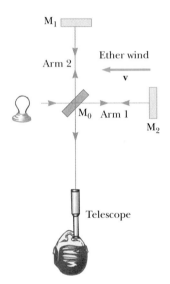

FIGURE 39.5 According to the ether wind theory, the speed of light should be $c - v$ as the beam approaches mirror M_2 and $c + v$ after reflection.

---

[2] From an Earth observer's point of view, changes in the Earth's speed and direction of motion in the course of a year are viewed as ether wind shifts. Even if the speed of the Earth with respect to the ether were zero at some time, six months later the speed of the Earth would be 60 km/s with respect to the ether and a clear fringe shift should be found. None has ever been observed, however.

blowing in the opposite direction and its speed relative to the Earth is $c - v$. On its return journey, as the light beam moves to the left along arm 1, its direction is the same as that of the ether wind and so its speed relative to the Earth is $c + v$. Thus, the time of travel to the right is $L/(c - v)$ and the time of travel to the left is $L/(c + v)$. The total time of travel for the round trip along the horizontal path is

$$t_1 = \frac{L}{c + v} + \frac{L}{c - v} = \frac{2Lc}{c^2 - v^2} = \frac{2L}{c}\left(1 - \frac{v^2}{c^2}\right)^{-1}$$

Now consider the light beam traveling along arm 2, perpendicular to the ether wind. Because the speed of the beam relative to the Earth is $(c^2 - v^2)^{1/2}$ in this case (see Fig. 39.5), then the time of travel for each half of this trip is given by $L/(c^2 - v^2)^{1/2}$, and the total time of travel for the round trip is

$$t_2 = \frac{2L}{(c^2 - v^2)^{1/2}} = \frac{2L}{c}\left(1 - \frac{v^2}{c^2}\right)^{-1/2}$$

Thus, the time difference between the light beam traveling horizontally (arm 1) and the beam traveling vertically (arm 2) is

$$\Delta t = t_1 - t_2 = \frac{2L}{c}\left[\left(1 - \frac{v^2}{c^2}\right)^{-1} - \left(1 - \frac{v^2}{c^2}\right)^{-1/2}\right]$$

Because $v^2/c^2 \ll 1$, this expression can be simplified by using the following binomial expansion after dropping all terms higher than second order:

$$(1 - x)^n \approx 1 - nx \quad \text{(for } x \ll 1\text{)}$$

In our case, $x = v^2/c^2$, and we find

$$\Delta t = t_1 - t_2 \approx \frac{Lv^2}{c^3} \tag{39.3}$$

The two light beams in Figure 39.5 start out in phase and return to form an interference pattern. Let us assume that the interferometer is adjusted for parallel fringes and that a telescope is focused on one of these fringes. The time difference between the two instants at which the light beams arrive at the telescope gives rise to a phase difference between the beams, producing an interference pattern when they combine at the position of the telescope. A difference in the pattern should be detected by rotating the interferometer through 90° in a horizontal plane, so that the two arms of the interferometer exchange positions. This results in a net time difference of twice that given by Equation 39.3. Thus, the path difference that corresponds to this time difference is

$$\Delta d = c(2\,\Delta t) = \frac{2Lv^2}{c^2}$$

The corresponding fringe shift is equal to this path difference divided by the wavelength of light, since a change in path length of one wavelength corresponds to a shift one fringe:

$$\text{Shift} = \frac{2Lv^2}{\lambda c^2} \tag{39.4}$$

In the experiments by Michelson and Morley, each light beam was reflected by mirrors many times to give an effective path length L of approximately 11 m. Using this value, and taking v to be equal to 3.0×10^4 m/s, the speed of the Earth

Albert A. Michelson (1852–1931). A German-American physicist, Michelson invented the interferometer and spent much of his life making accurate measurements of the speed of light. In 1907 he was the first American to be awarded the Nobel prize, which he received for his work in optics. His most famous experiment, conducted with Edward Morley in 1887, implied that it was impossible to measure the absolute velocity of the Earth with respect to the ether. Subsequent work by Einstein in his special theory of relativity eliminated the ether concept by assuming that the speed of light has the same value in all inertial reference frames. *(AIP Emilio Segrè Visual Archives, Michelson Collection)*

around the Sun, we get a path difference of

$$\Delta d = \frac{2(11 \text{ m})(3.0 \times 10^4 \text{ m/s})^2}{(3.0 \times 10^8 \text{ m/s})^2} = 2.2 \times 10^{-7} \text{ m}$$

This extra travel distance should produce a noticeable shift in the fringe pattern. Specifically, using light of wavelength 500 nm, we would expect a fringe shift for rotation through 90° of

$$\text{Shift} = \frac{\Delta d}{\lambda} = \frac{2.2 \times 10^{-7}}{5.0 \times 10^{-7}} \cong 0.40$$

The instrument used by Michelson and Morley had the capability of detecting shifts as small as 0.01 fringe. However, they *reported a shift that is less than one fourth of a fringe*. Since then, the experiment has been repeated many times by various scientists under various conditions, and no fringe shift has ever been detected. Thus, it was concluded that the motion of the Earth with respect to the ether cannot be detected.

Many efforts were made to explain the null results of the Michelson-Morley experiment and to save the ether frame concept and the Galilean addition law for the velocity of light. Since all of these proposals have been shown to be wrong, they will not be considered here. In the 1890s, G. F. Fitzgerald and Hendrik A. Lorentz independently tried to explain the null results by making the following ad hoc assumption. They proposed that the length of an object moving at speed v would contract along the direction of motion by a factor of $\sqrt{1 - v^2/c^2}$. The net result of this contraction would be a change in length of one of the arms of the interferometer such that no path difference would occur as the apparatus was rotated. Such a physical contraction would fully explain the original Michelson-Morley experiment but would be inconsistent with the same experiment when the two arms of the interferometer have different lengths.

No experiment in the history of physics has received such valiant efforts to explain the absence of an expected result as did the Michelson-Morley experiment. The stage was set for the brilliant Albert Einstein, who solved the problem in 1905 with his special theory of relativity.

39.3 EINSTEIN'S PRINCIPLE OF RELATIVITY

In the previous section we noted the impossibility of measuring the speed of the ether with respect to the Earth and the failure of the Galilean addition law for velocities in the case of light. Albert Einstein proposed a theory that boldly removed these difficulties and at the same time completely altered our notion of space and time.[3] Einstein based his special theory of relativity on two postulates:

1. **The Principle of Relativity:** All the laws of physics are the same in all inertial reference frames.
2. **The Constancy of the Speed of Light:** The speed of light in vacuum has the same value, $c = 3.00 \times 10^8$ m/s, in all inertial frames, regardless of the velocity of the observer or the velocity of the source emitting the light.

The postulates of the special theory of relativity

---

[3] A. Einstein, "On the Electrodynamics of Moving Bodies," *Ann. Physik* 17:891 (1905). For an English translation of this article and other publications by Einstein, see the book by H. Lorentz, A. Einstein, H. Minkowski, and H. Weyl, *The Principle of Relativity*, Dover, 1958.

The first postulate asserts that *all* the laws of physics, those dealing with mechanics, electricity and magnetism, optics, thermodynamics, and so on, are the same in all reference frames moving with constant velocity relative to each other. This postulate is a sweeping generalization of the principle of Newtonian relativity that only refers to the laws of mechanics. From an experimental point of view, Einstein's principle of relativity means that any kind of experiment (measuring the speed of light for example), performed in a laboratory at rest, must give the same result when performed in a laboratory moving a constant velocity past the first one. Hence, no preferred inertial reference frame exists and it is impossible to detect absolute motion.

Note that postulate 2, the principle of the constancy of the speed of light, is required by postulate 1: If the speed of light were not the same in all inertial frames, it would be possible to distinguish between inertial frames and a preferred, absolute frame could be identified, in contradiction to postulate 1. Postulate 2 also eliminates the problem of measuring the speed of the ether by denying the existence of the ether and boldly asserting that light always moves with speed c relative to all inertial observers.

Although the Michelson-Morley experiment was performed before Einstein published his work on relativity, it is not clear whether or not Einstein was aware of the details of the experiment. Nonetheless, the null result of the experiment can be readily understood within the framework of Einstein's theory. According to his principle of relativity, the premises of the Michelson-Morley experiment were incorrect. In the process of trying to explain the expected results, we stated that when light traveled against the ether wind its speed was $c - v$, in accordance with the Galilean addition law for velocities. However, if the state of motion of the observer or of the source has no influence on the value found for the speed of light, one will always measure the value to be c. Likewise, the light makes the return trip after reflection from the mirror at speed c, not at speed $c + v$. Thus, the motion of the Earth does not influence the fringe pattern observed in the Michelson-Morley experiment, and a null result should be expected.

If we accept Einstein's theory of relativity, we must conclude that relative motion is unimportant when measuring the speed of light. At the same time, we must alter our common-sense notion of space and time and be prepared for some rather bizarre consequences.

39.4 CONSEQUENCES OF SPECIAL RELATIVITY

Before we discuss the consequences of special relativity, we must first understand how an observer located in an inertial reference frame describes an event. As mentioned earlier, an event is an occurrence describable by three space coordinates and one time coordinate. Different observers in different inertial frames usually describe the same event with different space-time coordinates.

The reference frame used to describe an event consists of a coordinate grid and a set of synchronized clocks located at the grid intersections as shown in Figure 39.6 in two dimensions. The clocks can be synchronized in many ways with the help of light signals. For example, suppose the observer is located at the origin with his master clock and sends out a pulse of light at $t = 0$. The light pulse takes a time r/c to reach a second clock located a distance r from the origin. Hence, the second clock is synchronized with the clock at the origin if the second clock reads a time r/c at the instant the pulse reaches it. This procedure of synchronization assumes that the speed of light has the same value in all directions and in all inertial frames.

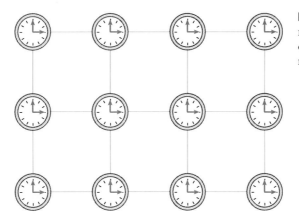

FIGURE 39.6 In relativity, we use a reference frame consisting of a coordinate grid and a set of synchronized clocks.

Furthermore, the procedure concerns an event recorded by an observer in a specific inertial reference frame. An observer in some other inertial frame would assign different space-time coordinates to events being observed by using another coordinate grid and another array of clocks.

Almost everyone who has dabbled even superficially with science is aware of some of the startling predictions that arise because of Einstein's approach to relative motion. As we examine some of the consequences of relativity in the following three sections, we see that they conflict with our basic notions of space and time. We restrict our discussion to the concepts of length, time, and simultaneity, which are quite different in relativistic mechanics than in Newtonian mechanics. For example, *the distance between two points and the time interval between two events depend on the frame of reference in which they are measured.* That is, *there is no such thing as absolute length or absolute time in relativity.* Furthermore, *events at different locations that occur simultaneously in one frame are not simultaneous in another frame moving uniformly past the first.*

Simultaneity and the Relativity of Time

A basic premise of Newtonian mechanics is that a universal time scale exists that is the same for all observers. In fact, Newton wrote that "Absolute, true, and mathematical time, of itself, and from its own nature, flows equably without relation to anything external." Thus, Newton and his followers simply took simultaneity for granted. In his special theory of relativity, Einstein abandoned this assumption. According to Einstein, *a time interval measurement depends on the reference frame in which the measurement is made.*

Einstein devised the following thought experiment to illustrate this point. A boxcar moves with uniform velocity, and two lightning bolts strike its ends, as in Figure 39.7a, leaving marks on the boxcar and on the ground. The marks on the boxcar are labeled A' and B', while those on the ground are labeled A and B. An observer at O' moving with the boxcar is midway between A' and B', while a ground observer at O is midway between A and B. The events recorded by the observers are the light signals from the lightning bolts.

The two light signals reach the observer at O at the same time, as indicated in Figure 39.7b. This observer realizes that the light signals have traveled at the same speed over equal distances, and so rightly concludes that the events at A and B occurred simultaneously. Now consider the same events as viewed by the observer on the boxcar at O'. By the time the light has reached observer O, observer O' has moved as indicated in Figure 39.7b. Thus, the light signal from B' has already

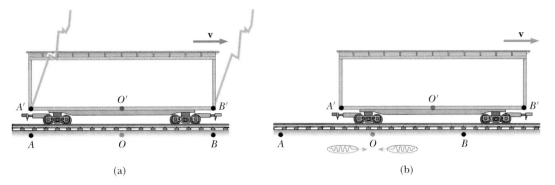

FIGURE 39.7 Two lightning bolts strike the end of a moving boxcar. (a) The events appear to be simultaneous to the stationary observer at O, who is midway between A and B. (b) The events do not appear to be simultaneous to the observer at O', who claims that the front of the train is struck before the rear.

swept past O', while the light from A' has not yet reached O'. According to Einstein, *observer O' must find that light travels at the same speed as that measured by observer O.* Therefore, the observer O' concludes that the lightning struck the front of the boxcar before it struck the back. This thought experiment clearly demonstrates that the two events, which appear to be simultaneous to observer O, do not appear to be simultaneous to observer O'. In other words,

> two events that are simultaneous in one reference frame are in general not simultaneous in a second frame moving relative to the first. That is, simultaneity is not an absolute concept but one that depends upon the state of motion of the observer.

At this point, you might wonder which observer is right concerning the two events. The answer is that *both are correct*, because the principle of relativity states that *there is no preferred inertial frame of reference*. Although the two observers reach different conclusions, both are correct in their own reference frame because the concept of simultaneity is not absolute. This, in fact, is the central point of relativity—any uniformly moving frame of reference can be used to describe events and do physics. However, observers in different inertial frames of reference always measure different time intervals with their clocks and different distances with their meter sticks. Nevertheless, all observers agree on the forms of the laws of physics in their respective frames, because these laws must be the same for all observers in uniform motion. It is the alteration of time and space that allows the laws of physics (including Maxwell's equations) to be the same for all observers in uniform motion.

Time Dilation

The fact that observers in different inertial frames always measure different time intervals between a pair of events can be illustrated by considering a vehicle moving to the right with a speed v as in Figure 39.8a. A mirror is fixed to the ceiling of the vehicle and observer O' at rest in this system holds a laser a distance d below the mirror. At some instant, the laser emits a pulse of light directed towards the mirror (event 1), and at some later time after reflecting from the mirror, the pulse

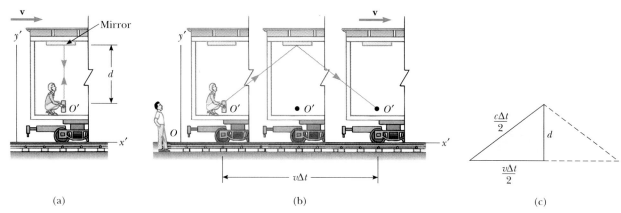

FIGURE 39.8 (a) A mirror is fixed to a moving vehicle, and a light pulse leaves O' at rest in the vehicle. (b) Relative to a stationary observer on Earth, the mirror and O' move with a speed v. Note that the distance the pulse travels is greater than $2d$ as measured by the stationary observer. (c) The right triangle for calculating the relationship between Δt and $\Delta t'$.

arrives back at the laser (event 2). Observer O' carries a clock C' that she uses to measure the time interval $\Delta t'$ between these two events. Because the light pulse has a speed c, the time it takes the pulse to travel from O' to the mirror and back to O' can be found from the definition of speed:

$$\Delta t' = \frac{\text{distance traveled}}{\text{speed}} = \frac{2d}{c} \quad (39.5)$$

This time interval $\Delta t'$ measured by O', who is at rest in the moving vehicle, requires only a single clock C' located at the same place in this frame.

Now consider the same pair of events as viewed by observer O in a second frame as in Figure 39.8b. According to this observer, the mirror and laser are moving to the right with a speed v. The sequence of events appears entirely different as viewed by this observer. By the time the light from the laser reaches the mirror, the mirror has moved a distance $v\,\Delta t/2$, where Δt is the time it takes the light to travel from O' to the mirror and back to O' as measured by observer O. In other words, the second observer concludes that, because of the motion of the vehicle, if the light is to hit the mirror, it must leave the laser at an angle with respect to the vertical direction. Comparing Figures 39.8a and 39.8b, we see that the light must travel farther in the second frame than in the first frame.

According to the second postulate of special relativity, both observers must measure c for the speed of light. Because the light travels farther in the second frame, it follows that the time interval Δt measured by the observer in the second frame is longer than the time interval $\Delta t'$ measured by the observer in the first frame. To obtain a relationship between these two time intervals, it is convenient to use the right triangle shown in Figure 39.8c. The Pythagorean theorem gives

$$\left(\frac{c\,\Delta t}{2}\right)^2 = \left(\frac{v\,\Delta t}{2}\right)^2 + d^2$$

Solving for Δt gives

$$\Delta t = \frac{2d}{\sqrt{c^2 - v^2}} = \frac{2d}{c\sqrt{1 - \dfrac{v^2}{c^2}}} \quad (39.6)$$

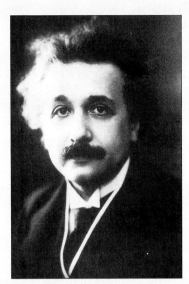

Albert Einstein
1879–1955

Albert Einstein, one of the greatest physicists of all times, was born in Ulm, Germany. He left the highly disciplined German school system after one teacher stated, "You will never amount to anything, Einstein." Following a vacation in Italy, he completed his education at the Swiss Federal Polytechnic School in 1901. Although Einstein attended very few lectures, he was able to pass the courses with the help of excellent lecture notes taken by a friend. Unable to find an academic position, Einstein accepted a position as a junior official in the Swiss Patent Office in Berne. In this setting, and during his "spare time," he continued his independent studies in theoretical physics. In 1905, at the age of 26, he published four scientific papers that revolutionized physics. (In that same year, he earned his Ph.D.) One of these papers, for which he was awarded the 1921 Nobel prize in physics, dealt with the photoelectric effect. Another was concerned with Brownian motion, the irregular motion of small particles suspended in a liquid. The remaining two papers were concerned with what is now considered his most important contribution of all, the special theory of relativity. In 1916, Einstein published his work on the general theory of relativity, which relates gravity to the structure of space and time. The most dramatic prediction of this theory is the degree to which light is deflected by a gravitational field. Measurements made by astronomers on bright stars in the vicinity of the eclipsed Sun in 1919 confirmed Einstein's prediction, and Einstein suddenly became a world celebrity.

In 1913, following academic appointments in Switzerland and Czechoslovakia, Einstein accepted a special position created for him at the Kaiser Wilhelm Institute in Berlin. This made it possible for him to devote all of his time to research free of financial troubles and routine duties. Einstein left Germany in 1933, which was then under Hitler's power, thereby escaping the fate of millions of other European Jews. In the same year he accepted a special position at the Institute for Advanced Study in Princeton where he remained for the rest of his life. He became an American citizen in 1940. Although he was a pacifist, Einstein was persuaded by Leo Szilard to write a letter to President Franklin D. Roosevelt urging him to initiate a program to develop a nuclear bomb. The result was the successful six-year Manhattan project and two nuclear explosions in Japan that ended World War II in 1945.

Einstein made many important contributions to the development of modern physics, including the concept of the light quantum and the idea of stimulated emission of radiation, which led to the invention of the laser 40 years later. He was deeply disturbed by the development of quantum mechanics in the 1920s despite his own role as a scientific revolutionary. In particular, he could never accept the probabilistic view of events in nature that is a central feature of the Copenhagen interpretation of quantum theory. He once said, "God does not play dice with nature." The last few decades of his life were devoted to an unsuccessful search for a unified theory that would combine gravitation and electromagnetism into one picture.

Photo credit: (Photograph courtesy of AIP Niels Bohr Library)

Because $\Delta t' = 2d/c$, we can express Equation 39.6 as

$$\Delta t = \frac{\Delta t'}{\sqrt{1 - \frac{v^2}{c^2}}} = \gamma \, \Delta t' \qquad (39.7)$$

Time dilation

where $\gamma = (1 - v^2/c^2)^{-1/2}$. This result says that *the time interval Δt measured by an observer moving with respect to the clock is longer than the time interval $\Delta t'$ measured by the observer at rest with respect to the clock* because γ is always greater than unity. That is, $\Delta t > \Delta t'$. This effect is known as **time dilation**.

The time interval $\Delta t'$ in Equation 39.7 is called the **proper time**. In general, proper time is defined as *the time interval between two events measured by an observer who sees the events occur at the same point in space*. In our case, observer O' measures the proper time. That is, *proper time is always the time measured with a single clock at rest in the frame in which the event takes place*.

Because the time between ticks of a moving clock, $\gamma(2d/c)$, is observed to be longer than the time between ticks of an identical clock at rest, $2d/c$, it is often said "*A moving clock runs slower than a clock at rest by a factor γ.*" This is true for ordinary mechanical clocks as well as for the light clock just described. In fact, we can generalize these results by stating that *all physical, chemical, and biological processes slow down relative to a stationary clock when those processes occur in a moving frame*. For example, the heartbeat of an astronaut moving through space would keep time with a clock inside the spaceship. Both the astronaut's clock and heartbeat are slowed down relative to a stationary clock. The astronaut would not have any sensation of life slowing down in the spaceship. For the astronaut, it is the clock on Earth and the companions at Mission Control that are moving and therefore keep a slow time.

Time dilation is a verifiable phenomenon. For example, muons are unstable elementary particles that have a charge equal to that of the electron and a mass 207 times that of the electron. Muons can be produced by the collision of cosmic radiation with atoms high in the atmosphere. Muons have a lifetime of only 2.2 μs when measured in a reference frame at rest relative to them. If we take 2.2 μs as the average lifetime of a muon and assume that its speed is close to the speed of light, we find that these particles can travel a distance of only approximately 600 m before they decay (Fig. 39.9a). Hence, they cannot reach the Earth from the upper atmosphere where they are produced. However, experiments show that a large number of muons do reach the Earth. The phenomenon of time dilation explains this effect. Relative to an observer on Earth, the muons have a lifetime equal to $\gamma\tau$, where $\tau = 2.2$ μs is the lifetime in a frame of reference traveling with the muons. For example, for $v = 0.99c$, $\gamma \approx 7.1$ and $\gamma\tau \approx 16$ μs. Hence, the average distance traveled as measured by an observer on Earth is $\gamma v\tau \approx 4800$ m, as indicated in Figure 39.9b.

In 1976, at the laboratory of the European Council for Nuclear Research (CERN) in Geneva, muons injected into a large storage ring reached speeds of approximately $0.9994c$. Electrons produced by the decaying muons were detected by counters around the ring, enabling scientists to measure decay rate and, hence, the muon lifetime. The lifetime of the moving muons was measured to be approximately 30 times as long as that of the stationary muon (Fig. 39.10), in agreement with the prediction of relativity to within two parts in a thousand.

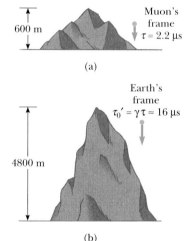

FIGURE 39.9 (a) Muons traveling with a speed of $0.99c$ travel only approximately 600 m as measured in the muons' reference frame, where their lifetime is about 2.2 μs. (b) The muons travel approximately 4800 m as measured by an observer on Earth. Because of time dilation, the muons' lifetime is longer as measured by the Earth observer.

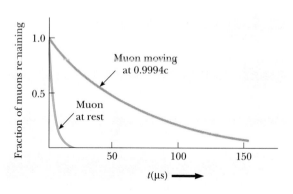

FIGURE 39.10 Decay curves for muons traveling at a speed of $0.9994c$ and for muons at rest.

EXAMPLE 39.1 What Is the Period of the Pendulum?

The period of a pendulum is measured to be 3.0 s in the rest frame of the pendulum. What is the period when measured by an observer moving at a speed of $0.95c$ relative to the pendulum?

Reasoning Instead of the observer moving at $0.95c$, we can take the equivalent point of view that the observer is at rest and that the pendulum is moving at $0.95c$ past the stationary observer. Hence the pendulum is an example of a moving clock.

Solution The proper time is 3.0 s. Because a moving clock runs slower than a stationary clock by γ, Equation 39.7 gives

$$T = \gamma T' = \frac{1}{\sqrt{1 - \frac{(0.95c)^2}{c^2}}} T' = (3.2)(3.0 \text{ s}) = 9.6 \text{ s}$$

That is, a moving pendulum slows down or takes longer to complete a period compared with one at rest.

Length Contraction

The measured distance between two points also depends on the frame of reference. The **proper length** of an object is defined as *the length of the object measured by someone who is at rest relative to the object*. The length of an object measured by someone in a reference frame that is moving with respect to the object is always less than the proper length. This effect is known as **length contraction**.

Consider a spaceship traveling with a speed v from one star to another. There are two observers, one on Earth and the other in the spaceship. The observer at rest on Earth (and also assumed to be at rest with respect to the two stars) measures the distance between the stars to be L_p, the proper length. According to this observer, the time it takes the spaceship to complete the voyage is $\Delta t = L_p/v$. What does an observer in the moving spaceship measure for the distance between the stars? Because of time dilation, the space traveler measures a smaller time of travel: $\Delta t' = \Delta t/\gamma$. The space traveler claims to be at rest and sees the destination star moving toward the spaceship with speed v. Because the space traveler reaches the star in the time $\Delta t'$, he concludes that the distance, L, between the stars is shorter than L_p. This distance measured by the space traveler is

$$L = v\,\Delta t' = v\,\frac{\Delta t}{\gamma}$$

Since $L_p = v\,\Delta t$, we see that $L = L_p/\gamma$ or

$$L = L_p\left(1 - \frac{v^2}{c^2}\right)^{1/2} \qquad (39.8)$$

where $(1 - v^2/c^2)^{1/2}$ is a factor less than one. This result may be interpreted as follows:

If an object has a length L_p when it is at rest, then when it moves with speed v in a direction parallel to its length, it contracts to the length L, where $L = L_p(1 - v^2/c^2)^{1/2}$.

Note that *length contraction takes place only along the direction of motion*. For example, suppose a stick moves past a stationary Earth observer with speed v as in Figure 39.11. The length of the stick as measured by an observer in a frame attached to

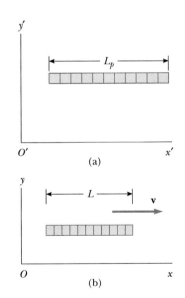

FIGURE 39.11 (a) A stick as viewed by an observer in a frame attached to the stick (i.e., both have the same velocity). (b) The stick as seen by an observer in a frame in which the stick has a velocity **v** relative to the frame. The length is shorter than the proper length, L_p, by a factor $(1 - v^2/c^2)^{1/2}$.

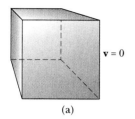

(a)

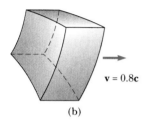

(b)

FIGURE 39.12 Computer-simulated photographs of a box (a) at rest relative to the camera and (b) moving at a velocity $v = 0.8c$ relative to the camera.

the stick is the proper length, L_p, as in Figure 39.11a. The length of the stick, L, measured by the Earth observer is shorter than L_p by the factor $(1 - v^2/c^2)^{1/2}$. Furthermore, length contraction is a symmetrical effect: If the stick is at rest on Earth, an observer in the moving frame would also measure its length to be shorter by the same factor $(1 - v^2/c^2)^{1/2}$.

It is important to emphasize that proper length and proper time are measured in different reference frames. As an example of this point, let us return to the decaying muons moving at speeds close to the speed of light. An observer in the muon's reference frame would measure the proper lifetime, while an Earth-based observer would measure the proper height of the mountain in Figure 39.9. In the muon's reference frame, there is no time dilation but the distance of travel is observed to be shorter when measured in this frame. Likewise, in the Earth observer's reference frame, there is time dilation, but the distance of travel is measured to be the actual height of the mountain. Thus, when calculations on the muon are performed in both frames, the effect of "offsetting penalties" is seen and the outcome of the experiment in one frame is the same as the outcome in the other frame!

If a box passing by at a speed close to c could be photographed, its image would show length contraction, but its shape would also be distorted. This is illustrated in Figure 39.12 for a box moving past a camera with a speed $v = 0.8c$. When the shutter of the camera is opened, it records the shape of the box at a given instant of time. Since light from different parts of the box must arrive at the camera at the same instant (the instant at which the photograph is taken), light from the more distant parts must start its journey earlier than light from closer parts. Hence, the photograph records different parts of the box at different times. This results in a highly distorted image, which shows horizontal length contraction, vertical curvature, and image rotation.

EXAMPLE 39.2 The Contraction of a Spaceship

A spaceship is measured to be 120 m long while at rest relative to an observer. If this spaceship now flies by the observer with a speed $0.99c$, what length does the observer measure?

Solution From Equation 39.8, the length measured by the observer is

$$L = L_p\sqrt{1 - \frac{v^2}{c^2}} = (120 \text{ m})\sqrt{1 - \frac{(0.99c)^2}{c^2}} = 17 \text{ m}$$

Exercise If the ship moves past the observer with a speed of $0.01000c$, what length will the observer measure?

Answer 119.40 m.

EXAMPLE 39.3 The Triangular Spaceship

A spaceship in the form of a triangle flies by an observer with a speed of $0.95c$. When the ship is at rest relative to the observer (Fig. 39.13a), the distances x and y are measured to be 52 m and 25 m, respectively. What is the shape of the ship as seen by an observer at rest when the ship is in motion along the direction shown in Figure 39.13b?

FIGURE 39.13 (Example 39.4) (a) When the spaceshift is at rest, its shape is as shown. (b) The spaceship appears to look like this when it moves to the right with a speed v. Note that only its x dimension is contracted in this case.

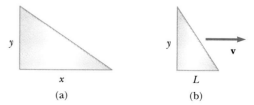

Solution The observer sees the horizontal length of the ship to be contracted to a length

$$L = L_p\sqrt{1 - \frac{v^2}{c^2}} = (52 \text{ m})\sqrt{1 - \frac{(0.95\,c)^2}{c^2}} = 16 \text{ m}$$

The 25-m vertical height is unchanged because it is perpendicular to the direction of relative motion between observer and spaceship. Figure 39.13b represents the size of the spaceship as seen by the observer at rest.

CONCEPTUAL EXAMPLE 39.4   A Voyage to Sirius

An astronaut takes a trip to Sirius, located 8 lightyears from Earth. The astronaut measures the time of the one-way journey to be 6 y. If the spaceship moved at a constant speed of $0.8\,c$, how can the 8-lightyear distance be reconciled with the 6-y duration measured by the astronaut?

Reasoning The 8 lightyears (ly) represents the proper length (distance) from Earth to Sirius measured by an observer seeing both nearly at rest. The astronaut sees Sirius approaching her at $0.8\,c$, but she also sees the distance contracted to

$$\frac{8 \text{ ly}}{\gamma} = (8 \text{ ly})\sqrt{1 - \frac{v^2}{c^2}} = (8 \text{ ly})\sqrt{1 - \frac{(0.8\,c)^2}{c^2}} = 5 \text{ ly}$$

So the travel time measured on her clocks is

$$t = \frac{d}{v} = \frac{5 \text{ ly}}{0.8\,c} = 6 \text{ y}$$

---

*The Twins Paradox

An intriguing consequence of time dilation is the so-called twins paradox. Consider an experiment involving a set of twins named Speedo and Goslo who are, say, 21 years old. The twins carry with them identical clocks that have been synchronized (Fig. 39.14). Speedo, the more adventuresome of the two, sets out on an epic journey to Planet X, located 10 lightyears from Earth. Furthermore, his spaceship is capable of reaching a speed of $0.500\,c$ relative to the inertial frame of his twin brother on Earth. After reaching Planet X, Speedo becomes homesick and immediately returns to Earth at the same high speed he had attained on the outbound journey. Upon his return, Speedo is shocked to discover that many things have changed in his absence. To Speedo, the most significant change to have occurred is that his twin brother Goslo has aged 40 years and is now 61 years old. Speedo, on the other hand, has aged by only 34.6 years.

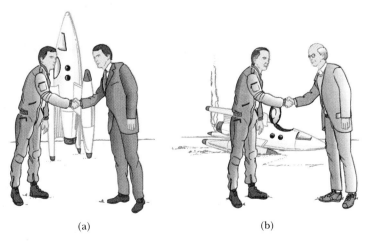

(a)   (b)

FIGURE 39.14 (a) As the twins depart, they are the same age. (b) When Speedo returns from his journey to Planet X, he is younger than his twin Goslo, who remained on Earth.

At this point, it is fair to raise the following question—which twin is the traveler and which twin is really younger as a result of this experiment? From Goslo's frame of reference, he was at rest while his brother traveled at a high speed. From Speedo's perspective, it is he who is at rest while Goslo is on the high-speed space journey. According to Speedo, it is Goslo and the Earth that have raced away on a 17.3-year journey and then headed back for another 17.3 years. This leads to an apparent contradiction. Which twin has developed signs of excess aging?

To resolve this apparent paradox, recall that the special theory of relativity deals with inertial frames of reference moving relative to each other at uniform speed. However, the trip situation is not symmetrical. Speedo, the space traveler, must experience a series of accelerations during his journey. As a result, his speed is not always uniform and consequently Speedo is not in an inertial frame. He cannot be regarded as always being at rest and Goslo to be in uniform motion, because to do so would be an incorrect application of the special theory of relativity. Therefore there is no paradox.

The conclusion that Speedo is in a noninertial frame is inescapable. The time required to accelerate and decelerate Speedo's spaceship may be made very small using large rockets, so that Speedo can claim that he spends most of his time traveling to Planet X at $0.500c$ in an inertial frame. However, Speedo must slow down, reverse his motion, and return to Earth in an altogether different inertial frame. At the very best, Speedo is in two different inertial frames during his journey. Only Goslo, who is in a single inertial frame, can apply the simple time dilation formula to Speedo's trip. Thus, Goslo finds that instead of aging 40 years, Speedo ages only $(1 - v^2/c^2)^{1/2}(40 \text{ years}) = 34.6$ years. On the other hand, Speedo spends 17.3 years traveling to Planet X and 17.3 years returning, for a total travel time of 34.6 years, in agreement with our earlier statement.

CONCEPTUAL EXAMPLE 39.5

Suppose astronauts were paid according to the time spent traveling in space. After a long voyage traveling at a speed near that of light, a crew of astronauts return to Earth and open their pay envelopes. What will their reaction be?

Reasoning Assuming that their on-duty time was kept on Earth, they will be pleasantly surprised with a large paycheck. Less time will have passed for the astronauts in their frame of reference than for their employer back on Earth.

An experiment reported by Hafele and Keating provided direct evidence of time dilation.[4] Time intervals measured with four cesium beam atomic clocks in jet flight were compared with time intervals measured by Earth-based reference atomic clocks. In order to compare these results with the theory, many factors had to be considered, including periods of acceleration and deceleration relative to the Earth, variations in direction of travel, and the weaker gravitational field experienced by the flying clocks compared with the Earth-based clock. Their results were in good agreement with the predictions of the special theory of relativity and can be explained in terms of the relative motion between the Earth's rotation and the jet aircraft. In their paper, Hafele and Keating say: "Relative to the atomic time scale of the U.S. Naval Observatory, the flying clocks lost 59 ± 10 ns during the

---

[4] J. C. Hafele and R. E. Keating, "Around the World Atomic Clocks: Relativistic Time Gains Observed," *Science*, July 14, 1972, p. 168.

eastward trip and gained 273 ± 7 ns during the westward trip. . . . These results provide an unambiguous empirical resolution of the famous clock paradox with macroscopic clocks."

39.5 THE LORENTZ TRANSFORMATION EQUATIONS

We have seen that the Galilean transformation is not valid when v approaches the speed of light. In this section, we state the correct transformation equations that apply for all speeds in the range $0 \leq v < c$.

Suppose an event that occurs at some point P is reported by two observers, one at rest in a frame S and another in a frame S′ that is moving to the right with speed v as in Figure 39.15. The observer in S reports the event with space-time coordinates (x, y, z, t), while the observer in S′ reports the same event using the coordinates (x', y', z', t'). We would like to find a relationship between these coordinates that is valid for all speeds. In Section 39.1, we found that the Galilean transformation of coordinates, given by Equation 39.1, does not agree with experiment at speeds comparable to the speed of light.

The equations that are valid from $v = 0$ to $v = c$ and enable us to transform coordinates from S to S′ are given by the **Lorentz transformation equations**:

$$x' = \gamma(x - vt)$$
$$y' = y$$
$$z' = z$$
$$t' = \gamma\left(t - \frac{v}{c^2}x\right)$$

(39.9) Lorentz transformation for S → S′

This transformation, known as the Lorentz transformation, was developed by Hendrik A. Lorentz (1853–1928) in 1890 in connection with electromagnetism. However, it was Einstein who recognized their physical significance and took the bold step of interpreting them within the framework of the theory of relativity.

We see that the value for t' assigned to an event by an observer standing at O' depends both on the time t and on the coordinate x as measured by an observer at O. This is consistent with the notion that an event is characterized by four space-time coordinates (x, y, z, t). In other words, in relativity, space and time are not separate concepts but rather are closely interwoven with each other. This is unlike the case of the Galilean transformation in which $t = t'$.

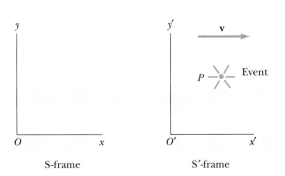

FIGURE 39.15 Representation of an event that occurs at some point P as observed by an observer at rest in the S frame and another in the S′ frame, which is moving to the right with a speed v.

If we wish to transform coordinates in the S' frame to coordinates in the S frame, we simply replace v by $-v$ and interchange the primed and unprimed coordinates in Equation 39.9:

Inverse Lorentz transformation for S' → S

$$x = \gamma(x' + vt')$$
$$y = y'$$
$$z = z' \quad (39.10)$$
$$t = \gamma\left(t' + \frac{v}{c^2} x'\right)$$

When $v \ll c$, the Lorentz transformation should reduce to the Galilean transformation. To check this, note that as $v \to 0$, $v/c^2 \ll 1$ and $v^2/c^2 \ll 1$, so that $\gamma = 1$ and Equation 39.9 reduces in this limit to the Galilean coordinate transformation equations

$$x' = x - vt \quad y' = y \quad z' = z \quad t' = t$$

In many situations, we would like to know the difference in coordinates between two events or the time interval between two events as seen by observers at O and O'. This can be accomplished by writing the Lorentz equations in a form suitable for describing pairs of events. From Equations 39.9 and 39.10, we can express the differences between the four variables x, x', t, and t' in the form

$$\left.\begin{array}{l} \Delta x' = \gamma(\Delta x - v\,\Delta t) \\ \Delta t' = \gamma\left(\Delta t - \dfrac{v}{c^2}\Delta x\right) \end{array}\right\} \text{S} \to \text{S}' \quad (39.11)$$

$$\left.\begin{array}{l} \Delta x = \gamma(\Delta x' + v\,\Delta t') \\ \Delta t = \gamma\left(\Delta t' + \dfrac{v}{c^2}\Delta x'\right) \end{array}\right\} \text{S}' \to \text{S} \quad (39.12)$$

where $\Delta x' = x'_2 - x'_1$ and $\Delta t' = t'_2 - t'_1$ are the differences measured by the observer at O', while $\Delta x = x_2 - x_1$ and $\Delta t = t_2 - t_1$ are the differences measured by the observer at O. We have not included the expressions for relating the y and z coordinates because they are unaffected by motion along the x direction.[5]

EXAMPLE 39.6 Simultaneity and Time Dilation Revisited

Use the Lorentz transformation equations in difference form to show that (a) simultaneity is not an absolute concept and (b) moving clocks run slower than stationary clocks.

Solution (a) Suppose that two events are simultaneous according to a moving observer at O', so that $\Delta t' = 0$. From the expression for Δt given in Equation 39.12, we see that in this case, $\Delta t = \gamma v\,\Delta x'/c^2$. That is, the time interval for the same two events as measured by an observer at O is nonzero, and so they do not appear to be simultaneous in O.

(b) Suppose that an observer at O' finds that two events occur at the same place ($\Delta x' = 0$), but at different times ($\Delta t' \neq 0$). In this situation, the expression for Δt given in Equation 39.12 becomes $\Delta t = \gamma\,\Delta t'$. This is the equation for time dilation found earlier, Equation 39.7, where $\Delta t' = \Delta t$ is the proper time measured by the single clock located at O'.

Exercise Use the Lorentz transformation equations in difference form to confirm that $L = L_p/\gamma$.

---

[5] Although motion along x does not change y and z coordinates, it does change velocity components along y and z.

Lorentz Velocity Transformation

Let us now derive the Lorentz velocity transformation, which is the relativistic counterpart of the Galilean velocity transformation. Once again S is our stationary frame of reference and S' is our frame of reference that moves at a speed v relative to S. Suppose that an object is observed in the S' frame with an instantaneous speed u'_x measured in S' given by

$$u'_x = \frac{dx'}{dt'} \quad (39.13)$$

Using Equations 39.9, we have

$$dx' = \gamma(dx - v\,dt)$$

$$dt' = \gamma\left(dt - \frac{v}{c^2}dx\right)$$

Substituting these values into Equation 39.13 gives

$$u'_x = \frac{dx'}{dt'} = \frac{dx - v\,dt}{dt - \frac{v}{c^2}dx} = \frac{\frac{dx}{dt} - v}{1 - \frac{v}{c^2}\frac{dx}{dt}}$$

But dx/dt is just the velocity component u_x of the object measured in S, and so this expression becomes

$$u'_x = \frac{u_x - v}{1 - \frac{u_x v}{c^2}} \quad (39.14)$$

Lorentz velocity transformation for S → S'

Similarly, if the object has velocity components along y and z, the components in S' are

$$u'_y = \frac{u_y}{\gamma\left(1 - \frac{u_x v}{c^2}\right)} \quad \text{and} \quad u'_z = \frac{u_z}{\gamma\left(1 - \frac{u_x v}{c^2}\right)} \quad (39.15)$$

When u_x and v are both much smaller than c (the nonrelativistic case), the denominator of Equation 39.14 approaches unity and so $u'_x \approx u_x - v$. This corresponds to the Galilean velocity transformations. In the other extreme, when $u_x = c$, Equation 39.14 becomes

$$u'_x = \frac{c - v}{1 - \frac{cv}{c^2}} = \frac{c\left(1 - \frac{v}{c}\right)}{1 - \frac{v}{c}} = c$$

From this result, we see that an object moving with a speed c relative to an observer in S also has a speed c relative to an observer in S'—independent of the relative motion of S and S'. Note that this conclusion is consistent with Einstein's second postulate, namely, that the speed of light must be c relative to all inertial frames of reference. Furthermore, the speed of an object can never exceed c. That is, the speed of light is the ultimate speed. We return to this point later when we consider the energy of a particle.

To obtain u_x in terms of u'_x, we replace v by $-v$ in Equation 39.15 and interchange the roles of u_x and u'_x:

Inverse Lorentz velocity transformation for $S' \rightarrow S$

$$u_x = \frac{u'_x + v}{1 + \dfrac{u'_x v}{c^2}} \qquad (39.16)$$

EXAMPLE 39.7 Relative Velocity of Spaceships

Two spaceships A and B are moving in opposite directions, as in Figure 39.16. An observer on Earth measures the speed of A to be $0.750c$ and the speed of B to be $0.850c$. Find the velocity of B with respect to A.

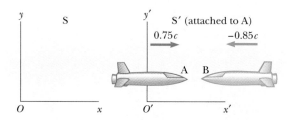

FIGURE 39.16 (Example 39.7) Two spaceships A and B move in opposite directions. The speed of B relative to A is *less* than c and is obtained by using the relativistic velocity transformation.

Solution This problem can be solved by taking the S' frame as being attached to A, so that $v = 0.750c$ relative to the Earth observer (the S frame). Spaceship B can be considered as an object moving with a velocity $u_x = -0.850c$ relative to the Earth observer. Hence, the velocity of B with respect to A can be obtained using Equation 39.14:

$$u'_x = \frac{u_x - v}{1 - \dfrac{u_x v}{c^2}} = \frac{-0.850c - 0.750c}{1 - \dfrac{(-0.850c)(0.750c)}{c^2}} = -0.980c$$

The negative sign indicates that spaceship B is moving in the negative x direction as observed by A. Note that the result is less than c. That is, a body whose speed is less than c in one frame of reference must have a speed less than c in any other frame. (If the Galilean velocity transformation were used in this example, we would find that $u'_x = u_x - v = -0.850c - 0.750c = -1.60c$, which is greater than c. The Galilean transformation does not work in relativistic situations.)

EXAMPLE 39.8 The Speeding Motorcycle

Imagine a motorcycle rider moving with a speed $0.80c$ past a stationary observer, as shown in Figure 39.17. If the rider tosses a ball in the forward direction with a speed of $0.70c$ relative to himself, what is the speed of the ball as seen by the stationary observer?

Solution In this situation, the velocity of the motorcycle with respect to the stationary observer is $v = 0.80c$. The velocity of the ball in the frame of reference of the motorcyclist is $0.70c$. Therefore, the velocity, u_x, of the ball relative to the stationary observer is

$$u_x = \frac{u'_x + v}{1 + \dfrac{u'_x v}{c^2}} = \frac{0.70c + 0.80c}{1 + \dfrac{(0.70c)(0.80c)}{c^2}} = 0.96c$$

Exercise Suppose that the motorcyclist turns on a beam of light that moves away from him with a speed c in the forward direction. What does the stationary observer measure for the speed of the light?

Answer c.

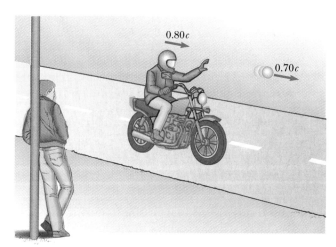

FIGURE 39.17 (Example 39.8) A motorcyclist moves past a stationary observer with a speed of $0.80c$ and throws a ball in the direction of motion with a speed of $0.70c$ relative to himself.

EXAMPLE 39.9 Relativistic Leaders of the Pack

Two motorcycle pack leaders named David and Emily are racing at relativistic speeds along perpendicular paths as in Figure 39.18. How fast does Emily recede over David's right shoulder as seen by David?

Solution Figure 39.18 represents the situation as seen by a police person at rest in frame S, who observes the following:

David: $u_x = 0.75\,c$ $u_y = 0$

Emily: $u_x = 0$ $u_y = -0.90\,c$

To get Emily's speed of recession as seen by David, we take S' to move along with David and we calculate u'_x and u'_y for Emily using Equations 39.14 and 39.15:

$$u'_x = \frac{u_x - v}{1 - \frac{u_x v}{c^2}} = \frac{0 - 0.75\,c}{1 - \frac{(0)(0.75\,c)}{c^2}} = -0.75\,c$$

$$u'_y = \frac{u_y}{\gamma\left(1 - \frac{u_x v}{c^2}\right)} = \frac{\sqrt{1 - \frac{(0.75\,c)^2}{c^2}}\,(-0.90\,c)}{\left(1 - \frac{(0)(0.75\,c)}{c^2}\right)}$$

$$= -0.60\,c$$

Thus, the speed of Emily as observed by David is

$$u' = \sqrt{(u'_x)^2 + (u'_y)^2} = \sqrt{(-0.75\,c)^2 + (-0.60\,c)^2} = \boxed{0.96\,c}$$

Note that this speed is less than c as required by special relativity.

Exercise Calculate the classical speed of recession for Emily as observed by David using a Galilean transformation.

Answer $1.2\,c$.

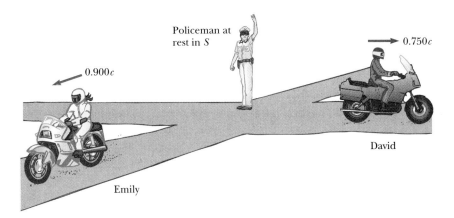

FIGURE 39.18 (Example 39.9) David moves to the east with a speed $0.750\,c$ relative to the policeman, while Emily travels south at a speed $0.900\,c$.

39.6 RELATIVISTIC MOMENTUM AND THE RELATIVISTIC FORM OF NEWTON'S LAWS

We have seen that to properly describe the motion of particles within the framework of special relativity, the Galilean transformation must be replaced by the Lorentz transformation. Because the laws of physics must remain unchanged under the Lorentz transformation, we must generalize Newton's laws and the definitions of momentum and energy to conform to the Lorentz transform and the principle of relativity. These generalized definitions should reduce to the classical (nonrelativistic) definitions for $v \ll c$.

First, recall that the law of conservation of momentum states that when two bodies collide, the total momentum remains constant, assuming the bodies are isolated. Suppose the collision is described in a reference frame S in which momentum is conserved. If the velocities in a second reference frame S' are calculated using the Lorentz transformation and the classical definition of momentum, **p** = m**u** (where **u** is the velocity of the particle), it is found that momentum is *not* conserved in the second reference frame. However, because the laws of physics are the same in all inertial frames, the momentum must be conserved in all systems. In view of this condition and assuming the Lorentz transformation is correct, we must modify the definition of momentum to satisfy the following conditions:

- **p** must be conserved in all collisions
- **p** must approach the classical value $m\mathbf{u}$ as $\mathbf{u} \to 0$

The correct relativistic equation for momentum that satisfies these conditions is

| Definition of relativistic momentum |

$$\mathbf{p} \equiv \frac{m\mathbf{u}}{\sqrt{1 - \dfrac{u^2}{c^2}}} \qquad (39.17)$$

where **u** is the velocity of the particle. (We use the symbol **u** for particle velocity rather than **v**, which is used for the relative velocity of two reference frames.) When u is much less than c, the denominator of Equation 39.17 approaches unity, so that **p** approaches $m\mathbf{u}$. Therefore, the relativistic equation for **p** reduces to the classical expression when u is small compared with c. Because it is simpler, Equation 39.17 is often written as

$$\mathbf{p} = \gamma m\mathbf{u} \qquad (39.18)$$

where $\gamma = (1 - u^2/c^2)^{-1/2}$. Note that γ has the same functional form as the γ in the Lorentz transformation. The transformation is from that of the particle to the frame of the observer moving at speed u relative to the particle.

The relativistic force **F** on a particle whose momentum is **p** is defined as

$$\mathbf{F} \equiv \frac{d\mathbf{p}}{dt} \qquad (39.19)$$

where **p** is given by Equation 39.17. This expression is reasonable because it preserves classical mechanics in the limit of low velocities and requires conservation of momentum for an isolated system ($\mathbf{F} = 0$) both relativistically and classically.

It is left as an end-of-chapter problem (Problem 55) to show that the acceleration **a** of a particle decreases under the action of a constant force, in which case $a \propto (1 - u^2/c^2)^{3/2}$. From this formula note that as the particle's speed approaches

---

EXAMPLE 39.10 Momentum of an Electron

An electron, which has a mass of 9.11×10^{-31} kg, moves with a speed of $0.750c$. Find its relativistic momentum and compare this with the momentum calculated from the classical expression.

Solution Using Equation 39.17 with $u = 0.75c$, we have

$$p = \frac{mu}{\sqrt{1 - \dfrac{u^2}{c^2}}}$$

$$p = \frac{(9.11 \times 10^{-31} \text{ kg})(0.750 \times 3.00 \times 10^8 \text{ m/s})}{\sqrt{1 - \dfrac{(0.750c)^2}{c^2}}}$$

$$= 3.10 \times 10^{-22} \text{ kg} \cdot \text{m/s}$$

The incorrect classical expression gives

$$\text{Momentum} = mu = 2.05 \times 10^{-22} \text{ kg} \cdot \text{m/s}$$

Hence, the correct relativistic result is 50% greater than the classical result!

c, the acceleration caused by any finite force approaches zero. Hence, it is impossible to accelerate a particle from rest to a speed $u \geq c$.

39.7 RELATIVISTIC ENERGY

We have seen that the definition of momentum and the laws of motion require generalization to make them compatible with the principle of relativity. This implies that the definition of kinetic energy must also be modified.

To derive the relativistic form of the work-energy theorem, let us start with the definition of the work done on a particle by a force F and use the definition of relativistic force, Equation 39.19:

$$W = \int_{x_1}^{x_2} F\, dx = \int_{x_1}^{x_2} \frac{dp}{dt}\, dx \tag{39.20}$$

for force and motion both along the x axis. In order to perform this integration, and find the work done on a particle, or the relativistic kinetic energy as a function of u, we first evaluate dp/dt:

$$\frac{dp}{dt} = \frac{d}{dt} \frac{mu}{\sqrt{1 - \dfrac{u^2}{c^2}}} = \frac{m(du/dt)}{\left(1 - \dfrac{u^2}{c^2}\right)^{3/2}}$$

Substituting this expression for dp/dt and $dx = u\, dt$ into Equation 39.20 gives

$$W = \int_{x_1}^{x_2} \frac{m(du/dt)\, u\, dt}{\left(1 - \dfrac{u^2}{c^2}\right)^{3/2}} = m \int_0^u \frac{u}{\left(1 - \dfrac{u^2}{c^2}\right)^{3/2}} du$$

where we have assumed that the particle is accelerated from rest to some final speed u. Evaluating the integral, we find that

$$W = \frac{mc^2}{\sqrt{1 - \dfrac{u^2}{c^2}}} - mc^2 \tag{39.21}$$

Recall from Chapter 7 that the work done by a force acting on a particle equals the change in kinetic energy of the particle. Because the initial kinetic energy is zero, we conclude that the work W is equivalent to the relativistic kinetic energy K:

$$K = \frac{mc^2}{\sqrt{1 - \dfrac{u^2}{c^2}}} - mc^2 = \gamma mc^2 - mc^2 \tag{39.22}$$

Relativistic kinetic energy

This equation is routinely confirmed by experiments using high-energy particle accelerators.

At low speeds, where $u/c \ll 1$, Equation 39.22 should reduce to the classical expression $K = \frac{1}{2}mu^2$. We can check this by using the binomial expansion $(1 - x^2)^{-1/2} \approx 1 + \frac{1}{2}x^2 + \cdots$ for $x \ll 1$, where the higher-order powers of x are neglected in the expansion. In our case, $x = u/c$, so that

$$\frac{1}{\sqrt{1 - \frac{u^2}{c^2}}} = \left(1 - \frac{u^2}{c^2}\right)^{-1/2} \approx 1 + \frac{1}{2}\frac{u^2}{c^2} + \cdots$$

Substituting this into Equation 39.22 gives

$$K \approx mc^2\left(1 + \frac{1}{2}\frac{u^2}{c^2} + \cdots\right) - mc^2 = \frac{1}{2}mu^2$$

which agrees with the classical result. A graph comparing the relativistic and nonrelativistic expressions is given in Figure 39.19. In the relativistic case, the particle speed never exceeds c, regardless of the kinetic energy. The two curves are in good agreement when $u \ll c$.

The constant term mc^2 in Equation 39.22, which is independent of the speed, is called the **rest energy** of the free particle E_R. The term γmc^2, which depends on the particle speed, is therefore the sum of the kinetic and rest energies. We define γmc^2 to be the **total energy** E, that is,

Total energy = kinetic energy + rest energy

Definition of total energy

$$E = \gamma mc^2 = K + mc^2 \tag{39.23}$$

or, when γ is replaced by its equivalent,

Conservation of mass-energy

$$E = \frac{mc^2}{\sqrt{1 - \frac{u^2}{c^2}}} \tag{39.24}$$

This, of course, is Einstein's famous mass-energy equivalence equation. The relation $E = \gamma mc^2 = \gamma E_R$ shows that *mass is a property of energy*. Furthermore, this result shows that a small mass corresponds to an enormous amount of energy. This concept is fundamental to much of the field of nuclear physics.

In many situations, the momentum or energy of a particle is measured rather than its speed. It is therefore useful to have an expression relating the total energy E to the relativistic momentum p. This is accomplished by using the expressions $E = \gamma mc^2$ and $p = \gamma mu$. By squaring these equations and subtracting, we can eliminate u (Problem 23). The result, after some algebra, is

Energy-momentum relationship

$$E^2 = p^2c^2 + (mc^2)^2 \tag{39.25}$$

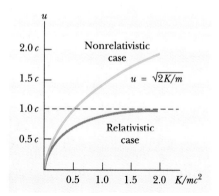

FIGURE 39.19 A graph comparing relativistic and nonrelativistic kinetic energy. The energies are plotted versus speed. In the relativistic case, u is always less than c.

When the particle is at rest, $p = 0$, and so $E = E_R = mc^2$. That is, the total energy equals the rest energy. For the case of particles that have zero mass, such as photons (massless, chargeless particles of light) and neutrinos (massless, chargeless particles associated with beta decay of a nucleus), we set $m = 0$ in Equation 39.25, and we see that

$$E = pc \tag{39.26}$$

This equation is an exact expression relating energy and momentum for photons and neutrinos, which always travel at the speed of light.

Finally, note that since the mass m of a particle is independent of its motion, m must have the same value in all reference frames. For this reason, m is often called the *invariant mass*. On the other hand, the total energy and momentum of a particle depend on the reference frame in which they are measured, since they both depend on velocity. Since m is a constant, then according to Equation 39.26 the quantity $E^2 - p^2c^2$ must have the same value in all reference frames. That is, $E^2 - p^2c^2$ is invariant under a Lorentz transformation. These equations do not yet make provision for potential energy.

When dealing with subatomic particles, it is convenient to express their energy in electron volts (eV), because the particles are usually given this energy by acceleration through a potential difference. The conversion factor is

$$1 \text{ eV} = 1.60 \times 10^{-19} \text{ J}$$

For example, the mass of an electron is 9.11×10^{-31} kg. Hence, the rest energy of the electron is

$$mc^2 = (9.11 \times 10^{-31} \text{ kg})(3.00 \times 10^8 \text{ m/s})^2 = 8.20 \times 10^{-14} \text{ J}$$

Converting this to eV, we have

$$mc^2 = (8.20 \times 10^{-14} \text{ J})(1 \text{ eV}/1.60 \times 10^{-19} \text{ J}) = 0.511 \text{ MeV}$$

EXAMPLE 39.11 The Energy of a Speedy Electron

An electron moves with a speed $u = 0.850c$. Find its total energy and kinetic energy in electron volts.

Solution Using the fact that the rest energy of the electron is 0.511 MeV together with Equation 39.24 gives

$$E = \frac{mc^2}{\sqrt{1 - \dfrac{u^2}{c^2}}} = \frac{0.511 \text{ MeV}}{\sqrt{1 - \dfrac{(0.850\,c)^2}{c^2}}}$$

$$= 1.90(0.511 \text{ MeV}) = 0.970 \text{ MeV}$$

The kinetic energy is obtained by subtracting the rest energy from the total energy:

$$K = E - mc^2 = 0.970 \text{ MeV} - 0.511 \text{ MeV} = 0.459 \text{ MeV}$$

EXAMPLE 39.12 The Energy of a Speedy Proton

The total energy of a proton is three times its rest energy.
(a) Find the proton's rest energy in electron volts.

Solution

Rest energy $= mc^2 = (1.67 \times 10^{-27} \text{ kg})(3.00 \times 10^8 \text{ m/s})^2$

$= (1.50 \times 10^{-10} \text{ J})(1.00 \text{ eV}/1.60 \times 10^{-19} \text{ J})$

$= 938 \text{ MeV}$

(b) With what speed is the proton moving?

Solution Since the total energy E is three times the rest energy, $E = \gamma mc^2$ (Eq. 39.24) gives

$$E = 3mc^2 = \frac{mc^2}{\sqrt{1 - \dfrac{u^2}{c^2}}}$$

$$3 = \frac{1}{\sqrt{1 - \dfrac{u^2}{c^2}}}$$

Solving for u gives

$$\left(1 - \frac{u^2}{c^2}\right) = \frac{1}{9} \quad \text{or} \quad \frac{u^2}{c^2} = \frac{8}{9}$$

$$u = \frac{\sqrt{8}}{3}\,c = 2.83 \times 10^8 \text{ m/s}$$

(c) Determine the kinetic energy of the proton in electron volts.

Solution

$$K = E - mc^2 = 3mc^2 - mc^2 = 2mc^2$$

Since $mc^2 = 938$ MeV

$$K = 1876 \text{ MeV}$$

(d) What is the proton's momentum?

Solution We can use Equation 39.25 to calculate the momentum with $E = 3mc^2$:

$$E^2 = p^2c^2 + (mc^2)^2 = (3mc^2)^2$$

$$p^2c^2 = 9(mc^2)^2 - (mc^2)^2 = 8(mc^2)^2$$

$$p = \sqrt{8}\,\frac{mc^2}{c} = \sqrt{8}\,\frac{(938 \text{ MeV})}{c} = 2650\,\frac{\text{MeV}}{c}$$

The unit of momentum is written MeV/c for convenience.

39.8 EQUIVALENCE OF MASS AND ENERGY

To understand the equivalence of mass and energy, consider the following "thought experiment" proposed by Einstein in developing his famous equation $E = mc^2$. Imagine a box of mass M and length L initially at rest as in Figure 39.20a. Suppose that a pulse of light is emitted from the left side of the box as in Figure 39.20b. From Equation 39.26, we know that the light of energy E carries momentum $p = E/c$. Hence, the box must recoil to the left with a speed v to conserve momentum. Assuming the box is very massive, the recoil speed is small compared with the speed of light, and conservation of momentum gives $Mv = E/c$, or

$$v = \frac{E}{Mc}$$

The time it takes the light to move the length of the box is approximately $\Delta t = L/c$ (where, again, we assume that $v \ll c$). In this time interval, the box moves a small distance Δx to the left, where

$$\Delta x = v\,\Delta t = \frac{EL}{Mc^2}$$

The light then strikes the right end of the box, transfers its momentum to the box, causing the box to stop. With the box in its new position, it appears as if its center of mass has moved to the left. However, its center of mass cannot move because the box is an isolated system. Einstein resolved this perplexing situation by assuming that in addition to energy and momentum, light also carries mass. If m is the equivalent mass carried by the pulse of light, and the center of mass of the box is to remain fixed, then

$$mL = M\,\Delta x$$

Solving for m, and using the previous expression for Δx, we get

$$m = \frac{M\,\Delta x}{L} = \frac{M}{L}\,\frac{EL}{Mc^2} = \frac{E}{c^2}$$

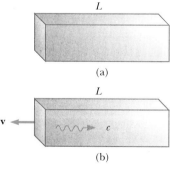

FIGURE 39.20 (a) A box of length L at rest. (b) When a light pulse is emitted at the left end of the box, the box recoils to the left as the pulse strikes the right end.

or
$$E = mc^2$$

Thus, Einstein reached the profound conclusion "If a body gives off the energy E in the form of radiation, its mass diminishes by E/c^2, ... The mass of a body is a measure of its energy content."

It follows that mass varies with speed (relative to the observer). We must therefore distinguish between the **rest mass,** m_o, which is the mass measured by an observer at rest relative to the particle (and at the same location), and the mass measured in real experiments. For a free particle, $m = \gamma m_o$ provides an adequate description. For a larger object, whose center of mass is at rest relative to the observer, $m = \Sigma_i \gamma m_{oi}$ (if we neglect the energy of particle interactions). In every case, the real mass is provided by the total energy, E, divided by the square of the speed of light.

Although we derived the relationship $E = mc^2$ for light energy, the equivalence of mass and energy is universal. Equation 39.24, $E = \gamma m_o c^2$, which represents the total energy of any particle, suggests that even when a particle is at rest ($\gamma = 1$) it still possesses enormous energy through its mass. Probably the clearest experimental proof of the equivalence of mass and energy occurs in nuclear and elementary particle interactions in which there are large amounts of energy released, accompanied by the release of mass. Because energy and mass are related, we see that the laws of conservation of energy and conservation of mass are one and the same. The left side of $E = mc^2$ is constant for an isolated system; therefore the right side must be constant, and thus m (but not m_o) is also constant.

Simply put, this law states that

the energy, and therefore the mass, of a system of particles before interaction must equal the energy, and therefore the mass, of the system after interaction, where the mass is defined as $m = E/c^2$. For free particles, $m = \gamma m_o$, so

$$E_i = \frac{m_{oi} c^2}{\sqrt{1 - \dfrac{u_i^2}{c^2}}}$$

Conservation of mass-energy

The release of enormous energy, accompanied by the change in masses of particles after they have lost their excess energy as they are brought to rest, is the basis of atomic and hydrogen bombs. In fact, whenever energy is released, as in chemical reactions, the residual mass (and energy) are decreased. In a conventional nuclear reactor, the uranium nucleus undergoes fission, a reaction that results in several lighter fragments having considerable kinetic energy. In the case of ^{235}U (the parent nucleus), which undergoes spontaneous fission, the fragments are two lighter nuclei and two neutrons. The total rest mass of the fragments is less than that of the parent nucleus by an amount Δm. The corresponding energy Δmc^2 associated with this mass difference is exactly equal to the total kinetic energy of the fragments. This kinetic energy is then used to produce heat and steam for the generation of electrical power.

Next, consider the basic fusion reaction in which two deuterium atoms combine to form one helium atom. This reaction is of major importance in current research and development of controlled-fusion reactors. The decrease in rest mass that results from the creation of one helium atom from two deuterium atoms is $\Delta m = 4.25 \times 10^{-29}$ kg. Hence, the corresponding excess energy that results from one fusion reaction is $\Delta mc^2 = 3.83 \times 10^{-12}$ J = 23.9 MeV. To appreciate the magnitude of this result, if 1 g of deuterium is converted to helium, the energy released is about 10^{12} J! At the 1995 cost of electrical energy, this would be worth about $60 000.

CONCEPTUAL EXAMPLE 39.13

Because mass is a measure of energy, can we conclude that a compressed spring has more mass than the same spring when it is not compressed?

Reasoning   Recall that when a spring of force constant k is compressed (or stretched) from its equilibrium position by a distance x, it stores elastic potential energy $U = kx^2/2$. According to the theory of special relativity, any change in the total energy of a system is equivalent to a change in mass of the system. Therefore, the mass of a compressed (or stretched) spring is greater than the mass of the spring in its equilibrium position by an amount U/c^2.

EXAMPLE 39.14   Binding Energy of the Deuteron

The mass of the deuteron, which is the nucleus of "heavy hydrogen," is not equal to the sum of the masses of its constituents, which are the proton and neutron. Calculate this mass difference and determine its energy equivalence.

Solution   Using atomic mass units (u), we have

$$m_p = \text{mass of proton} = 1.007276 \text{ u}$$

$$m_n = \text{mass of neutron} = 1.008665 \text{ u}$$

$$m_p + m_n = 2.015941 \text{ u}$$

Since the mass of the deuteron is 2.013553 u (Appendix A), we see that the mass difference Δm is 0.002388 u. By definition, 1 u = 1.66×10^{-27} kg, and therefore

$$\Delta m = 0.002388 \text{ u} = \boxed{3.96 \times 10^{-30} \text{ kg}}$$

Using $E = \Delta mc^2$, we find that

$$E = \Delta mc^2 = (3.96 \times 10^{-30} \text{ kg})(3.00 \times 10^8 \text{ m/s})^2$$

$$= 3.56 \times 10^{-13} \text{ J} = \boxed{2.23 \text{ MeV}}$$

Therefore, the minimum energy required to separate the proton from the neutron of the deuterium nucleus (the binding energy) is 2.23 MeV.

39.9   RELATIVITY AND ELECTROMAGNETISM

Relativity requires that the laws of physics must be the same in all inertial frames. The intimate connection between relativity and electromagnetism was recognized in his first paper on relativity entitled "On the Electrodynamics of Moving Bodies." As we have seen, a light wave in one frame must be a light wave in any other frame, and its speed must be c relative to any observer. This is consistent with Maxwell's theory of electromagnetism. That is, Maxwell's equations require no modification in relativity.

To understand the relation between relativity and electromagnetism, we now describe a situation that shows how an electric field in one frame of reference is viewed as a magnetic field in another frame of reference. Consider a wire carrying a current, and suppose that a positive test charge q is moving parallel to the wire with velocity **u** as in Figure 39.21a. We assume that the net charge on the wire is

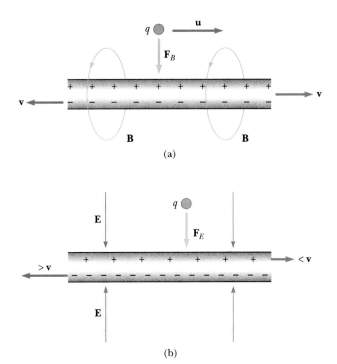

FIGURE 39.21 (a) A positive test charge moving to the right with a velocity **u** near a wire carrying a current. In the frame of the wire, positive and negative charges of equal densities move in opposite directions, the net charge is zero, and **E** = 0. A magnetic field **B** surrounds the wire, and the charge experiences a magnetic force toward the wire. (b) In the rest frame of the test charge, the negative charges in the wire are contracted more than the positive charges, so the wire has a net negative charge, creating an electric field **E** directed toward the wire. Hence, the test charge experiences an electric force toward the wire.

zero. The current in the wire produces a magnetic field that forms circles around the wire and is directed out of the page at the moving test charge. This results in a magnetic force $\mathbf{F}_B = q\mathbf{u} \times \mathbf{B}$ on the test charge acting toward the wire, but no electric force because the net charge on the wire is zero when viewed in this frame.

Now consider the same situation as viewed from the frame of the test charge as in Figure 39.21b. In this frame, the positive charges in the wire move slower relative to the test charge than the negative charges in the wire. Because of length contraction, distances between positive charges in the wire are smaller than distances between negative charges. Hence, there is a net negative charge on the wire when viewed in this frame. The net negative charge produces an electric field pointing toward the wire, and our positive test charge experiences an electric force toward the wire. Thus, what was viewed as a magnetic field in the frame of the wire transforms into an electric field in the frame of the test charge.

*39.10 GENERAL RELATIVITY

Up to this point, we have sidestepped a curious puzzle. Mass has two seemingly different properties: a *gravitational attraction* for other masses and an *inertial* property that resists acceleration. To designate these two attributes, we use the subscripts g and i and write

$$\text{Gravitational property} \quad F_g = m_g g$$

$$\text{Inertial property} \quad F = m_i a$$

The value for the gravitational constant G was chosen to make the magnitudes of m_g and m_i numerically equal. Regardless of how G is chosen, however, the strict equality of m_g and m_i has been measured to an extremely high degree: a few parts in 10^{12}. Thus, it appears that gravitational mass and inertial mass may indeed be exactly equal.

But why? They seem to involve two entirely different concepts: a force of mutual gravitational attraction between two masses and the resistance of a single mass to being accelerated. This question, which puzzled Newton and many other physicists over the years, was answered when Einstein published his theory of gravitation, known as *general relativity,* in 1916. Because it is a mathematically complex theory, we merely offer a hint of its elegance and insight.

In Einstein's view, the remarkable coincidence that m_g and m_i seemed to be exactly proportional was evidence for a very intimate and basic connection between the two concepts. He pointed out that no mechanical experiment (such as dropping a mass) could distinguish between the two situations illustrated in Figures 39.22a and 39.22b. In each case, a mass released by the observer undergoes a downward acceleration of g relative to the floor.

Einstein carried this idea further and proposed that *no* experiment, mechanical or otherwise, could distinguish between the two cases. This extension to include all phenomena (not just mechanical ones) has interesting consequences. For example, suppose that a light pulse is sent horizontally across the box, as in Figure 39.22c. The trajectory of the light pulse bends downward as the box accelerates upward to meet it. Therefore, Einstein proposed that a beam of light should

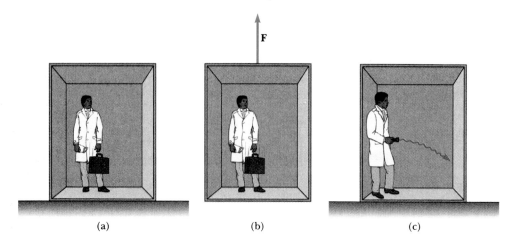

FIGURE 39.22 (a) The observer is at rest in a uniform gravitational field **g**. (b) The observer is in a region where gravity is negligible, but the frame of reference is accelerated by an external force **F** that produces an acceleration **g**. According to Einstein, the frames of reference in parts (a) and (b) are equivalent in every way. No local experiment could distinguish any difference between the two frames. (c) If parts (a) and (b) are truly equivalent, as Einstein proposed, then a ray of light would bend in a gravitational field.

also be bent downward by a gravitational field. (No such bending is predicted in Newton's theory of gravitation.)

The two postulates of Einstein's **general relativity** are as follows:

- All the laws of nature have the same form for observers in any frame of reference, whether accelerated or not.
- In the vicinity of any given point, a gravitational field is equivalent to an accelerated frame of reference in the absence of gravitational effects. (This is the *principle of equivalence*.)

The second postulate implies that gravitational mass and inertial mass are completely equivalent, not just proportional. What were thought to be two different types of mass are actually identical.

One interesting effect predicted by general relativity is that time scales are altered by gravity. A clock in the presence of gravity runs more slowly than one where gravity is negligible. Consequently, the frequencies of radiation emitted by atoms in the presence of a strong gravitational field are *red-shifted* to lower frequencies when compared with the same emissions in a weak field. This gravitational red shift has been detected in spectral lines emitted by atoms in massive stars. It has also been verified on the Earth by comparing the frequencies of gamma rays (a high energy form of electromagnetic radiation) emitted from nuclei separated vertically by about 20 m.

The second postulate suggests that a gravitational field may be "transformed away" at any point if we choose an appropriate accelerated frame of reference—a freely falling one. Einstein developed an ingenious method of describing the acceleration necessary to make the gravitational field "disappear." He specified a certain quantity, the *curvature of space-time,* that describes the gravitational effect at every point. In fact, the curvature of space-time completely replaces Newton's gravitational theory. According to Einstein, there is no such thing as a gravitational force. Rather, the presence of a mass causes a curvature of space-time in the vicinity of the mass, and this curvature dictates the space-time path that all freely moving objects must follow. As one physicist says: "Mass tells space-time how to curve; curved space-time tells mass how to move." One important test of general relativity is the prediction that a light ray passing near the Sun should be deflected by some angle. This prediction was confirmed by astronomers as bending of starlight during a total solar eclipse shortly following World War I (Fig. 39.23).

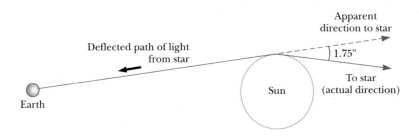

FIGURE 39.23 Deflection of starlight passing near the Sun. Because of this effect, the Sun and other remote objects can act as a *gravitational lens*. In his general theory of relativity, Einstein calculated that starlight just grazing the Sun's surface should be deflected by an angle of 1.75″.

If the concentration of mass becomes very great, as is believed to occur when a large star exhausts its nuclear fuel and collapses to a very small volume, a **black hole** may form. Here the curvature of space-time is so extreme that, within a certain distance from the center of the black hole, all matter and light become trapped.

SUMMARY

The two basic postulates of the special theory of relativity are

- All the laws of physics are the same in all inertial reference frames.
- The speed of light in vacuum has the same value, $c = 3.00 \times 10^8$ m/s, in all inertial frames, regardless of the velocity of the observer or the velocity of the source emitting the light.

Three consequences of the special theory of relativity are as follows:

- Events that are simultaneous for one observer are not simultaneous for another observer who is in motion relative to the first.
- Clocks in motion relative to an observer appear to be slowed down by a factor γ. This is known as **time dilation.**
- Lengths of objects in motion appear to be contracted in the direction of motion.

To satisfy the postulates of special relativity, the Galilean transformations must be replaced by the **Lorentz transformations:**

$$x' = \gamma(x - vt)$$
$$y' = y$$
$$z' = z \qquad (39.9)$$
$$t' = \gamma\left(t - \frac{v}{c^2}x\right)$$

where $\gamma = (1 - v^2/c^2)^{-1/2}$.

The relativistic form of the **velocity transformation** is

$$u'_x = \frac{u_x - v}{1 - \dfrac{u_x v}{c^2}} \qquad (39.14)$$

where u_x is the speed of an object as measured in the S frame and u'_x is its speed measured in the S' frame.

The relativistic expression for the **momentum** of a particle moving with a velocity **u** is

$$\mathbf{p} \equiv \frac{m\mathbf{u}}{\sqrt{1 - \dfrac{u^2}{c^2}}} = \gamma m\mathbf{u} \qquad (39.18)$$

The relativistic expression for the **kinetic energy** of a particle is

$$K = \gamma mc^2 - mc^2 \qquad (39.22)$$

where mc^2 is called the **rest energy** of the particle.

The total energy E of a particle is related to the **mass** through the famous **energy-mass** equivalence expression:

$$E = \gamma mc^2 = \frac{mc^2}{\sqrt{1 - \frac{u^2}{c^2}}} \qquad (39.24)$$

The relativistic momentum is related to the total energy through the equation

$$E^2 = p^2 c^2 + (mc^2)^2 \qquad (39.25)$$

QUESTIONS

1. What two speed measurements do two observers in relative motion always agree on?
2. A spaceship in the shape of a sphere moves past an observer on Earth with a speed $0.5c$. What shape does the observer see as the spaceship moves past?
3. An astronaut moves away from the Earth at a speed close to the speed of light. If an observer on Earth measures the astronaut's size and pulse rate, what changes (if any) would the observer measure? Would the astronaut measure any changes?
4. Two identical clocks are synchronized. One is put in orbit directed eastward around the Earth while the other remains on Earth. Which clock runs slower? When the moving clock returns to Earth, are the two still synchronized?
5. Two lasers situated on a moving spacecraft are triggered simultaneously. An observer on the spacecraft claims to see the pulses of light simultaneously. What condition is necessary so that a secondary observer agrees?
6. When we say that a moving clock runs slower than a stationary one, does this imply that there is something physically unusual about the moving clock?
7. List some ways our day-to-day lives would change if the speed of light were only 50 m/s.
8. Give a physical argument that shows that it is impossible to accelerate an object of mass m to the speed of light, even with a continuous force acting on it.
9. It is said that Einstein, in his teenage years, asked the question, "What would I see in a mirror if I carried it in my hands and ran at the speed of light?" How would you answer this question?
10. What happens to the density of an object as its speed increases? Note that relativistic density is $m/V = E/c^2V$.
11. Some of the distant stars, called quasars, are receding from us at half the speed of light (or greater). What is the speed of the light we receive from these quasars?
12. How is it possible that photons of light, which have zero rest mass, have momentum?
13. With regard to reference frames, how does general relativity differ from special relativity?

PROBLEMS

Section 39.1 The Principle of Newtonian Relativity

1. In a laboratory frame of reference, an observer notes that Newton's second law is valid. Show that it is also valid for an observer moving at a constant speed relative to the laboratory frame.
2. Show that Newton's second law is not valid in a reference frame moving past the laboratory frame of Problem 1 with a constant acceleration.
3. A 2000-kg car moving at 20 m/s collides with and sticks to a 1500-kg car at rest at a stop sign. Show that momentum is conserved in a reference frame moving at 10 m/s in the direction of the moving car.
4. A billiard ball of mass of 0.30 kg moving at 5.0 m/s collides elastically with a ball of mass 0.20 kg moving in the opposite direction at 3.0 m/s. Show that momentum is conserved in a frame of reference moving with a speed of 2.0 m/s in the direction of the second ball.
5. A ball is thrown at 20 m/s inside a boxcar moving along the tracks at 40 m/s. What is the speed of the ball relative to the ground if the ball is thrown (a) forward, (b) backward, (c) out the side door?
5A. A ball is thrown at a speed v_b inside a boxcar moving along the tracks at a speed v. What is the speed of the ball relative to the ground if the ball is thrown (a) forward, (b) backward, (c) out the side door?

□ indicates problems that have full solutions available in the Student Solutions Manual and Study Guide.

Section 39.4 Consequences of Special Relativity

6. At what speed does a clock have to move in order to run at a rate that is one-half the rate of a clock at rest?

7. In 1962, when Scott Carpenter orbited Earth 22 times, the press stated that for each orbit he aged 2.0×10^{-6} s less than he would have had he remained on Earth. (a) Assuming he was 160 km above Earth in a circular orbit, determine the time difference between someone on Earth and Carpenter for the 22 orbits. (*Hint:* Use the approximation $\sqrt{1-x} \approx 1 - x/2$ for small x.) (b) Was the press information accurate? Explain.

8. The proper length of one spaceship is three times that of another. The two ships are traveling in the same direction and, while both are passing overhead, an Earth observer measures them to have the same length. If the slower ship is moving at $0.35c$, determine the speed of the faster one.

8A. The proper length of one spaceship is N times that of another. The two ships are traveling in the same direction and, while both are passing overhead, an Earth observer measures them to have the same length. If the slower ship is moving with speed v, determine the speed of the faster spaceship.

9. A spaceship of proper length 300 m takes 0.75 μs to pass an Earth observer. Determine its speed as measured by the Earth observer.

9A. A spaceship of proper length L_p takes t seconds to pass an Earth observer. Determine its speed as measured by the Earth observer.

10. Muons move in circular orbits at a speed of $0.9994c$ in a storage ring of radius 500 m. If a muon at rest decays into other particles after $T = 2.20$ μs, how many trips around the storage ring do we expect the muons to make before they decay?

11. A spacecraft moves at $0.90c$. If its length is L_0 when measured from inside the spacecraft, what is its length measured by a ground observer?

12. The cosmic rays of highest energy are protons, having kinetic energy of 10^{13} MeV. (a) How long would it take a proton of this energy to travel across the Milky Way galaxy, of diameter 10^5 lightyears, as measured in the proton's frame? (b) From the point of view of the proton, how many kilometers across is the galaxy?

13. The pion has an average lifetime of 26.0 ns when at rest. In order for it to travel 10.0 m, how fast must it move?

14. If astronauts could travel at $v = 0.95c$, we on Earth would say it takes $(4.2/0.95) = 4.4$ years to reach Alpha Centauri, 4.2 lightyears away. The astronauts disagree. (a) How much time passes on the astronauts' clocks? (b) What distance to Alpha Centauri do the astronauts measure?

Section 39.5 The Lorentz Transformation Equations

15. A spaceship travels at $0.75c$ relative to Earth. If the spaceship fires a small rocket in the forward direction, what initial speed (relative to the ship) must the rocket have in order for it to travel at $0.95c$ relative to Earth?

16. A certain quasar recedes from the Earth at $v = 0.87c$. A jet of material ejected from the quasar toward the Earth moves at $0.55c$ relative to the quasar. Find the speed of the ejected material relative to the Earth.

17. Two jets of material from the center of a radio galaxy fly away in opposite directions. Both jets move at $0.75c$ relative to the galaxy. Determine the speed of one jet relative to the other.

18. A Klingon spaceship moves away from the Earth at a speed of $0.80c$ (Fig. P39.18). The Starship Enterprise pursues at a speed of $0.90c$ relative to the Earth. Observers on Earth see the Enterprise overtaking the Klingon ship at a relative speed of $0.10c$. With what speed is the Enterprise overtaking the Klingon ship as seen by the crew of the Enterprise?

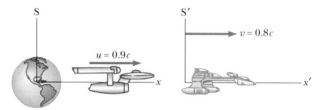

FIGURE P39.18 The Earth is frame S. The Klingon ship is frame S'. The Enterprise is the object whose motion is followed from S and S'.

19. A cube of steel has a volume of 1.0 cm³ and a mass of 8.0 g when at rest on the Earth. If this cube is now given a speed $v = 0.90c$, what is its density as measured by a stationary observer? Note that relativistic density is $m/V = E/c^2V$.

19A. A cube of steel has a volume V and a mass m when at rest on the Earth. If this cube is now given a speed v, what is its density as measured by a stationary observer? Note that relativistic density is $m/V = E/c^2V$.

Section 39.6 Relativistic Momentum and Relativistic Form of Newton's Laws

20. Calculate the momentum of a proton moving at (a) $0.0100c$, (b) $0.500c$, (c) $0.900c$.

21. Find the momentum of a proton in MeV/c units if its total energy is twice its rest energy.

22. Show that the speed of an object having momentum p and mass m is

$$v = \frac{c}{\sqrt{1 + (mc/p)^2}}$$

Section 39.7 Relativistic Energy

23. Show that the energy-momentum relationship $E^2 = p^2c^2 + (mc^2)^2$ follows from the expressions $E = \gamma mc^2$ and $p = \gamma mu$.

24. Cherenkov radiation is given off when an electron travels faster than the speed of light in a medium, the relativistic equivalent of a sonic boom. Consider an electron traveling 10 percent faster than light in water. Determine (a) the electron's total energy and kinetic energy in electron volts and (b) its momentum in MeV/c.

25. A proton moves at $0.95c$. Calculate its (a) rest energy, (b) total energy, and (c) kinetic energy.

26. The total volume of water in the oceans is approximately 1.4×10^9 km^3. The density of sea water is 1030 kg/m^3, and its heat capacity is 4200 J/kg·°C. Estimate the increase in mass of sea water corresponding to an increase in temperature of 10°C.

27. Find the speed of a particle whose total energy is twice its rest energy.

28. A proton in a high-energy accelerator is given a kinetic energy of 50 GeV. Determine (a) its momentum and (b) its speed.

29. Determine the energy required to accelerate an electron from (a) $0.50c$ to $0.90c$ and (b) $0.90c$ to $0.99c$.

30. In a typical color television tube, the electrons are accelerated through a potential difference of 25 000 V. (a) What speed do the electrons have when they strike the screen? (b) What is their kinetic energy in joules?

31. Electrons are accelerated to an energy of 20 GeV in the 3.0-km-long Stanford Linear Accelerator. (a) What is the γ factor for the electrons? (b) What is their speed? (c) How long does the accelerator appear to them?

32. A spaceship of mass 1.0×10^6 kg is to be accelerated to $0.60c$. (a) How much energy does this require? (b) How many kilograms of mass (apart from its fuel) does the spaceship gain from burning its fuel?

33. A pion at rest ($m_\pi = 270 m_e$) decays to a muon ($m_\mu = 206 m_e$) and an antineutrino ($m_\nu = 0$): $\pi^- \rightarrow \mu^- + \overline{\nu}$. Find the kinetic energy of the muon and the antineutrino in electron volts. (*Hint:* Relativistic momentum is conserved.)

Section 39.8 Equivalence of Mass and Energy

34. In a nuclear power plant, the fuel rods last three years. If a 1.0-GW plant operates at 80 percent capacity for the three years, what is the loss of mass (to the steam) of the fuel?

35. Consider the decay $^{55}_{24}$Cr $\rightarrow$ $^{55}_{25}$Mn + e, where e is an electron. The ^{55}Cr nucleus has a mass of 54.9279 u, and the ^{55}Mn nucleus has a mass of 54.9244 u. (a) Calculate the mass difference between the two nuclei in electron volts. (b) What is the maximum kinetic energy of the emitted electron?

36. A ^{57}Fe nucleus at rest emits a 14-keV photon. Use the conservation of energy and momentum to deduce the kinetic energy of the recoiling nucleus in electron volts. (Use $Mc^2 = 8.6 \times 10^{-9}$ J for the final state of the ^{57}Fe nucleus.)

37. The power output of the Sun is 3.8×10^{26} W. How much rest mass is converted to kinetic energy in the Sun each second?

38. A gamma ray (a high-energy photon of light) can produce an electron (e$^-$) and a positron (e$^+$) when it enters the electric field of a heavy nucleus: ($\gamma \rightarrow$ e$^+$ + e$^-$). What minimum γ-ray energy is required to accomplish this task? (*Hint:* The masses of the electron and the positron are equal.)

ADDITIONAL PROBLEMS

39. A spaceship moves away from Earth at $0.50c$ and fires a shuttle craft that then moves in the forward direction at $0.50c$ relative to the ship. The pilot of the shuttle launches a probe forward at speed $0.50c$ relative to the shuttle. Determine (a) the speed of the shuttle relative to Earth and (b) the speed of the probe relative to Earth.

39A. A spaceship moves away from Earth at a speed v and fires a shuttle craft that then moves in the forward direction at a speed v relative to the ship. The pilot of the shuttle launches a probe at speed v relative to the shuttle. Determine (a) the speed of the shuttle relative to Earth and (b) the speed of the probe relative to Earth.

40. An astronaut wishes to visit the Andromeda galaxy (2 million lightyears away) in a one-way trip that will take 30 years in the spaceship's frame of reference. Assuming that his speed is constant, how fast must he travel relative to the Earth?

41. The net nuclear reaction inside the Sun is 4p $\rightarrow$ ^{4}He + ΔE. If the rest mass of each proton is 938.2 MeV and the rest mass of the ^{4}He nucleus is 3727 MeV, calculate the percentage of the starting mass that is released as energy.

42. The annual energy requirement for the United States is on the order of 10^{20} J. How many kilograms of matter would have to be released as energy to meet this requirement?

43. A rocket moves toward a mirror at $0.80c$ relative to reference frame S in Figure P39.43. The mirror is stationary relative to S. A light pulse emitted by the rocket travels to the mirror and is reflected back to the rocket. The front of the rocket is 1.8×10^{12} m from the mirror (as measured by observers in S) at the moment the light pulse leaves the rocket. What is

the total travel time of the pulse as measured by observers in (a) the S frame and (b) the front of the rocket?

43A. A rocket moves toward a mirror at speed v relative to the reference frame labeled by S in Figure P39.43. The mirror is stationary with respect to S. A light pulse emitted by the rocket travels toward the mirror and is reflected back to the rocket. The front of the rocket is a distance d from the mirror (as measured by observers in S) at the moment the light pulse leaves the rocket. What is the total travel time of the pulse as measured by observers in (a) the S frame and (b) the front of the rocket?

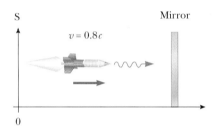

FIGURE P39.43

44. How fast would a motorist have to be going to make a red light appear green? ($\lambda_{red} = 650$ nm, $\lambda_{green} = 550$ nm.) In computing this, use the correct relativistic formula for Doppler shift:

$$\frac{\Delta \lambda}{\lambda} + 1 = \sqrt{\frac{c-v}{c+v}}$$

where v is the approach speed and λ is the source wavelength.

45. A physics professor on Earth gives an exam to students who are on a rocket ship traveling at speed v relative to Earth. The moment the ship passes the professor, she signals the start of the exam. If she wishes her students to have time T_0 (rocket time) to complete the exam, show that she should wait an Earth time

$$T = T_0 \sqrt{\frac{1 - v/c}{1 + v/c}}$$

before sending a light signal telling them to stop. (*Hint:* Remember that it takes some time for the second light signal to travel from the professor to the students.)

46. Ted and Mary are playing catch in frame S', which is moving at $0.60c$ relative to frame S, while Jim in frame S watches the action (Fig. P39.46). Ted throws the ball to Mary at $0.80c$ (according to Ted) and their separation (measured in S') is 1.8×10^{12} m. (a) According to Mary, how fast is the ball moving? (b) According to Mary, how long does it take the ball to reach her? (c) According to Jim, how far apart are Ted and Mary, and how fast is the ball moving? (d) According to Jim, how long does it take the ball to reach Mary?

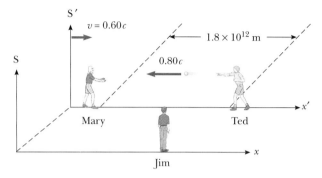

FIGURE P39.46

47. Spaceship I, which contains students taking a physics exam, approaches Earth with a speed of $0.60c$ (relative to Earth), while spaceship II, which contains professors proctoring the exam, moves at $0.28c$ (relative to Earth) directly toward the students. If the professors stop the exam after 50 min have passed on their clock, how long does the exam last as measured by (a) the students and (b) an observer on Earth?

47A. Spaceship I, which contains students taking a physics exam, approaches Earth with a speed v_I (relative to Earth), while spaceship II, which contains professors proctoring the exam, moves at a speed v_{II} (relative to Earth) directly toward the students. If the professors stop the exam after t_{II} minutes have passed on their clock, how long does the exam last as measured by (a) the students and (b) an observer on Earth?

48. A rod of length L_0 moving with a speed v along the horizontal direction makes an angle of θ_0 with respect to the x' axis. (a) Show that the length of the rod as measured by a stationary observer is $L = L_0[1 - (v^2/c^2)\cos^2\theta_0]^{1/2}$. (b) Show that the angle that the rod makes with the x axis is $\tan\theta = \gamma \tan\theta_0$. These results show that the rod is both contracted and rotated. (Take the lower end of the rod to be at the origin of the primed coordinate system.)

49. Imagine a spacecraft that starts from Earth moving at constant speed to the yet-to-be-discovered planet Retah, which is 20 lighthours away from Earth. It takes 25 h (according to an Earth observer) for the spacecraft to reach this planet. Assuming that clocks on Earth and in the spacecraft are synchronized at the beginning of the journey, compare the time elapsed in the spacecraft's frame for the one-way journey with the time elapsed in Earth's frame.

50. If the number of muons at $t = 0$ is N_0, the number at time t is $N = N_0 e^{-t/\tau}$ where τ is the mean lifetime,

equal to 2.2 μs. Suppose muons move at $0.95c$ and there are 5.0×10^4 of them at $t = 0$. (a) What is the observed lifetime of the muons? (b) How many remain after traveling 3.0 km?

51. Consider two inertial reference frames S and S′, where S′ is moving to the right with a constant speed of $0.60c$ relative to S. A stick of proper length 1.0 m moves to the left toward the origins of both S and S′, and the length of the stick is 50 cm as measured by an observer in S′. (a) Determine the speed of the stick as measured by observers in S and S′. (b) What is the length of the stick as measured by an observer in S?

51A. Consider two inertial reference frames S and S′, where S′ is moving to the right with a constant speed v relative to S. A stick of proper length L_p moves to the left toward the origins of both S and S′, and the length of the stick is L' as measured by an observer in S′. (a) Determine the speed of the stick as measured by observers in S and S′. (b) What is the length of the stick as measured by an observer in S?

52. Suppose our Sun is about to explode. In an effort to escape, we depart in a spaceship at $v = 0.80c$ and head toward the star Tau Ceti, 12 lightyears away. When we reach the midpoint of our journey from the Earth, we see our Sun explode and, unfortunately, at the same instant we see Tau Ceti explode as well. (a) In the spaceship's frame of reference, should we conclude that the two explosions occurred simultaneously? If not, which occurred first? (b) In a frame of reference in which the Sun and Tau Ceti are at rest, did they explode simultaneously? If not, which exploded first?

53. Two rockets are on a collision course. They are moving at $0.800c$ and $0.600c$ and are initially 2.52×10^{12} m apart as measured by Liz, the Earth observer in Figure P39.53. Both rockets are 50.0 m in length as measured by Liz. (a) What are their respective proper lengths? (b) What is the length of each rocket as measured by an observer in the other rocket? (c) According to Liz, how long before the rockets collide? (d) According to rocket 1, how long before they collide? (e) According to rocket 2, how long before they collide? (f) If both rocket crews are capable of total evacuation within 90 min (their own time), will there be any casualties?

54. *The red shift.* A light source recedes from an observer with a speed v_s, which is small compared with c. (a) Show that the fractional shift in the measured wavelength is given by the approximate expression

$$\frac{\Delta \lambda}{\lambda} \approx \frac{v_s}{c}$$

This result is known as the red shift, because the visible light is shifted toward the red. (b) Spectroscopic measurements of light at $\lambda = 397$ nm coming from a galaxy in Ursa Major reveal a red shift of 20 nm. What is the recessional speed of the galaxy?

55. A particle having charge q moves at speed v along a straight line in a uniform electric field E. If the motion and the electric field are both in the x direction, (a) show that the acceleration of the particle in the x direction is

$$a = \frac{dv}{dt} = \frac{qE}{m}\left(1 - \frac{v^2}{c^2}\right)^{3/2}$$

(b) Discuss the significance of the fact that the acceleration depends on the speed. (c) If the particle starts from rest at $x = 0$ at $t = 0$, how would you find its speed and position after a time t has elapsed?

56. Consider two inertial reference frames S and S′, where S′ is moving to the right with constant speed $0.60c$ relative to S. Jennifer is located 1.8×10^{11} m to the right of the origin of S and is fixed in S (as measured by an observer in S), while Matt is fixed in S′ at the origin (as measured by an observer in S′). At the instant their origins coincide, Matt throws a ball toward Jennifer at $0.80c$ as measured by Matt (Fig. P39.56). (a) What is the speed of the ball as measured by Jennifer? How long before Jennifer catches the ball, as measured by (b) Jennifer, (c) the ball, and (d) Matt?

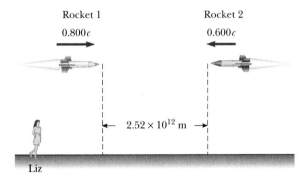

FIGURE P39.53

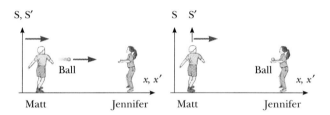

FIGURE P39.56

57. As measured by observers in a reference frame S, a particle having charge q moves with velocity **v** in a magnetic field **B** and an electric field **E**. The resulting force on the particle is $\mathbf{F} = q(\mathbf{E} + \mathbf{v} \times \mathbf{B})$. Another observer moves along with the particle and measures its charge to be q also but its electric field to be **E**'. If both observers are to measure the same force **F**, show that $\mathbf{E}' = \mathbf{E} + \mathbf{v} \times \mathbf{B}$.

SPREADSHEET PROBLEMS

S1. Astronomers use the Doppler shift in the Balmer series of the hydrogen spectrum to determine the radial speed of a galaxy. The fractional change in the wavelength of the spectral line is given by

$$Z = \frac{\Delta\lambda}{\lambda_0} = \frac{\lambda - \lambda_0}{\lambda_0} = \sqrt{\frac{1 + v/c}{1 - v/c}} - 1$$

Once this quantity is measured for a particular receding galaxy, the speed of recession can be found by solving for v/c in terms of Z. Spreadsheet 39.1 calculates the speed of recession for a range of Z values. (a) What is the speed for galaxies having $Z = 0.2$, 0.5, 1.0, and 2.0? (b) The largest Z values, $Z \approx 3.8$, have been measured for several quasars (quasi-stellar radio sources). How fast are these quasars moving away from us?

S2. Astronauts in a starship traveling at a speed v relative to a starbase are given instructions from mission control to call back in 1 h as measured by the starship clocks. Spreadsheet 39.2 calculates how long mission control has to wait for the call for different starship speeds. How long does mission control have to wait if the ship is traveling at $v = 0.1c$, $0.2c$, $0.4c$, $0.6c$, $0.8c$, $0.9c$, $0.95c$, $0.995c$, $0.9995c$?

S3. Most astronomers believe that the Universe began at some instant with an explosion called the Big Bang and that the observed recession of the galaxies is a direct result of this explosion. If the galaxies recede from each other at a constant rate, then we expect that the galaxies moving fastest are now farthest away from Earth. This result, called Hubble's law after Edwin Hubble, can be written $v = Hr$, where H can be determined from observation, v is the speed of recession of the galaxy, and r is its distance from Earth. Current estimates of H range from 15 to 30 km/s/Mly, where Mly is the distance light travels in one million years, and a conservative estimate is $H = 20$ km/s/Mly. Use Spreadsheet 39.1 to calculate the distance from Earth for each galaxy in Problem S1.

S4. Design a spreadsheet program to calculate and plot the relativistic kinetic energy (Eq. 39.22) and the classical kinetic energy ($\frac{1}{2}mu^2$) of a macroscopic object. Plot the relativistic and classical energies versus speed on the same graph. (a) For an object of rest mass $m_0 = 3$ kg, at what speed does the classical kinetic energy underestimate the relativistic value by 1 percent? 5 percent? 50 percent? What is the relativistic kinetic energy at these speeds? Repeat part (a) for (b) an electron and (c) a proton.

APPENDICES

Appendix A
Tables

Appendix B
Mathematics Review

Appendix C
Periodic Table of the Elements

Appendix D
SI Units

Appendix E
Nobel Prizes

Appendix F
Spreadsheet Problems

APPENDIX A

TABLE A.1 Conversion Factors

Length

	m	cm	km	in.	ft	mi
1 meter	1	10^2	10^{-3}	39.37	3.281	6.214×10^{-4}
1 centimeter	10^{-2}	1	10^{-5}	0.3937	3.281×10^{-2}	6.214×10^{-6}
1 kilometer	10^3	10^5	1	3.937×10^4	3.281×10^3	0.6214
1 inch	2.540×10^{-2}	2.540	2.540×10^{-5}	1	8.333×10^{-2}	1.578×10^{-5}
1 foot	0.3048	30.48	3.048×10^{-4}	12	1	1.894×10^{-4}
1 mile	1609	1.609×10^5	1.609	6.336×10^4	5280	1

Mass

	kg	g	slug	u
1 kilogram	1	10^3	6.852×10^{-2}	6.024×10^{26}
1 gram	10^{-3}	1	6.852×10^{-5}	6.024×10^{23}
1 slug	14.59	1.459×10^4	1	8.789×10^{27}
1 atomic mass unit	1.660×10^{-27}	1.660×10^{-24}	1.137×10^{-28}	1

Time

	s	min	h	day	year
1 second	1	1.667×10^{-2}	2.778×10^{-4}	1.157×10^{-5}	3.169×10^{-8}
1 minute	60	1	1.667×10^{-2}	6.994×10^{-4}	1.901×10^{-6}
1 hour	3600	60	1	4.167×10^{-2}	1.141×10^{-4}
1 day	8.640×10^4	1440	24	1	2.738×10^{-3}
1 year	3.156×10^7	5.259×10^5	8.766×10^3	365.2	1

Speed

	m/s	cm/s	ft/s	mi/h
1 meter/second	1	10^2	3.281	2.237
1 centimeter/second	10^{-2}	1	3.281×10^{-2}	2.237×10^{-2}
1 foot/second	0.3048	30.48	1	0.6818
1 mile/hour	0.4470	44.70	1.467	1

Note: 1 mi/min = 60 mi/h = 88 ft/s.

Force

	N	dyn	lb
1 newton	1	10^5	0.2248
1 dyne	10^{-5}	1	2.248×10^{-6}
1 pound	4.448	4.448×10^5	1

TABLE A.1 Continued

Work, Energy, Heat

	J	erg	ft·lb
1 joule	1	10^7	0.7376
1 erg	10^{-7}	1	7.376×10^{-8}
1 ft·lb	1.356	1.356×10^7	1
1 eV	1.602×10^{-19}	1.602×10^{-12}	1.182×10^{-19}
1 cal	4.186	4.186×10^7	3.087
1 Btu	1.055×10^3	1.055×10^{10}	7.779×10^2
1 kWh	3.600×10^6	3.600×10^{13}	2.655×10^6

	eV	cal	Btu	kWh
1 joule	6.242×10^{18}	0.2389	9.481×10^{-4}	2.778×10^{-7}
1 erg	6.242×10^{11}	2.389×10^{-8}	9.481×10^{-11}	2.778×10^{-14}
1 ft·lb	8.464×10^{18}	0.3239	1.285×10^{-3}	3.766×10^{-7}
1 eV	1	3.827×10^{-20}	1.519×10^{-22}	4.450×10^{-26}
1 cal	2.613×10^{19}	1	3.968×10^{-3}	1.163×10^{-6}
1 Btu	6.585×10^{21}	2.520×10^2	1	2.930×10^{-4}
1 kWh	2.247×10^{25}	8.601×10^5	3.413×10^2	1

Pressure

	Pa	dyn/cm²	atm
1 pascal	1	10	9.869×10^{-6}
1 dyne/centimeter²	10^{-1}	1	9.869×10^{-7}
1 atmosphere	1.013×10^5	1.013×10^6	1
1 centimeter mercury*	1.333×10^3	1.333×10^4	1.316×10^{-2}
1 pound/inch²	6.895×10^3	6.895×10^4	6.805×10^{-2}
1 pound/foot²	47.88	4.788×10^2	4.725×10^{-4}

	cm Hg	lb/in.²	lb/ft²
1 newton/meter²	7.501×10^{-4}	1.450×10^{-4}	2.089×10^{-2}
1 dyne/centimeter²	7.501×10^{-5}	1.450×10^{-5}	2.089×10^{-3}
1 atmosphere	76	14.70	2.116×10^3
1 centimeter mercury*	1	0.1943	27.85
1 pound/inch²	5.171	1	144
1 pound/foot²	3.591×10^{-2}	6.944×10^{-3}	1

*At 0°C and at a location where the acceleration due to gravity has its "standard" value, 9.80665 m/s².

TABLE A.2 Symbols, Dimensions, and Units of Physical Quantities

Quantity	Common Symbol	Unit*	Dimensions†	Unit in Terms of Base SI Units
Acceleration	a	m/s²	L/T^2	m/s²
Amount of substance	n	mole		mol
Angle	θ, ϕ	radian (rad)	1	
Angular acceleration	α	rad/s²	T^{-2}	s⁻²
Angular frequency	ω	rad/s	T^{-1}	s⁻¹
Angular momentum	L	kg·m²/s	ML^2/T	kg·m²/s
Angular velocity	ω	rad/s	T^{-1}	s⁻¹
Area	A	m²	L^2	m²
Atomic number	Z			
Capacitance	C	farad (F) (= Q/V)	Q^2T^2/ML^2	$A^2 \cdot s^4/kg \cdot m^2$
Charge	q, Q, e	coulomb (C)	Q	A·s
Charge density				
Line	λ	C/m	Q/L	A·s/m
Surface	σ	C/m²	Q/L^2	A·s/m²
Volume	ρ	C/m³	Q/L^3	A·s/m³
Conductivity	σ	1/Ω·m	Q^2T/ML^3	$A^2 \cdot s^3/kg \cdot m^3$
Current	I	AMPERE	Q/T	A
Current density	J	A/m²	Q/T^2	A/m²
Density	ρ	kg/m³	M/L^3	kg/m³
Dielectric constant	κ			
Displacement	s	METER	L	m
Distance	d, h			
Length	ℓ, L			
Electric dipole moment	p	C·m	QL	A·s·m
Electric field	E	V/m	ML/QT^2	kg·m/A·s³
Electric flux	Φ	V·m	ML^3/QT^2	kg·m³/A·s³
Electromotive force	$\mathcal{E}$	volt (V)	ML^2/QT^2	kg·m²/A·s³
Energy	E, U, K	joule (J)	ML^2/T^2	kg·m²/s²
Entropy	S	J/K	$ML^2/T^2 \cdot K$	kg·m²/s²·K
Force	F	newton (N)	ML/T^2	kg·m/s²
Frequency	f, ν	hertz (Hz)	T^{-1}	s⁻¹
Heat	Q	joule (J)	ML^2/T^2	kg·m²/s²
Inductance	L	henry (H)	ML^2/Q^2	kg·m²/A²·s²
Magnetic dipole moment	μ	N·m/T	QL^2/T	A·m²
Magnetic field	B	tesla (T) (= Wb/m²)	M/QT	kg/A·s²
Magnetic flux	Φ_m	weber (Wb)	ML^2/QT	kg·m²/A·s²
Mass	m, M	KILOGRAM	M	kg
Molar specific heat	C	J/mol·K		kg·m²/s²·mol·K
Moment of inertia	I	kg·m²	ML^2	kg·m²
Momentum	p	kg·m/s	ML/T	kg·m/s
Period	T	s	T	s
Permeability of space	μ_0	N/A² (= H/m)	ML/Q^2T	kg·m/A²·s²
Permittivity of space	ϵ_0	C²/N·m² (= F/m)	Q^2T^2/ML^3	$A^2 \cdot s^4/kg \cdot m^3$
Potential (voltage)	V	volt (V) (= J/C)	ML^2/QT^2	kg·m²/A·s³
Power	P	watt (W) (= J/s)	ML^2/T^3	kg·m²/s³

continued

TABLE A.2 Continued

Quantity	Common Symbol	Unit*	Dimensions†	Unit in Terms of Base SI Units
Pressure	P, p	pascal (Pa) = (N/m^2)	M/LT^2	kg/m·s^2
Resistance	R	ohm (Ω) (= V/A)	ML^2/Q^{2}T	kg·m^2/A^2·s^3
Specific heat	c	J/kg·K	L^2/T^2·K	m^2/s^2·K
Temperature	T	KELVIN	K	K
Time	t	SECOND	T	s
Torque	τ	N·m	ML^2/T^2	kg·m^2/s^2
Speed	v	m/s	L/T	m/s
Volume	V	m^3	L^3	m^3
Wavelength	λ	m	L	m
Work	W	joule (J) (= N·m)	ML^2/T^2	kg·m^2/s^2

* The base SI units are given in upper case letters.

† The symbols M, L, T, and Q denote mass, length, time, and charge, respectively.

TABLE A.3 Table of Atomic Masses[a]

Z	Element	Symbol	Chemical Atomic Mass (u)	Mass Number (* Indicates Radioactive) A	Atomic Mass (u)	Percent Abundance	Half-Life (if Radioactive) $T_{1/2}$
0	(Neutron)	n		1*	1.008 665		10.4 m
1	Hydrogen	H	1.0079	1	1.007 825	99.985	
	Deuterium	D		2	2.014 102	0.015	
	Tritium	T		3*	3.016 049		12.33 y
2	Helium	He	4.00260	3	3.016 029	0.00014	
				4	4.002 602	99.99986	
				6*	6.018 886		0.81 s
3	Lithium	Li	6.941	6	6.015 121	7.5	
				7	7.016 003	92.5	
				8*	8.022 486		0.84 s
4	Beryllium	Be	9.0122	7*	7.016 928		53.3 d
				9	9.012 174	100	
				10*	10.013 534		1.5 × 10^6 y
5	Boron	B	10.81	10	10.012 936	19.9	
				11	11.009 305	80.1	
				12*	12.014 352		0.0202 s
6	Carbon	C	12.011	10*	10.016 854		19.3 s
				11*	11.011 433		20.4 m
				12	12.000 000	98.90	
				13	13.003 355	1.10	
				14*	14.003 242		5730 y
				15*	15.010 599		2.45 s
7	Nitrogen	N	14.0067	12*	12.018 613		0.0110 s
				13*	13.005 738		9.96 m
				14	14.003 074	99.63	
				15	15.000 108	0.37	
				16*	16.006 100		7.13 s
				17*	17.008 450		4.17 s

TABLE A.3 *Continued*

Z	Element	Symbol	Chemical Atomic Mass (u)	Mass Number (* Indicates Radioactive) A	Atomic Mass (u)	Percent Abundance	Half-Life (if Radioactive) $T_{1/2}$
8	Oxygen	O	15.9994	14*	14.008 595		70.6 s
				15*	15.003 065		122
				16	15.994 915	99.761	
				17	16.999 132	0.039	
				18	17.999 160	0.20	
				19*	19.003 577		26.9 s
9	Fluorine	F	18.99840	17*	17.002 094		64.5 s
				18*	18.000 937		109.8 m
				19	18.998 404	100	
				20*	19.999 982		11.0 s
				21*	20.999 950		4.2 s
10	Neon	Ne	20.180	18*	18.005 710		1.67 s
				19*	19.001 880		17.2 s
				20	19.992 435	90.48	
(10)	(Neon)			21	20.993 841	0.27	
				22	21.991 383	9.25	
				23*	22.994 465		37.2 s
11	Sodium	Na	22.98987	21*	20.997 650		22.5 s
				22*	21.994 434		2.61 y
				24*	23.990 961		14.96 h
12	Magnesium	Mg	24.305	23*	22.994 124		11.3 s
				24	23.985 042	78.99	
				25	24.985 838	10.00	
				26	25.982 594	11.01	
				27*	26.984 341		9.46 m
13	Aluminum	Al	26.98154	26*	25.986 892		7.4×10^5 y
				27	26.981 538	100	
				28*	27.981 910		2.24 m
14	Silicon	Si	28.086	28	27.976 927	92.23	
				29	28.976 495	4.67	
				30	29.973 770	3.10	
				31*	30.975 362		2.62 h
				32*	31.974 148		172 y
15	Phosphorus	P	30.97376	30*	29.978 307		2.50 m
				31	30.973 762	100	
				32*	31.973 908		14.26 d
				33*	32.971 725		25.3 d
16	Sulfur	S	32.066	32	31.972 071	95.02	
				33	32.971 459	0.75	
				34	33.967 867	4.21	
				35*	34.969 033		87.5 d
				36	35.967 081	0.02	
17	Chlorine	Cl	35.453	35	34.968 853	75.77	
				36*	35.968 307		3.0×10^5 y
				37	36.965 903	24.23	

continued

TABLE A.3 Continued

Z	Element	Symbol	Chemical Atomic Mass (u)	Mass Number (* Indicates Radioactive) A	Atomic Mass (u)	Percent Abundance	Half-Life (if Radioactive) $T_{1/2}$
18	Argon	Ar	39.948	36	35.967 547	0.337	
				37*	36.966 776		35.04 d
				38	37.962 732	0.063	
				39*	38.964 314		269 y
				40	39.962 384	99.600	
				42*	41.963 049		33 y
19	Potassium	K	39.0983	39	38.963 708	93.2581	
				40*	39.964 000	0.0117	1.28×10^9 y
				41	40.961 827	6.7302	
20	Calcium	Ca	40.08	40	39.962 591	96.941	
				41*	40.962 279		1.0×10^5 y
				42	41.958 618	0.647	
				43	42.958 767	0.135	
				44	43.955 481	2.086	
				46	45.953 687	0.004	
				48	47.952 534	0.187	
21	Scandium	Sc	44.9559	41*	40.969 250		0.596 s
				45	44.955 911	100	
22	Titanium	Ti	47.88	44*	43.959 691		49 y
				46	45.952 630	8.0	
				47	46.951 765	7.3	
				48	47.947 947	73.8	
(22)	(Titanium)			49	48.947 871	5.5	
				50	49.944 792	5.4	
23	Vanadium	V	50.9415	48*	47.952 255		15.97 d
				50*	49.947 161	0.25	1.5×10^{17} y
				51	50.943 962	99.75	
24	Chromium	Cr	51.996	48*	47.954 033		21.6 h
				50	49.946 047	4.345	
				52	51.940 511	83.79	
				53	52.940 652	9.50	
				54	53.938 883	2.365	
25	Manganese	Mn	54.93805	54*	53.940 361		312.1 d
				55	54.938 048	100	
26	Iron	Fe	55.847	54	53.939 613	5.9	
				55*	54.938 297		2.7 y
				56	55.934 940	91.72	
				57	56.935 396	2.1	
				58	57.933 278	0.28	
				60*	59.934 078		1.5×10^6 y
27	Cobalt	Co	58.93320	59	58.933 198	100	
				60*	59.933 820		5.27 y
28	Nickel	Ni	58.693	58	57.935 346	68.077	
				59*	58.934 350		7.5×10^4 y
				60	59.930 789	26.223	
				61	60.931 058	1.140	
				62	61.928 346	3.634	
				63*	62.929 670		100 y
				64	63.927 967	0.926	

TABLE A.3 *Continued*

Z	Element	Symbol	Chemical Atomic Mass (u)	Mass Number (* Indicates Radioactive) A	Atomic Mass (u)	Percent Abundance	Half-Life (if Radioactive) $T_{1/2}$
29	Copper	Cu	63.54	63	62.929 599	69.17	
				65	64.927 791	30.83	
30	Zinc	Zn	65.39	64	63.929 144	48.6	
				66	65.926 035	27.9	
				67	66.927 129	4.1	
				68	67.924 845	18.8	
				70	69.925 323	0.6	
31	Gallium	Ga	69.723	69	68.925 580	60.108	
				71	70.924 703	39.892	
32	Germanium	Ge	72.61	70	69.924 250	21.23	
				72	71.922 079	27.66	
				73	72.923 462	7.73	
				74	73.921 177	35.94	
				76	75.921 402	7.44	
33	Arsenic	As	74.9216	75	74.921 594	100	
34	Selenium	Se	78.96	74	73.922 474	0.89	
				76	75.919 212	9.36	
				77	76.919 913	7.63	
				78	77.917 307	23.78	
				79*	78.918 497		$\leq 6.5 \times 10^4$ y
				80	79.916 519	49.61	
				82*	81.916 697	8.73	1.4×10^{20} y
35	Bromine	Br	79.904	79	78.918 336	50.69	
				81	80.916 287	49.31	
36	Krypton	Kr	83.80	78	77.920 400	0.35	
				80	79.916 377	2.25	
				81*	80.916 589		2.1×10^5 y
(36)	(Krypton)			82	81.913 481	11.6	
				83	82.914 136	11.5	
				84	83.911 508	57.0	
				85*	84.912 531		10.76 y
				86	85.910 615	17.3	
37	Rubidium	Rb	85.468	85	84.911 793	72.17	
				87*	86.909 186	27.83	4.75×10^{10} y
38	Strontium	Sr	87.62	84	83.913 428	0.56	
				86	85.909 266	9.86	
				87	86.908 883	7.00	
				88	87.905 618	82.58	
				90*	89.907 737		29.1 y
39	Yttrium	Y	88.9058	89	88.905 847	100	
40	Zirconium	Zr	91.224	90	89.904 702	51.45	
				91	90.905 643	11.22	
				92	91.905 038	17.15	
				93*	92.906 473		1.5×10^6 y
				94	93.906 314	17.38	
				96	95.908 274	2.80	

continued

TABLE A.3 *Continued*

Z	Element	Symbol	Chemical Atomic Mass (u)	Mass Number (* Indicates Radioactive) A	Atomic Mass (u)	Percent Abundance	Half-Life (if Radioactive) $T_{1/2}$
41	Niobium	Nb	92.9064	91*	90.906 988		6.8×10^2 y
				92*	91.907 191		3.5×10^7 y
				93	92.906 376	100	
				94*	93.907 280		2×10^4 y
42	Molybdenum	Mo	95.94	92	91.906 807	14.84	
				93*	92.906 811		3.5×10^3 y
				94	93.905 085	9.25	
				95	94.905 841	15.92	
				96	95.904 678	16.68	
				97	96.906 020	9.55	
				98	97.905 407	24.13	
				100	99.907 476	9.63	
43	Technetium	Tc		97*	96.906 363		2.6×10^6 y
				98*	97.907 215		4.2×10^6 y
				99*	98.906 254		2.1×10^5 y
44	Ruthenium	Ru	101.07	96	95.907 597	5.54	
				98	97.905 287	1.86	
				99	98.905 939	12.7	
				100	99.904 219	12.6	
				101	100.905 558	17.1	
				102	101.904 348	31.6	
				104	103.905 428	18.6	
45	Rhodium	Rh	102.9055	103	102.905 502	100	
46	Palladium	Pd	106.42	102	101.905 616	1.02	
				104	103.904 033	11.14	
				105	104.905 082	22.33	
				106	105.903 481	27.33	
				107*	106.905 126		6.5×10^6 y
				108	107.903 893	26.46	
				110	109.905 158	11.72	
47	Silver	Ag	107.868	107	106.905 091	51.84	
				109	108.904 754	48.16	
48	Cadmium	Cd	112.41	106	105.906 457	1.25	
				108	107.904 183	0.89	
				109*	108.904 984		462 d
(48)	(Cadmium)			110	109.903 004	12.49	
				111	110.904 182	12.80	
				112	111.902 760	24.13	
				113*	112.904 401	12.22	9.3×10^{15} y
				114	113.903 359	28.73	
				116	115.904 755	7.49	
49	Indium	In	114.82	113	112.904 060	4.3	
				115*	114.903 876	95.7	4.4×10^{14} y
50	Tin	Sn	118.71	112	111.904 822	0.97	
				114	113.902 780	0.65	
				115	114.903 345	0.36	
				116	115.901 743	14.53	
				117	116.902 953	7.68	

TABLE A.3 *Continued*

Z	Element	Symbol	Chemical Atomic Mass (u)	Mass Number (* Indicates Radioactive) A	Atomic Mass (u)	Percent Abundance	Half-Life (if Radioactive) $T_{1/2}$
				118	117.901 605	24.22	
				119	118.903 308	8.58	
				120	119.902 197	32.59	
				121*	120.904 237		55 y
				122	121.903 439	4.63	
				124	123.905 274	5.79	
51	Antimony	Sb	121.76	121	120.903 820	57.36	
				123	122.904 215	42.64	
				125*	124.905 251		2.7 y
52	Tellurium	Te	127.60	120	119.904 040	0.095	
				122	121.903 052	2.59	
				123*	122.904 271	0.905	1.3×10^{13} y
				124	123.902 817	4.79	
				125	124.904 429	7.12	
				126	125.903 309	18.93	
				128*	127.904 463	31.70	$> 8 \times 10^{24}$ y
				130*	129.906 228	33.87	$\leq 1.25 \times 10^{21}$ y
53	Iodine	I	126.9045	127	126.904 474	100	
				129*	128.904 984		1.6×10^7 y
54	Xenon	Xe	131.29	124	123.905 894	0.10	
				126	125.904 268	0.09	
				128	127.903 531	1.91	
				129	128.904 779	26.4	
				130	129.903 509	4.1	
				131	130.905 069	21.2	
				132	131.904 141	26.9	
				134	133.905 394	10.4	
				136*	135.907 215	8.9	$\geq 2.36 \times 10^{21}$ y
55	Cesium	Cs	132.9054	133	132.905 436	100	
				134*	133.906 703		2.1 y
				135*	134.905 891		2×10^6 y
				137*	136.907 078		30 y
56	Barium	Ba	137.33	130	129.906 289	0.106	
				132	131.905 048	0.101	
				133*	132.905 990		10.5 y
				134	133.904 492	2.42	
				135	134.905 671	6.593	
				136	135.904 559	7.85	
				137	136.905 816	11.23	
				138	137.905 236	71.70	
57	Lanthanum	La	138.905	137*	136.906 462		6×10^4 y
(57)	(Lanthanum)			138*	137.907 105	0.0902	1.05×10^{11} y
				139	138.906 346	99.9098	
58	Cerium	Ce	140.12	136	135.907 139	0.19	
				138	137.905 986	0.25	
				140	139.905 434	88.43	
				142*	141.909 241	11.13	$> 5 \times 10^{16}$ y
59	Praseodymium	Pr	140.9076	141	140.907 647	100	

continued

TABLE A.3 Continued

Z	Element	Symbol	Chemical Atomic Mass (u)	Mass Number (* Indicates Radioactive) A	Atomic Mass (u)	Percent Abundance	Half-Life (if Radioactive) $T_{1/2}$
60	Neodymium	Nd	144.24	142	141.907 718	27.13	
				143	142.909 809	12.18	
				144*	143.910 082	23.80	2.3×10^{15} y
				145	144.912 568	8.30	
				146	145.913 113	17.19	
				148	147.916 888	5.76	
				150*	149.920 887	5.64	$>1 \times 10^{18}$ y
61	Promethium	Pm		143*	142.910 928		265 d
				145*	144.912 745		17.7 y
				146*	145.914 698		5.5 y
				147*	146.915 134		2.623 y
62	Samarium	Sm	150.36	144	143.911 996	3.1	
				146*	145.913 043		1.0×10^{8} y
				147*	146.914 894	15.0	1.06×10^{11} y
				148*	147.914 819	11.3	7×10^{15} y
				149*	148.917 180	13.8	$>2 \times 10^{15}$ y
				150	149.917 273	7.4	
				151*	150.919 928		90 y
				152	151.919 728	26.7	
				154	153.922 206	22.7	
63	Europium	Eu	151.96	151	150.919 846	47.8	
				152*	151.921 740		13.5 y
				153	152.921 226	52.2	
				154*	153.922 975		8.59 y
				155*	154.922 888		4.7 y
64	Gadolinium	Gd	157.25	148*	147.918 112		75 y
				150*	149.918 657		1.8×10^{6} y
				152*	151.919 787	0.20	1.1×10^{14} y
				154	153.920 862	2.18	
				155	154.922 618	14.80	
				156	155.922 119	20.47	
				157	156.923 957	15.65	
				158	157.924 099	24.84	
				160	159.927 050	21.86	
65	Terbium	Tb	158.9253	159	158.925 345	100	
66	Dysprosium	Dy	162.50	156	155.924 277	0.06	
				158	157.924 403	0.10	
				160	159.925 193	2.34	
				161	160.926 930	18.9	
				162	161.926 796	25.5	
				163	162.928 729	24.9	
				164	163.929 172	28.2	
67	Holmium	Ho	164.9303	165	164.930 316	100	
				166*	165.932 282		1.2×10^{3} y
68	Erbium	Er	167.26	162	161.928 775	0.14	
				164	163.929 198	1.61	
				166	165.930 292	33.6	

TABLE A.3 Continued

Z	Element	Symbol	Chemical Atomic Mass (u)	Mass Number (* Indicates Radioactive) A	Atomic Mass (u)	Percent Abundance	Half-Life (if Radioactive) $T_{1/2}$
(68)	(Erbium)			167	166.932 047	22.95	
				168	167.932 369	27.8	
				170	169.935 462	14.9	
69	Thulium	Tm	168.9342	169	168.934 213	100	
				171*	170.936 428		1.92 y
70	Ytterbium	Yb	173.04	168	167.933 897	0.13	
				170	169.934 761	3.05	
				171	170.936 324	14.3	
				172	171.936 380	21.9	
				173	172.938 209	16.12	
				174	173.938 861	31.8	
				176	175.942 564	12.7	
71	Lutecium	Lu	174.967	173*	172.938 930		1.37 y
				175	174.940 772	97.41	
				176*	175.942 679	2.59	3.78×10^{10} y
72	Hafnium	Hf	178.49	174*	173.940 042	0.162	2.0×10^{15} y
				176	175.941 404	5.206	
				177	176.943 218	18.606	
				178	177.943 697	27.297	
				179	178.945 813	13.629	
				180	179.946 547	35.100	
73	Tantalum	Ta	180.9479	180	179.947 542	0.012	
				181	180.947 993	99.988	
74	Tungsten (Wolfram)	W	183.85	180	179.946 702	0.12	
				182	181.948 202	26.3	
				183	182.950 221	14.28	
				184	183.950 929	30.7	
				186	185.954 358	28.6	
75	Rhenium	Re	186.207	185	184.952 951	37.40	
				187*	186.955 746	62.60	4.4×10^{10} y
76	Osmium	Os	190.2	184	183.952 486	0.02	
				186*	185.953 834	1.58	2.0×10^{15} y
				187	186.955 744	1.6	
				188	187.955 832	13.3	
				189	188.958 139	16.1	
				190	189.958 439	26.4	
				192	191.961 468	41.0	
				194*	193.965 172		6.0 y
77	Iridium	Ir	192.2	191	190.960 585	37.3	
				193	192.962 916	62.7	
78	Platinum	Pt	195.08	190*	189.959 926	0.01	6.5×10^{11} y
				192	191.961 027	0.79	
				194	193.962 655	32.9	
				195	194.964 765	33.8	
				196	195.964 926	25.3	
				198	197.967 867	7.2	
79	Gold	Au	196.9665	197	196.966 543	100	

continued

TABLE A.3 Continued

Z	Element	Symbol	Chemical Atomic Mass (u)	Mass Number (* Indicates Radioactive) A	Atomic Mass (u)	Percent Abundance	Half-Life (if Radioactive) $T_{1/2}$
80	Mercury	Hg	200.59	196	195.965 806	0.15	
				198	197.966 743	9.97	
				199	198.968 253	16.87	
				200	199.968 299	23.10	
				201	200.970 276	13.10	
				202	201.970 617	29.86	
				204	203.973 466	6.87	
81	Thallium	Tl	204.383	203	202.972 320	29.524	
				204*	203.973 839		3.78 y
				205	204.974 400	70.476	
		(Ra E″)		206*	205.976 084		4.2 m
		(Ac C″)		207*	206.977 403		4.77 m
		(Th C″)		208*	207.981 992		3.053 m
		(Ra C″)		210*	209.990 057		1.30 m
82	Lead	Pb	207.2	202*	201.972 134		5×10^4 y
				204*	203.973 020	1.4	$\geq 1.4 \times 10^{17}$ y
				205*	204.974 457		1.5×10^7 y
				206	205.974 440	24.1	
				207	206.975 871	22.1	
				208	207.976 627	52.4	
		(Ra D)		210*	209.984 163		22.3 y
		(Ac B)		211*	210.988 734		36.1 m
		(Th B)		212*	211.991 872		10.64 h
		(Ra B)		214*	213.999 798		26.8 m
83	Bismuth	Bi	208.9803	207*	206.978 444		32.2 y
				208*	207.979 717		3.7×10^5 y
				209	208.980 374	100	
		(Ra E)		210*	209.984 096		5.01 d
		(Th C)		211*	210.987 254		2.14 m
				212*	211.991 259		60.6 m
		(Ra C)		214*	213.998 692		19.9 m
				215*	215.001 836		7.4 m
84	Polonium	Po		209*	208.982 405		102 y
		(Ra F)		210*	209.982 848		138.38 d
		(Ac C′)		211*	210.986 627		0.52 s
		(Th C′)		212*	211.988 842		0.30 μs
		(Ra C′)		214*	213.995 177		164 μs
		(Ac A)		215*	214.999 418		0.0018 s
		(Th A)		216*	216.001 889		0.145 s
		(Ra A)		218*	218.008 965		3.10 m
85	Astatine	At		215*	214.998 638		≈ 100 μs
				218*	218.008 685		1.6 s
				219*	219.011 294		0.9 m
86	Radon	Rn					
		(An)		219*	219.009 477		3.96 s
		(Tn)		220*	220.011 369		55.6 s
		(Rn)		222*	222.017 571		3.823 d
87	Francium	Fr					
		(Ac K)		223*	223.019 733		22 m

TABLE A.3 Continued

Z	Element	Symbol	Chemical Atomic Mass (u)	Mass Number (* Indicates Radioactive) A	Atomic Mass (u)	Percent Abundance	Half-Life (if Radioactive) $T_{1/2}$
88	Radium	Ra					
		(Ac X)		223*	223.018 499		11.43 d
		(Th X)		224*	224.020 187		3.66 d
		(Ra)		226*	226.025 402		1600 y
		(Ms Th$_1$)		228*	228.031 064		5.75 y
89	Actinium	Ac		227*	227.027 749		21.77 y
		(Ms Th$_2$)		228*	228.031 015		6.15 h
90	Thorium	Th	232.0381				
		(Rd Ac)		227*	227.027 701		18.72 d
		(Rd Th)		228*	228.028 716		1.913 y
				229*	229.031 757		7300 y
		(Io)		230*	230.033 127		75.000 y
(90)	(Thorium)	(UY)		231*	231.036 299		25.52 h
		(Th)		232*	232.038 051	100	1.40×10^{10} y
		(UX$_1$)		234*	234.043 593		24.1 d
91	Protactinium	Pa		231*	231.035 880		32.760 y
		(Uz)		234*	234.043 300		6.7 h
92	Uranium	U	238.0289	232*	232.037 131		69 y
				233*	233.039 630		1.59×10^5 y
				234*	234.040 946	0.0055	2.45×10^5 y
		(Ac U)		235*	235.043 924	0.720	7.04×10^8 y
				236*	236.045 562		2.34×10^7 y
		(UI)		238*	238.050 784	99.2745	4.47×10^9 y
93	Neptunium	Np		235*	235.044 057		396 d
				236*	236.046 560		115,000 y
				237*	237.048 168		2.14×10^6 y
94	Plutonium	Pu		236*	236.046 033		2.87 y
				238*	238.049 555		87.7 y
				239*	239.052 157		24,120 y
				240*	240.053 808		6560 y
				241*	241.056 846		14.4 y
				242*	242.058 737		3.73×10^5 y
				244*	244.064 200		8.1×10^7 y

[a] The masses in the sixth column are atomic masses, which include the mass of Z electrons. Data are from the National Nuclear Data Center, Brookhaven National Laboratory, prepared by Jagdish K. Tuli, July 1990. The data are based on experimental results reported in *Nuclear Data Sheets* and *Nuclear Physics* and also from *Chart of the Nuclides,* 14th ed. Atomic masses are based on those by A. H. Wapstra, G. Audi, and R. Hoekstra. Isotopic abundances are based on those by N. E. Holden.

APPENDIX B

Mathematics Review

These appendices in mathematics are intended as a brief review of operations and methods. Early in this course, you should be totally familiar with basic algebraic techniques, analytic geometry, and trigonometry. The appendices on differential and integral calculus are more detailed and are intended for those students who have difficulty applying calculus concepts to physical situations.

B.1 SCIENTIFIC NOTATION

Many quantities that scientists deal with often have very large or very small values. For example, the speed of light is about 300 000 000 m/s and the ink required to make the dot over an i in this textbook has a mass of about 0.000 000 001 kg. Obviously, it is very cumbersome to read, write, and keep track of numbers such as these. We avoid this problem by using a method dealing with powers of the number 10:

$$10^0 = 1$$
$$10^1 = 10$$
$$10^2 = 10 \times 10 = 100$$
$$10^3 = 10 \times 10 \times 10 = 1000$$
$$10^4 = 10 \times 10 \times 10 \times 10 = 10\,000$$
$$10^5 = 10 \times 10 \times 10 \times 10 \times 10 = 100\,000$$

and so on. The number of zeros corresponds to the power to which 10 is raised, called the **exponent** of 10. For example, the speed of light, 300 000 000 m/s, can be expressed as 3×10^8 m/s.

In this method, some representative numbers smaller than unity are

$$10^{-1} = \frac{1}{10} = 0.1$$

$$10^{-2} = \frac{1}{10 \times 10} = 0.01$$

$$10^{-3} = \frac{1}{10 \times 10 \times 10} = 0.001$$

$$10^{-4} = \frac{1}{10 \times 10 \times 10 \times 10} = 0.0001$$

$$10^{-5} = \frac{1}{10 \times 10 \times 10 \times 10 \times 10} = 0.00001$$

In these cases, the number of places the decimal point is to the left of the digit 1 equals the value of the (negative) exponent. Numbers expressed as some power of 10 multiplied by another number between 1 and 10 are said to be in **scientific notation**. For example, the scientific notation for 5 943 000 000 is 5.943×10^9 and that for 0.0000832 is 8.32×10^{-5}.

When numbers expressed in scientific notation are being multiplied, the following general rule is very useful:

$$10^n \times 10^m = 10^{n+m} \tag{B.1}$$

where n and m can be *any* numbers (not necessarily integers). For example, $10^2 \times 10^5 = 10^7$. The rule also applies if one of the exponents is negative: $10^3 \times 10^{-8} = 10^{-5}$.

When dividing numbers expressed in scientific notation, note that

$$\frac{10^n}{10^m} = 10^n \times 10^{-m} = 10^{n-m} \tag{B.2}$$

EXERCISES

With help from the above rules, verify the answers to the following:

1. $86\,400 = 8.64 \times 10^4$
2. $9\,816{,}762.5 = 9.8167625 \times 10^6$
3. $0.0000000398 = 3.98 \times 10^{-8}$
4. $(4 \times 10^8)(9 \times 10^9) = 3.6 \times 10^{18}$
5. $(3 \times 10^7)(6 \times 10^{-12}) = 1.8 \times 10^{-4}$
6. $\dfrac{75 \times 10^{-11}}{5 \times 10^{-3}} = 1.5 \times 10^{-7}$
7. $\dfrac{(3 \times 10^6)(8 \times 10^{-2})}{(2 \times 10^{17})(6 \times 10^5)} = 2 \times 10^{-18}$

B.2 ALGEBRA

Some Basic Rules

When algebraic operations are performed, the laws of arithmetic apply. Symbols such as x, y, and z are usually used to represent quantities that are not specified, what are called the **unknowns**.

First, consider the equation

$$8x = 32$$

If we wish to solve for x, we can divide (or multiply) each side of the equation by the same factor without destroying the equality. In this case, if we divide both sides by 8, we have

$$\frac{8x}{8} = \frac{32}{8}$$

$$x = 4$$

Next consider the equation

$$x + 2 = 8$$

In this type of expression, we can add or subtract the same quantity from each side. If we subtract 2 from each side, we get

$$x + 2 - 2 = 8 - 2$$
$$x = 6$$

In general, if $x + a = b$, then $x = b - a$.

Now consider the equation

$$\frac{x}{5} = 9$$

If we multiply each side by 5, we are left with x on the left by itself and 45 on the right:

$$\left(\frac{x}{5}\right)(5) = 9 \times 5$$
$$x = 45$$

In all cases, *whatever operation is performed on the left side of the equality must also be performed on the right side.*

The following rules for multiplying, dividing, adding, and subtracting fractions should be recalled, where a, b, and c are three numbers:

	Rule	Example
Multiplying	$\left(\dfrac{a}{b}\right)\left(\dfrac{c}{d}\right) = \dfrac{ac}{bd}$	$\left(\dfrac{2}{3}\right)\left(\dfrac{4}{5}\right) = \dfrac{8}{15}$
Dividing	$\dfrac{(a/b)}{(c/d)} = \dfrac{ad}{bc}$	$\dfrac{2/3}{4/5} = \dfrac{(2)(5)}{(4)(3)} = \dfrac{10}{12}$
Adding	$\dfrac{a}{b} \pm \dfrac{c}{d} = \dfrac{ad \pm bc}{bd}$	$\dfrac{2}{3} - \dfrac{4}{5} = \dfrac{(2)(5) - (4)(3)}{(3)(5)} = -\dfrac{2}{15}$

EXERCISES

In the following exercises, solve for x:

Answers

1. $a = \dfrac{1}{1 + x}$ $\quad\quad x = \dfrac{1 - a}{a}$

2. $3x - 5 = 13$ $\quad\quad x = 6$

3. $ax - 5 = bx + 2$ $\quad\quad x = \dfrac{7}{a - b}$

4. $\dfrac{5}{2x + 6} = \dfrac{3}{4x + 8}$ $\quad\quad x = -\dfrac{11}{7}$

Powers

When powers of a given quantity x are multiplied, the following rule applies:

$$x^n x^m = x^{n+m} \tag{B.3}$$

For example, $x^2 x^4 = x^{2+4} = x^6$.

When dividing the powers of a given quantity, the rule is

$$\frac{x^n}{x^m} = x^{n-m} \tag{B.4}$$

For example, $x^8/x^2 = x^{8-2} = x^6$.

A power that is a fraction, such as $\frac{1}{3}$, corresponds to a root as follows:

$$x^{1/n} = \sqrt[n]{x} \tag{B.5}$$

For example, $4^{1/3} = \sqrt[3]{4} = 1.5874$. (A scientific calculator is useful for such calculations.)

Finally, any quantity x^n raised to the mth power is

$$(x^n)^m = x^{nm} \tag{B.6}$$

TABLE B.1 Rules of Exponents

$x^0 = 1$
$x^1 = x$
$x^n x^m = x^{n+m}$
$x^n/x^m = x^{n-m}$
$x^{1/n} = \sqrt[n]{x}$
$(x^n)^m = x^{nm}$

Table B.1 summarizes the rules of exponents.

EXERCISES

Verify the following:

1. $3^2 \times 3^3 = 243$
2. $x^5 x^{-8} = x^{-3}$
3. $x^{10}/x^{-5} = x^{15}$
4. $5^{1/3} = 1.709975$ (Use your calculator.)
5. $60^{1/4} = 2.783158$ (Use your calculator.)
6. $(x^4)^3 = x^{12}$

Factoring

Some useful formulas for factoring an equation are

$ax + ay + az = a(x + y + x)$ common factor
$a^2 + 2ab + b^2 = (a + b)^2$ perfect square
$a^2 - b^2 = (a + b)(a - b)$ differences of squares

Quadratic Equations

The general form of a quadratic equation is

$$ax^2 + bx + c = 0 \tag{B.7}$$

where x is the unknown quantity and a, b, and c are numerical factors referred to as **coefficients** of the equation. This equation has two roots, given by

$$x = \frac{-b \pm \sqrt{b^2 - 4ac}}{2a} \tag{B.8}$$

If $b^2 \geq 4ac$, the roots are real.

EXAMPLE 1

The equation $x^2 + 5x + 4 = 0$ has the following roots corresponding to the two signs of the square-root term:

$$x = \frac{-5 \pm \sqrt{5^2 - (4)(1)(4)}}{2(1)} = \frac{-5 \pm \sqrt{9}}{2} = \frac{-5 \pm 3}{2}$$

$$x_+ = \frac{-5 + 3}{2} = -1 \quad x_- = \frac{-5 - 3}{2} = -4$$

where x_+ refers to the root corresponding to the positive sign and x_- refers to the root corresponding to the negative sign.

EXERCISES

Solve the following quadratic equations:

Answers

1. $x^2 + 2x - 3 = 0$ $x_+ = 1$ $x_- = -3$
2. $2x^2 - 5x + 2 = 0$ $x_+ = 2$ $x_- = \frac{1}{2}$
3. $2x^2 - 4x - 9 = 0$ $x_+ = 1 + \sqrt{22}/2$ $x_- = 1 - \sqrt{22}/2$

Linear Equations

A linear equation has the general form

$$y = mx + b \tag{B.9}$$

where m and b are constants. This equation is referred to as being linear because the graph of y versus x is a straight line, as shown in Figure B.1. The constant b, called the **y-intercept**, represents the value of y at which the straight line intersects the y axis. The constant m is equal to the **slope** of the straight line and is also equal to the tangent of the angle that the line makes with the x axis. If any two points on the straight line are specified by the coordinates (x_1, y_1) and (x_2, y_2), as in Figure B.1, then the **slope** of the straight line can be expressed as

$$\text{Slope} = \frac{y_2 - y_1}{x_2 - x_1} = \frac{\Delta y}{\Delta x} = \tan \theta \tag{B.10}$$

Note that m and b can have either positive or negative values. If $m > 0$, the straight line has a *positive* slope, as in Figure B.1. If $m < 0$, the straight line has a *negative* slope. In Figure B.1, both m and b are positive. Three other possible situations are shown in Figure B.2.

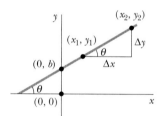

FIGURE B.1

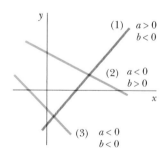

FIGURE B.2

EXERCISES

1. Draw graphs of the following straight lines:
(a) $y = 5x + 3$ (b) $y = -2x + 4$ (c) $y = -3x - 6$
2. Find the slopes of the straight lines described in Exercise 1.

Answers (a) 5   (b) -2 (c) -3

3. Find the slopes of the straight lines that pass through the following sets of points:
(a) (0, −4) and (4, 2), (b) (0, 0) and (2, −5), and (c) (−5, 2) and (4, −2)
Answers (a) 3/2   (b) −5/2   (c) −4/9

Solving Simultaneous Linear Equations

Consider the equation $3x + 5y = 15$, which has two unknowns, x and y. Such an equation does not have a unique solution. Instead, $(x = 0, y = 3)$, $(x = 5, y = 0)$, and $(x = 2, y = 9/5)$ are all solutions to this equation.

If a problem has two unknowns, a unique solution is possible only if we have *two* equations. In general, if a problem has n unknowns, its solution requires n equations. In order to solve two simultaneous equations involving two unknowns, x and y, we solve one of the equations for x in terms of y and substitute this expression into the other equation.

EXAMPLE 2

Solve the following two simultaneous equations:

(1) $5x + y = -8$

(2) $2x - 2y = 4$

Solution From (2), $x = y + 2$. Substitution of this into (1) gives

$$5(y + 2) + y = -8$$

$$6y = -18$$

$$y = -3$$

$$x = y + 2 = \boxed{-1}$$

Alternate Solution Multiply each term in (1) by the factor 2 and add the result to (2):

$$10x + 2y = -16$$

$$\underline{2x - 2y = 4}$$

$$12x = -12$$

$$x = -1$$

$$y = x - 2 = \boxed{-3}$$

Two linear equations containing two unknowns can also be solved by a graphical method. If the straight lines corresponding to the two equations are plotted in a conventional coordinate system, the intersection of the two lines represents the solution. For example, consider the two equations

$$x - y = 2$$

$$x - 2y = -1$$

These are plotted in Figure B.3. The intersection of the two lines has the coordinates $x = 5, y = 3$. This represents the solution to the equations. You should check this solution by the analytical technique discussed above.

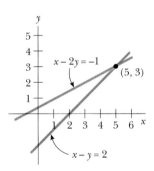

FIGURE B.3

EXERCISES

Solve the following pairs of simultaneous equations involving two unknowns:

Answers

1. $x + y = 8$ $x = 5, y = 3$
    $x - y = 2$
2. $98 - T = 10a$ $T = 65, a = 3.27$
    $T - 49 = 5a$
3. $6x + 2y = 6$ $x = 2, y = -3$
    $8x - 4y = 28$

Appendix B

Logarithms

Suppose that a quantity x is expressed as a power of some quantity a:

$$x = a^y \tag{B.11}$$

The number a is called the **base** number. The **logarithm** of x with respect to the base a is equal to the exponent to which the base must be raised in order to satisfy the expression $x = a^y$:

$$y = \log_a x \tag{B.12}$$

Conversely, the **antilogarithm** of y is the number x:

$$x = \text{antilog}_a y \tag{B.13}$$

In practice, the two bases most often used are base 10, called the *common* logarithm base, and base $e = 2.718\ldots$, called the *natural* logarithm base. When common logarithms are used,

$$y = \log_{10} x \quad (\text{or } x = 10^y) \tag{B.14}$$

When natural logarithms are used,

$$y = \ln_e x \quad (\text{or } x = e^y) \tag{B.15}$$

For example, $\log_{10} 52 = 1.716$, so that $\text{antilog}_{10} 1.716 = 10^{1.716} = 52$. Likewise, $\ln_e 52 = 3.951$, so $\text{antiln}_e 3.951 = e^{3.951} = 52$.

In general, note that you can convert between base 10 and base e with the equality

$$\ln_e x = (2.302585) \log_{10} x \tag{B.16}$$

Finally, some useful properties of logarithms are

$$\log(ab) = \log a + \log b$$
$$\log(a/b) = \log a - \log b$$
$$\log(a^n) = n \log a$$
$$\ln e = 1$$
$$\ln e^a = a$$
$$\ln\left(\frac{1}{a}\right) = -\ln a$$

B.3 GEOMETRY

The **distance** d between two points having coordinates (x_1, y_1) and (x_2, y_2) is

$$d = \sqrt{(x_2 - x_1)^2 + (y_2 - y_1)^2} \tag{B.17}$$

Radian measure: The arc length s of a circular arc (Fig. B.4) is proportional to the radius r for a fixed value of θ (in radians):

$$s = r\theta$$
$$\theta = \frac{s}{r} \tag{B.18}$$

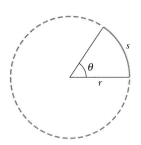

FIGURE B.4

Table B.2 gives the areas and volumes for several geometric shapes used throughout this text:

TABLE B.2 Useful Information for Geometry

Shape	Area or Volume	Shape	Area or Volume
Rectangle	Area = ℓw	Sphere	Surface area = $4\pi r^2$ Volume = $\frac{4\pi r^3}{3}$
Circle	Area = πr^2 (Circumference = $2\pi r$)	Cylinder	Volume = $\pi r^2 \ell$
Triangle	Area = $\frac{1}{2}bh$	Rectangular box	Area = $2(\ell h + \ell w + hw)$ Volume = $\ell w h$

The equation of a **straight line** (Fig. B.5) is

$$y = mx + b \tag{B.19}$$

where b is the y-intercept and m is the slope of the line.

The equation of a **circle** or radius R centered at the origin is

$$x^2 + y^2 = R^2 \tag{B.20}$$

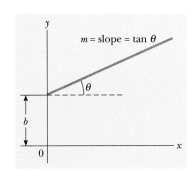

FIGURE B.5

The equation of an **ellipse** having the origin at its center (Fig. B.6) is

$$\frac{x^2}{a^2} + \frac{y^2}{b^2} = 1 \tag{B.21}$$

where a is the length of the semi-major axis (the longer one) and b is the length of the semi-minor axis (the shorter one).

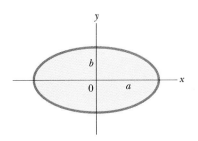

FIGURE B.6

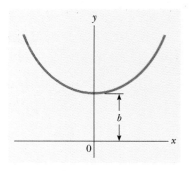

FIGURE B.7

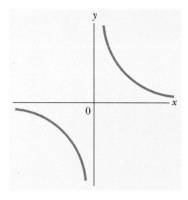

FIGURE B.8

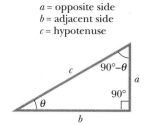

FIGURE B.9

The equation of a **parabola** the vertex of which is at $y = b$ (Fig. B.7) is

$$y = ax^2 + b \qquad (B.22)$$

The equation of a **rectangular hyperbola** (Fig. B.8) is

$$xy = \text{constant} \qquad (B.23)$$

B.4 TRIGONOMETRY

That portion of mathematics based on the special properties of the right triangle is called trigonometry. By definition, a right triangle is one containing at 90° angle. Consider the right triangle shown in Figure B.9, where side a is opposite the angle θ, side b is adjacent to the angle θ, and side c is the hypotenuse of the triangle. The three basic trigonometric functions defined by such a triangle are the sine (sin), cosine (cos), and tangent (tan) functions. In terms of the angle θ, these functions are defined by

$$\sin \theta \equiv \frac{\text{side opposite } \theta}{\text{hypotenuse}} = \frac{a}{c} \qquad (B.24)$$

$$\cos \theta \equiv \frac{\text{side adjacent to } \theta}{\text{hypotenuse}} = \frac{b}{c} \qquad (B.25)$$

$$\tan \theta \equiv \frac{\text{side opposite } \theta}{\text{side adjacent to } \theta} = \frac{a}{b} \qquad (B.26)$$

The Pythagorean theorem provides the following relationship between the sides of a right triangle:

$$c^2 = a^2 + b^2 \qquad (B.27)$$

From the above definitions and the Pythagorean theorem, it follows that

$$\sin^2 \theta + \cos^2 \theta = 1$$

$$\tan \theta = \frac{\sin \theta}{\cos \theta}$$

The cosecant, secant, and cotangent functions are defined by

$$\csc \theta \equiv \frac{1}{\sin \theta} \qquad \sec \theta \equiv \frac{1}{\cos \theta} \qquad \cot \theta \equiv \frac{1}{\tan \theta}$$

The relationship below follow directly from the right triangle shown in Figure B.9:

$$\sin \theta = \cos(90° - \theta)$$

$$\cos \theta = \sin(90° - \theta)$$

$$\cot \theta = \tan(90° - \theta)$$

Some properties of trigonometric functions are

$$\sin(-\theta) = -\sin \theta$$

$$\cos(-\theta) = \cos \theta$$

$$\tan(-\theta) = -\tan \theta$$

The following relationships apply to *any* triangle, as shown in Figure B.10:

$$\alpha + \beta + \gamma = 180°$$

Law of cosines
$$a^2 = b^2 + c^2 - 2bc \cos \alpha$$
$$b^2 = a^2 + c^2 - 2ac \cos \beta$$
$$c^2 = a^2 + b^2 - 2ab \cos \gamma$$

Law of sines
$$\frac{a}{\sin \alpha} = \frac{b}{\sin \beta} = \frac{c}{\sin \gamma}$$

Table B.3 lists a number of useful trigonometric identities.

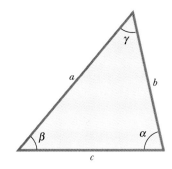

FIGURE B.10

TABLE B.3 Some Trigonometric Identities

$$\sin^2 \theta + \cos^2 \theta = 1 \qquad \csc^2 \theta = 1 + \cot^2 \theta$$

$$\sec^2 \theta = 1 + \tan^2 \theta \qquad \sin^2 \frac{\theta}{2} = \tfrac{1}{2}(1 - \cos \theta)$$

$$\sin 2\theta = 2 \sin \theta \cos \theta \qquad \cos^2 \frac{\theta}{2} = \tfrac{1}{2}(1 + \cos \theta)$$

$$\cos 2\theta = \cos^2 \theta - \sin^2 \theta \qquad 1 - \cos \theta = 2 \sin^2 \frac{\theta}{2}$$

$$\tan 2\theta = \frac{2 \tan \theta}{1 - \tan^2 \theta} \qquad \tan \frac{\theta}{2} = \sqrt{\frac{1 - \cos \theta}{1 + \cos \theta}}$$

$$\sin(A \pm B) = \sin A \cos B \pm \cos A \sin B$$
$$\cos(A \pm B) = \cos A \cos B \mp \sin A \sin B$$
$$\sin A \pm \sin B = 2 \sin[\tfrac{1}{2}(A \pm B)] \cos[\tfrac{1}{2}(A \mp B)]$$
$$\cos A + \cos B = 2 \cos[\tfrac{1}{2}(A + B)] \cos[\tfrac{1}{2}(A - B)]$$
$$\cos A - \cos B = 2 \sin[\tfrac{1}{2}(A + B)] \sin[\tfrac{1}{2}(B - A)]$$

EXAMPLE 3

Consider the right triangle in Figure B.11, in which $a = 2$, $b = 5$, and c is unknown. From the Pythagorean theorem, we have

$$c^2 = a^2 + b^2 = 2^2 + 5^2 = 4 + 25 = 29$$

$$c = \sqrt{29} = 5.39$$

To find the angle θ, note that

$$\tan \theta = \frac{a}{b} = \frac{2}{5} = 0.400$$

From a table of functions or from a calculator, we have

$$\theta = \tan^{-1}(0.400) = 21.8°$$

where $\tan^{-1}(0.400)$ is the notation for "angle whose tangent is 0.400," sometimes written as $\arctan(0.400)$.

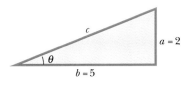

FIGURE B.11

EXERCISES

1. In Figure B.12, identify (a) the side opposite θ and (b) the side adjacent to ϕ and then find (c) $\cos \theta$, (d) $\sin \phi$, and (e) $\tan \phi$.

Answers (a) 3, (b) 3, (c) $\tfrac{4}{5}$, (d) $\tfrac{4}{5}$, and (e) $\tfrac{4}{3}$

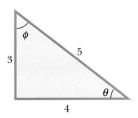

FIGURE B.12

2. In a certain right triangle, the two sides that are perpendicular to each other are 5 m and 7 m long. What is the length of the third side?

Answer 8.60 m

3. A right triangle has a hypotenuse of length 3 m, and one of its angles is 30°. What is the length of (a) the side opposite the 30° angle and (b) the side adjacent to the 30° angle?

Answers (a) 1.5 m, (b) 2.60 m

B.5 SERIES EXPANSIONS

$$(a + b)^n = a^n + \frac{n}{1!} a^{n-1} b + \frac{n(n-1)}{2!} a^{n-2} b^2 + \cdots$$

$$(1 + x)^n = 1 + nx + \frac{n(n-1)}{2!} x^2 + \cdots$$

$$e^x = 1 + x + \frac{x^2}{2!} + \frac{x^3}{3!} + \cdots$$

$$\ln(1 \pm x) = \pm x - \tfrac{1}{2} x^2 \pm \tfrac{1}{3} x^3 - \cdots$$

$$\left. \begin{array}{l} \sin x = x - \dfrac{x^3}{3!} + \dfrac{x^5}{5!} - \cdots \\[6pt] \cos x = 1 - \dfrac{x^2}{2!} + \dfrac{x^4}{4!} - \cdots \\[6pt] \tan x = x + \dfrac{x^3}{3} + \dfrac{2x^5}{15} + \cdots \quad |x| < \pi/2 \end{array} \right\} x \text{ in radians}$$

For $x \ll 1$, the following approximations can be used:

$$(1 + x)^n \approx 1 + nx \qquad \sin x \approx x$$
$$e^x \approx 1 + x \qquad \cos x \approx 1$$
$$\ln(1 \pm x) \approx \pm x \qquad \tan x \approx x$$

B.6 DIFFERENTIAL CALCULUS

In various branches of science, it is sometimes necessary to use the basic tools of calculus, invented by Newton, to describe physical phenomena. The use of calculus is fundamental in the treatment of various problems in newtonian mechanics, electricity, and magnetism. In this section, we simply state some basic properties and "rules of thumb" that should be a useful review to the student.

First, a **function** must be specified that relates one variable to another (such as a coordinate as a function of time). Suppose one of the variables is called y (the dependent variable), the other x (the independent variable). We might have a function relationship such as

$$y(x) = ax^3 + bx^2 + cx + d$$

If a, b, c, and d are specified constants, then y can be calculated for any value of x. We usually deal with continuous functions, that is, those for which y varies "smoothly" with x.

The **derivative** of y with respect to x is defined as the limit, as Δx approaches zero, of the slopes of chords drawn between two points on the y versus x curve. Mathematically, we write this definition as

$$\frac{dy}{dx} = \lim_{\Delta x \to 0} \frac{\Delta y}{\Delta x} = \lim_{\Delta x \to 0} \frac{y(x + \Delta x) - y(x)}{\Delta x} \quad \text{(B.28)}$$

where Δy and Δx are defined as $\Delta x = x_2 - x_1$ and $\Delta y = y_2 - y_1$ (Fig. B.13). It is important to note that dy/dx *does not* mean dy divided by dx, but is simply a notation of the limiting process of the derivative as defined by Equation B.28.

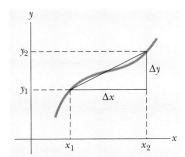

FIGURE B.13

A useful expression to remember when $y(x) = ax^n$, where a is a *constant* and n is *any* positive or negative number (integer or fraction), is

$$\frac{dy}{dx} = nax^{n-1} \quad \text{(B.29)}$$

If $y(x)$ is a polynomial or algebraic function of x, we apply Equation B.29 to *each* term in the polynomial and take $da/dx = 0$. In Examples 4 through 7, we evaluate the derivatives of several functions.

EXAMPLE 4

Suppose $y(x)$ (that is, y as a function of x) is given by

$$y(x) = ax^3 + bx + c$$

where a and b are constants. Then it follows that

$$y(x + \Delta x) = a(x + \Delta x)^3 + b(x + \Delta x) + c$$

$$y(x + \Delta x) = a(x^3 + 3x^2 \Delta x + 3x \Delta x^2 + \Delta x^3) + b(x + \Delta x) + c$$

so

$$\Delta y = y(x + \Delta x) - y(x) = a(3x^2 \Delta x + 3x \Delta x^2 + \Delta x^3) + b \Delta x$$

Substituting this into Equation B.28 gives

$$\frac{dy}{dx} = \lim_{\Delta x \to 0} \frac{\Delta y}{\Delta x} = \lim_{\Delta x \to 0} [3ax^2 + 3x \Delta x + \Delta x^2] + b$$

$$\frac{dy}{dx} = 3ax^2 + b$$

EXAMPLE 5

$$y(x) = 8x^5 + 4x^3 + 2x + 7$$

Solution Applying Equation B.29 to each term independently, and remembering that d/dx (constant) $= 0$, we have

$$\frac{dy}{dx} = 8(5)x^4 + 4(3)x^2 + 2(1)x^0 + 0$$

$$\frac{dy}{dx} = 40x^4 + 12x^2 + 2$$

Special Properties of the Derivative

A. **Derivative of the product of two functions** If a function $f(x)$ is given by the product of two functions, say, $g(x)$ and $h(x)$, then the derivative of $f(x)$ is defined as

$$\frac{d}{dx}f(x) = \frac{d}{dx}[g(x)h(x)] = g\frac{dh}{dx} + h\frac{dg}{dx} \qquad (B.30)$$

B. **Derivative of the sum of two functions** If a function $f(x)$ is equal to the sum of two functions, then the derivative of the sum is equal to the sum of the derivatives:

$$\frac{d}{dx}f(x) = \frac{d}{dx}[g(x) + h(x)] = \frac{dg}{dx} + \frac{dh}{dx} \qquad (B.31)$$

C. **Chain rule of differential calculus** If $y = f(x)$ and $x = f(z)$, then dy/dx can be written as the product of two derivatives:

$$\frac{dy}{dx} = \frac{dy}{dz}\frac{dz}{dx} \qquad (B.32)$$

D. **The second derivative** The second derivative of y with respect to x is defined as the derivative of the function dy/dx (the derivative of the derivative). It is usually written

$$\frac{d^2y}{dx^2} = \frac{d}{dx}\left(\frac{dy}{dx}\right) \qquad (B.33)$$

EXAMPLE 6

Find the derivative of $y(x) = x^3/(x + 1)^2$ with respect to x.

Solution We can rewrite this function as $y(x) = x^3(x + 1)^{-2}$ and apply Equation B.30:

$$\frac{dy}{dx} = (x + 1)^{-2}\frac{d}{dx}(x^3) + x^3\frac{d}{dx}(x + 1)^{-2}$$

$$= (x + 1)^{-2}3x^2 + x^3(-2)(x + 1)^{-3}$$

$$\frac{dy}{dx} = \frac{3x^2}{(x + 1)^2} - \frac{2x^3}{(x + 1)^3}$$

EXAMPLE 7

A useful formula that follows from Equation B.30 is the derivative of the quotient of two functions. Show that

$$\frac{d}{dx}\left[\frac{g(x)}{h(x)}\right] = \frac{h\dfrac{dg}{dx} - g\dfrac{dh}{dx}}{h^2}$$

Solution We can write the quotient as gh^{-1} and then apply Equations B.29 and B.30:

$$\frac{d}{dx}\left(\frac{g}{h}\right) = \frac{d}{dx}(gh^{-1}) = g\frac{d}{dx}(h^{-1}) + h^{-1}\frac{d}{dx}(g)$$

$$= -gh^{-2}\frac{dh}{dx} + h^{-1}\frac{dg}{dx}$$

$$= \frac{h\dfrac{dg}{dx} - g\dfrac{dh}{dx}}{h^2}$$

Some of the more commonly used derivatives of functions are listed in Table B.4.

B.7 INTEGRAL CALCULUS

We think of integration as the inverse of differentiation. As an example, consider the expression

$$f(x) = \frac{dy}{dx} = 3ax^2 + b \qquad (B.34)$$

which was the result of differentiating the function

$$y(x) = ax^3 + bx + c$$

in Example 4. We can write Equation B.34 as $dy = f(x)\,dx = (3ax^2 + b)\,dx$ and obtain $y(x)$ by "summing" over all values of x. Mathematically, we write this inverse operation

$$y(x) = \int f(x)\,dx$$

For the function $f(x)$ given by Equation B.34, we have

$$y(x) = \int (3ax^2 + b)\,dx = ax^3 + bx + c$$

where c is a constant of the integration. This type of integral is called an *indefinite integral* because its value depends on the choice of c.

A general **indefinite integral** $I(x)$ is defined as

$$I(x) = \int f(x)\,dx \qquad (B.35)$$

where $f(x)$ is called the *integrand* and $f(x) = \dfrac{dI(x)}{dx}$.

For a *general continuous* function $f(x)$, the integral can be described as the area under the curve bounded by $f(x)$ and the x axis, between two specified values of x, say, x_1 and x_2, as in Figure B.14.

The area of the blue element is approximately $f_i\,\Delta x_i$. If we sum all these area elements from x_1 to x_2 and take the limit of this sum as $\Delta x_i \to 0$, we obtain the *true* area under the curve bounded by $f(x)$ and x, between the limits x_1 and x_2:

$$\text{Area} = \lim_{\Delta x_i \to 0} \sum_i f(x_i)\,\Delta x_i = \int_{x_1}^{x_2} f(x)\,dx \qquad (B.36)$$

Integrals of the type defined by Equation B.36 are called **definite integrals**.

TABLE B.4 Derivatives for Several Functions

$$\frac{d}{dx}(a) = 0$$

$$\frac{d}{dx}(ax^n) = nax^{n-1}$$

$$\frac{d}{dx}(e^{ax}) = ae^{ax}$$

$$\frac{d}{dx}(\sin ax) = a\cos ax$$

$$\frac{d}{dx}(\cos ax) = -a\sin ax$$

$$\frac{d}{dx}(\tan ax) = a\sec^2 ax$$

$$\frac{d}{dx}(\cot ax) = -a\csc^2 ax$$

$$\frac{d}{dx}(\sec x) = \tan x \sec x$$

$$\frac{d}{dx}(\csc x) = -\cot x \csc x$$

$$\frac{d}{dx}(\ln ax) = \frac{1}{x}$$

Note: The letters a and n are constants.

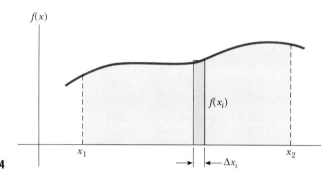

FIGURE B.14

One common integral that arises in practical situations has the form

$$\int x^n \, dx = \frac{x^{n+1}}{n+1} + c \qquad (n \neq -1) \tag{B.37}$$

This result is obvious, being that differentiation of the right-hand side with respect to x gives $f(x) = x^n$ directly. If the limits of the integration are known, this integral becomes a *definite integral* and is written

$$\int_{x_1}^{x_2} x^n \, dx = \frac{x_2^{n+1} - x_1^{n+1}}{n+1} \qquad (n \neq -1) \tag{B.38}$$

EXAMPLES

1. $\displaystyle\int_0^a x^2 \, dx = \left.\frac{x^3}{3}\right]_0^a = \frac{a^3}{3}$

2. $\displaystyle\int_0^b x^{3/2} \, dx = \left.\frac{x^{5/2}}{5/2}\right]_0^b = \frac{2}{5} b^{5/2}$

3. $\displaystyle\int_3^5 x \, dx = \left.\frac{x^2}{2}\right]_3^5 = \frac{5^2 - 3^2}{2} = 8$

Partial Integration

Sometimes it is useful to apply the method of *partial integration* (also called "integrating by parts") to evaluate certain integrals. The method uses the property that

$$\int u \, dv = uv - \int v \, du \tag{B.39}$$

where u and v are *carefully* chosen so as to reduce a complex integral to a simpler one. In many cases, several reductions have to be made. Consider the function

$$I(x) = \int x^2 e^x \, dx$$

This can be evaluated by integrating by parts twice. First, if we choose $u = x^2$, $v = e^x$, we get

$$\int x^2 e^x \, dx = \int x^2 \, d(e^x) = x^2 e^x - 2 \int e^x x \, dx + c_1$$

Now, in the second term, choose $u = x$, $v = e^x$, which gives

$$\int x^2 e^x \, dx = x^2 e^x - 2xe^x + 2 \int e^x \, dx + c_1$$

or

$$\int x^2 e^x \, dx = x^2 e^x - 2xe^x + 2e^x + c_2$$

The Perfect Differential

Another useful method to remember is the use of the *perfect differential*, in which we look for a change of variable such that the differential of the function is the differential of the independent variable appearing in the integrand. For example, consider the integral

$$I(x) = \int \cos^2 x \sin x \, dx$$

This becomes easy to evaluate if we rewrite the differential as $d(\cos x) = -\sin x \, dx$. The integral then becomes

$$\int \cos^2 x \sin x \, dx = -\int \cos^2 x \, d(\cos x)$$

If we now change variables, letting $y = \cos x$, we get

$$\int \cos^2 x \sin x \, dx = -\int y^2 \, dy = -\frac{y^3}{3} + c = -\frac{\cos^3 x}{3} + c$$

Table B.5 lists some useful indefinite integrals. Table B.6 gives Gauss's probability integral and other definite integrals. A more complete list can be found in various handbooks, such as *The Handbook of Chemistry and Physics,* CRC Press.

TABLE B.5 Some Indefinite Integrals (an arbitrary constant should be added to each of these integrals)

$$\int x^n \, dx = \frac{x^{n+1}}{n+1} \quad \text{(provided } n \neq -1)$$

$$\int \frac{dx}{x} = \int x^{-1} \, dx = \ln x$$

$$\int \frac{dx}{a + bx} = \frac{1}{b} \ln(a + bx)$$

$$\int \frac{dx}{(a + bx)^2} = -\frac{1}{b(a + bx)}$$

$$\int \frac{dx}{a^2 + x^2} = \frac{1}{a} \tan^{-1} \frac{x}{a}$$

$$\int \frac{dx}{a^2 - x^2} = \frac{1}{2a} \ln \frac{a + x}{a - x} \quad (a^2 - x^2 > 0)$$

$$\int \frac{dx}{x^2 - a^2} = \frac{1}{2a} \ln \frac{x - a}{x + a} \quad (x^2 - a^2 > 0)$$

$$\int \frac{x \, dx}{a^2 \pm x^2} = \pm \tfrac{1}{2} \ln(a^2 \pm x^2)$$

$$\int \frac{dx}{\sqrt{a^2 - x^2}} = \sin^{-1} \frac{x}{a} = -\cos^{-1} \frac{x}{a} \quad (a^2 - x^2 > 0)$$

$$\int \frac{dx}{\sqrt{x^2 \pm a^2}} = \ln(x + \sqrt{x^2 \pm a^2})$$

$$\int \frac{x \, dx}{\sqrt{a^2 - x^2}} = -\sqrt{a^2 - x^2}$$

$$\int \frac{x \, dx}{\sqrt{x^2 \pm a^2}} = \sqrt{x^2 \pm a^2}$$

$$\int \sqrt{a^2 - x^2} \, dx = \tfrac{1}{2}\left(x\sqrt{a^2 - x^2} + a^2 \sin^{-1} \frac{x}{a} \right)$$

$$\int x\sqrt{a^2 - x^2} \, dx = -\tfrac{1}{3}(a^2 - x^2)^{3/2}$$

$$\int \sqrt{x^2 \pm a^2} \, dx = \tfrac{1}{2}[x\sqrt{x^2 \pm a^2} \pm a^2 \ln(x + \sqrt{x^2 \pm a^2})]$$

$$\int x(\sqrt{x^2 \pm a^2}) \, dx = \tfrac{1}{3}(x^2 \pm a^2)^{3/2}$$

$$\int e^{ax} \, dx = \frac{1}{a} e^{ax}$$

$$\int \ln ax \, dx = (x \ln ax) - x$$

$$\int xe^{ax} \, dx = \frac{e^{ax}}{a^2}(ax - 1)$$

$$\int \frac{dx}{a + be^{cx}} = \frac{x}{a} - \frac{1}{ac} \ln(a + be^{cx})$$

$$\int \sin ax \, dx = -\frac{1}{a} \cos ax$$

$$\int \cos ax \, dx = \frac{1}{a} \sin ax$$

$$\int \tan ax \, dx = -\frac{1}{a} \ln(\cos ax) = \frac{1}{a} \ln(\sec ax)$$

$$\int \cot ax \, dx = \frac{1}{a} \ln(\sin ax)$$

$$\int \sec ax \, dx = \frac{1}{a} \ln(\sec ax + \tan ax) = \frac{1}{a} \ln\left[\tan\left(\frac{ax}{2} + \frac{\pi}{4}\right)\right]$$

$$\int \csc ax \, dx = \frac{1}{a} \ln(\csc ax - \cot ax) = \frac{1}{a} \ln\left(\tan \frac{ax}{2}\right)$$

$$\int \sin^2 ax \, dx = \frac{x}{2} - \frac{\sin 2ax}{4a}$$

$$\int \cos^2 ax \, dx = \frac{x}{2} + \frac{\sin 2ax}{4a}$$

$$\int \frac{dx}{\sin^2 ax} = -\frac{1}{a} \cot ax$$

$$\int \frac{dx}{\cos^2 ax} = \frac{1}{a} \tan ax$$

$$\int \tan^2 ax \, dx = \frac{1}{a}(\tan ax) - x$$

$$\int \cot^2 ax \, dx = -\frac{1}{a}(\cot ax) - x$$

$$\int \sin^{-1} ax \, dx = x(\sin^{-1} ax) + \frac{\sqrt{1 - a^2 x^2}}{a}$$

$$\int \cos^{-1} ax \, dx = x(\cos^{-1} ax) - \frac{\sqrt{1 - a^2 x^2}}{a}$$

$$\int \frac{dx}{(x^2 + a^2)^{3/2}} = \frac{x}{a^2 \sqrt{x^2 + a^2}}$$

$$\int \frac{x \, dx}{(x^2 + a^2)^{3/2}} = -\frac{1}{\sqrt{x^2 + a^2}}$$

TABLE B.6 Gauss's Probability Integral and Related Integrals

$$I_0 = \int_0^\infty e^{-\alpha x^2}\, dx = \tfrac{1}{2}\sqrt{\frac{\pi}{\alpha}} \quad \text{(Gauss's probability integral)}$$

$$I_1 = \int_0^\infty x e^{-\alpha x^2}\, dx = \frac{1}{2\alpha}$$

$$I_2 = \int_0^\infty x^2 e^{-\alpha x^2}\, dx = -\frac{dI_0}{d\alpha} = \tfrac{1}{4}\sqrt{\frac{\pi}{\alpha^3}}$$

$$I_3 = \int_0^\infty x^3 e^{-\alpha x^2}\, dx = -\frac{dI_1}{d\alpha} = \frac{1}{2\alpha^2}$$

$$I_4 = \int_0^\infty x^4 e^{-\alpha x^2}\, dx = \frac{d^2 I_0}{d\alpha^2} = \tfrac{3}{8}\sqrt{\frac{\pi}{\alpha^5}}$$

$$I_5 = \int_0^\infty x^5 e^{-\alpha x^2}\, dx = \frac{d^2 I_1}{d\alpha^2} = \frac{1}{\alpha^3}$$

$$\vdots$$

$$I_{2n} = (-1)^n \frac{d^n}{d\alpha^n} I_0$$

$$I_{2n+1} = (-1)^n \frac{d^n}{d\alpha^n} I_1$$

APPENDIX C

Periodic Table of the Elements*

Legend:
- Symbol — Ca
- Atomic mass † — 40.08
- Atomic number — 20
- Electron configuration — $4s^2$

Group I	Group II			Transition elements					
H 1 1.0080 $1s^1$									
Li 3 6.94 $2s^1$	**Be** 4 9.012 $2s^2$								
Na 11 22.99 $3s^1$	**Mg** 12 24.31 $3s^2$								
K 19 39.102 $4s^1$	**Ca** 20 40.08 $4s^2$	**Sc** 21 44.96 $3d^1 4s^2$	**Ti** 22 47.90 $3d^2 4s^2$	**V** 23 50.94 $3d^3 4s^2$	**Cr** 24 51.996 $3d^5 4s^1$	**Mn** 25 54.94 $3d^5 4s^2$	**Fe** 26 55.85 $3d^6 4s^2$	**Co** 27 58.93 $3d^7 4s^2$	
Rb 37 85.47 $5s^1$	**Sr** 38 87.62 $5s^2$	**Y** 39 88.906 $4d^1 5s^2$	**Zr** 40 91.22 $4d^2 5s^2$	**Nb** 41 92.91 $4d^4 5s^1$	**Mo** 42 95.94 $4d^5 5s^1$	**Tc** 43 (99) $4d^5 5s^2$	**Ru** 44 101.1 $4d^7 5s^1$	**Rh** 45 102.91 $4d^8 5s^1$	
Cs 55 132.91 $6s^1$	**Ba** 56 137.34 $6s^2$	57-71*	**Hf** 72 178.49 $5d^2 6s^2$	**Ta** 73 180.95 $5d^3 6s^2$	**W** 74 183.85 $5d^4 6s^2$	**Re** 75 186.2 $5d^5 6s^2$	**Os** 76 190.2 $5d^6 6s^2$	**Ir** 77 192.2 $5d^7 6s^2$	
Fr 87 (223) $7s^1$	**Ra** 88 (226) $7s^2$	89-103**	**Unq** 104 (261) $6d^2 7s^2$	**Unp** 105 (262) $6d^3 7s^2$	**Unh** 106 (263)	**Uns** 107 (262)	**Uno** 108 (265)	**Une** 109 (266)	

*Lanthanide series

La 57 138.91 $5d^1 6s^2$	**Ce** 58 140.12 $5d^1 4f^1 6s^2$	**Pr** 59 140.91 $4f^3 6s^2$	**Nd** 60 144.24 $4f^4 6s^2$	**Pm** 61 (147) $4f^5 6s^2$	**Sm** 62 150.4 $4f^6 6s^2$

**Actinide series

Ac 89 (227) $6d^1 7s^2$	**Th** 90 (232) $6d^2 7s^2$	**Pa** 91 (231) $5f^2 6d^1 7s^2$	**U** 92 (238) $5f^3 6d^1 7s^2$	**Np** 93 (239) $5f^4 6d^1 7s^2$	**Pu** 94 (239) $5f^6 6d^0 7s^2$

◻ Atomic mass values given are averaged over isotopes in the percentages in which they exist in nature.
† For an unstable element, mass number of the most stable known isotope is given in parentheses.

Appendix C   **A.33**

	Group III	Group IV	Group V	Group VI	Group VII	Group 0
					H 1 1.0080 $1s^1$	**He** 2 4.0026 $1s^2$
	B 5 10.81 $2p^1$	**C** 6 12.011 $2p^2$	**N** 7 14.007 $2p^3$	**O** 8 15.999 $2p^4$	**F** 9 18.998 $2p^5$	**Ne** 10 20.18 $2p^6$
	Al 13 26.98 $3p^1$	**Si** 14 28.09 $3p^2$	**P** 15 30.97 $3p^3$	**S** 16 32.06 $3p^4$	**Cl** 17 35.453 $3p^5$	**Ar** 18 39.948 $3p^6$

Ni 28 58.71 $3d^84s^2$	**Cu** 29 63.54 $3d^{10}4s^2$	**Zn** 30 65.37 $3d^{10}4s^2$	**Ga** 31 69.72 $4p^1$	**Ge** 32 72.59 $4p^2$	**As** 33 74.92 $4p^3$	**Se** 34 78.96 $4p^4$	**Br** 35 79.91 $4p^5$	**Kr** 36 83.80 $4p^6$
Pd 46 106.4 $4d^{10}$	**Ag** 47 107.87 $4d^{10}5s^1$	**Cd** 48 112.40 $4d^{10}5s^2$	**In** 49 114.82 $5p^1$	**Sn** 50 118.69 $5p^2$	**Sb** 51 121.75 $5p^3$	**Te** 52 127.60 $5p^4$	**I** 53 126.90 $5p^5$	**Xe** 54 131.30 $5p^6$
Pt 78 195.09 $5d^96s^1$	**Au** 79 196.97 $5d^{10}6s^1$	**Hg** 80 200.59 $5d^{10}6s^2$	**Tl** 81 204.37 $6p^1$	**Pb** 82 207.2 $6p^2$	**Bi** 83 208.98 $6p^3$	**Po** 84 (210) $6p^4$	**At** 85 (218) $6p^5$	**Rn** 86 (222) $6p^6$

Eu 63 152.0 $4f^76s^2$	**Gd** 64 157.25 $5d^14f^76s^2$	**Tb** 65 158.92 $5d^14f^86s^2$	**Dy** 66 162.50 $4f^{10}6s^2$	**Ho** 67 164.93 $4f^{11}6s^2$	**Er** 68 167.26 $4f^{12}6s^2$	**Tm** 69 168.93 $4f^{13}6s^2$	**Yb** 70 173.04 $4f^{14}6s^2$	**Lu** 71 174.97 $5d^14f^{14}6s^2$
Am 95 (243) $5f^76d^07s^2$	**Cm** 96 (245) $5f^76d^17s^2$	**Bk** 97 (247) $5f^86d^17s^2$	**Cf** 98 (249) $5f^{10}6d^07s^2$	**Es** 99 (254) $5f^{11}6d^07s^2$	**Fm** 100 (253) $5f^{12}6d^07s^2$	**Md** 101 (255) $5f^{13}6d^07s^2$	**No** 102 (255) $6d^07s^2$	**Lr** 103 (257) $6d^17s^2$

APPENDIX D

SI Units

TABLE D.1 SI Base Units

Base Quantity	SI Base Unit Name	Symbol
Length	Meter	m
Mass	Kilogram	kg
Time	Second	s
Electric current	Ampere	A
Temperature	Kelvin	K
Amount of substance	Mole	mol
Luminous intensity	Candela	cd

TABLE D.2 Some Derived SI Units

Quantity	Name	Symbol	Expression in Terms of Base Units	Expression in Terms of Other SI Units
Plane angle	Radian	rad	m/m	
Frequency	Hertz	Hz	s^{-1}	
Force	Newton	N	$kg \cdot m/s^2$	J/m
Pressure	Pascal	Pa	$kg/m \cdot s^2$	N/m^2
Energy: work	Joule	J	$kg \cdot m^2/s^2$	$N \cdot m$
Power	Watt	W	$kg \cdot m^2/s^3$	J/s
Electric charge	Coulomb	C	$A \cdot s$	
Electric potential (emf)	Volt	V	$kg \cdot m^2/A \cdot s^3$	W/A
Capacitance	Farad	F	$A^2 \cdot s^4/kg \cdot m^2$	C/V
Electric resistance	Ohm	Ω	$kg \cdot m^2/A^2 \cdot s^3$	V/A
Magnetic flux	Weber	Wb	$kg \cdot m^2/A \cdot s^2$	$V \cdot s$
Magnetic field intensity	Tesla	T	$kg/A \cdot s^2$	Wb/m^2
Inductance	Henry	H	$kg \cdot m^2/A^2 \cdot s^2$	Wb/A

APPENDIX E

Nobel Prizes

All Nobel Prizes in physics are listed (and marked with a P), as well as relevant Nobel Prizes in Chemistry (C). The key dates for some of the scientific work are supplied; they often antedate the prize considerably.

1901 (P) *Wilhelm Roentgen* for discovering x-rays (1895).
1902 (P) *Hendrik A. Lorentz* for predicting the Zeeman effect and *Pieter Zeeman* for discovering the Zeeman effect, the splitting of spectral lines in magnetic fields.
1903 (P) *Antoine-Henri Becquerel* for discovering radioactivity (1896) and *Pierre* and *Marie Curie* for studying radioactivity.
1904 (P) *Lord Rayleigh* for studying the density of gases and discovering argon.
 (C) *William Ramsay* for discovering the inert gas elements helium, neon, xenon, and krypton, and placing them in the periodic table.
1905 (P) *Philipp Lenard* for studying cathode rays, electrons (1898–1899).
1906 (P) *J. J. Thomson* for studying electrical discharge through gases and discovering the electron (1897).
1907 (P) *Albert A. Michelson* for inventing optical instruments and measuring the speed of light (1880s).
1908 (P) *Gabriel Lippmann* for making the first color photographic plate, using interference methods (1891).
 (C) *Ernest Rutherford* for discovering that atoms can be broken apart by alpha rays and for studying radioactivity.
1909 (P) *Guglielmo Marconi* and *Carl Ferdinand Braun* for developing wireless telegraphy.
1910 (P) *Johannes D. van der Waals* for studying the equation of state for gases and liquids (1881).
1911 (P) *Wilhelm Wien* for discovering Wien's law giving the peak of a blackbody spectrum (1893).
 (C) *Marie Curie* for discovering radium and polonium (1898) and isolating radium.
1912 (P) *Nils Dalén* for inventing automatic gas regulators for lighthouses.
1913 (P) *Heike Kamerlingh Onnes* for the discovery of superconductivity and liquefying helium (1908).
1914 (P) *Max T. F. von Laue* for studying x-rays from their diffraction by crystals, showing that x-rays are electromagnetic waves (1912).
 (C) *Theodore W. Richards* for determining the atomic weights of sixty elements, indicating the existence of isotopes.
1915 (P) *William Henry Bragg* and *William Lawrence Bragg*, his son, for studying the diffraction of x-rays in crystals.
1917 (P) *Charles Barkla* for studying atoms by x-ray scattering (1906).
1918 (P) *Max Planck* for discovering energy quanta (1900).
1919 (P) *Johannes Stark*, for discovering the Stark effect, the splitting of spectral lines in electric fields (1913).

1920 (P) *Charles-Édouard Guillaume* for discovering invar, a nickel-steel alloy with low coefficient of expansion.
(C) *Walther Nernst* for studying heat changes in chemical reactions and formulating the third law of thermodynamics (1918).

1921 (P) *Albert Einstein* for explaining the photoelectric effect and for his services to theoretical physics (1905).
(C) *Frederick Soddy* for studying the chemistry of radioactive substances and discovering isotopes (1912).

1922 (P) *Niels Bohr* for his model of the atom and its radiation (1913).
(C) *Francis W. Aston* for using the mass spectrograph to study atomic weights, thus discovering 212 of the 287 naturally occurring isotopes.

1923 (P) *Robert A. Millikan* for measuring the charge on an electron (1911) and for studying the photoelectric effect experimentally (1914).

1924 (P) *Karl M. G. Siegbahn* for his work in x-ray spectroscopy.

1925 (P) *James Franck* and *Gustav Hertz* for discovering the Franck-Hertz effect in electron-atom collisions.

1926 (P) *Jean-Baptiste Perrin* for studying Brownian motion to validate the discontinuous structure of matter and measure the size of atoms.

1927 (P) *Arthur Holly Compton* for discovering the Compton effect on x-rays, their change in wavelength when they collide with matter (1922), and *Charles T. R. Wilson* for inventing the cloud chamber, used to study charged particles (1906).

1928 (P) *Owen W. Richardson* for studying the thermionic effect and electrons emitted by hot metals (1911).

1929 (P) *Louis Victor de Broglie* for discovering the wave nature of electrons (1923).

1930 (P) *Chandrasekhara Venkata Raman* for studying Raman scattering, the scattering of light by atoms and molecules with a change in wavelength (1928).

1932 (P) *Werner Heisenberg* for creating quantum mechanics (1925).

1933 (P) *Erwin Schrödinger* and *Paul A. M. Dirac* for developing wave mechanics (1925) and relativistic quantum mechanics (1927).
(C) *Harold Urey* for discovering heavy hydrogen, deuterium (1931).

1935 (P) *James Chadwick* for discovering the neutron (1932).
(C) *Irène* and *Frédéric Joliot-Curie* for synthesizing new radioactive elements.

1936 (P) *Carl D. Anderson* for discovering the positron in particular and antimatter in general (1932) and *Victor F. Hess* for discovering cosmic rays.
(C) *Peter J. W. Debye* for studying dipole moments and diffraction of x-rays and electrons in gases.

1937 (P) *Clinton Davisson* and *George Thomson* for discovering the diffraction of electrons by crystals, confirming de Broglie's hypothesis (1927).

1938 (P) *Enrico Fermi* for producing the transuranic radioactive elements by neutron irradiation (1934–1937).

1939 (P) *Ernest O. Lawrence* for inventing the cyclotron.

1943 (P) *Otto Stern* for developing molecular-beam studies (1923), and using them to discover the magnetic moment of the proton (1933).

1944 (P) *Isidor I. Rabi* for discovering nuclear magnetic resonance in atomic and molecular beams.
(C) *Otto Hahn* for discovering nuclear fission (1938).

1945 (P) *Wolfgang Pauli* for discovering the exclusion principle (1924).

1946 (P) *Percy W. Bridgman* for studying physics at high pressures.

1947 (P) *Edward V. Appleton* for studying the ionosphere.

1948 (P) *Patrick M. S. Blackett* for studying nuclear physics with cloud-chamber photographs of cosmic-ray interactions.
1949 (P) *Hideki Yukawa* for predicting the existence of mesons (1935).
1950 (P) *Cecil F. Powell* for developing the method of studying cosmic rays with photographic emulsions and discovering new mesons.
1951 (P) *John D. Cockcroft* and *Ernest T. S. Walton* for transmuting nuclei in an accelerator (1932).
(C) *Edwin M. McMillan* for producing neptunium (1940) and *Glenn T. Seaborg* for producing plutonium (1941) and further transuranic elements.
1952 (P) *Felix Bloch* and *Edward Mills Purcell* for discovering nuclear magnetic resonance in liquids and gases (1946).
1953 (P) *Frits Zernike* for inventing the phase-contrast microscope, which uses interference to provide high contrast.
1954 (P) *Max Born* for interpreting the wave function as a probability (1926) and other quantum-mechanical discoveries and *Walther Bothe* for developing the coincidence method to study subatomic particles (1930–1931), producing, in particular, the particle interpreted by Chadwick as the neutron.
1955 (P) *Willis E. Lamb, Jr.*, for discovering the Lamb shift in the hydrogen spectrum (1947) and *Polykarp Kusch* for determining the magnetic moment of the electron (1947).
1956 (P) *John Bardeen, Walter H. Brattain,* and *William Shockley* for inventing the transistor (1956).
1957 (P) *T.-D. Lee* and *C.-N. Yang* for predicting that parity is not conserved in beta decay (1956).
1958 (P) *Pavel A. Čerenkov* for discovering Čerenkov radiation (1935) and *Ilya M. Frank* and *Igor Tamm* for interpreting it (1937).
1959 (P) *Emilio G. Segrè* and *Owen Chamberlain* for discovering the antiproton (1955).
1960 (P) *Donald A. Glaser* for inventing the bubble chamber to study elementary particles (1952).
(C) *Willard Libby* for developing radiocarbon dating (1947).
1961 (P) *Robert Hofstadter* for discovering internal structure in protons and neutrons and *Rudolf L. Mössbauer* for discovering the Mössbauer effect of recoilless gamma-ray emission (1957).
1962 (P) *Lev Davidovich Landau* for studying liquid helium and other condensed matter theoretically.
1963 (P) *Eugene P. Wigner* for applying symmetry principles to elementary-particle theory and *Maria Goeppert Mayer* and *J. Hans D. Jensen* for studying the shell model of nuclei (1947).
1964 (P) *Charles H. Townes, Nikolai G. Basov,* and *Alexandr M. Prokhorov* for developing masers (1951–1952) and lasers.
1965 (P) *Sin-itiro Tomonaga, Julian S. Schwinger,* and *Richard P. Feynman* for developing quantum electrodynamics (1948).
1966 (P) *Alfred Kastler* for his optical methods of studying atomic energy levels
1967 (P) *Hans Albrecht Bethe* for discovering the routes of energy production in stars (1939).
1968 (P) *Luis W. Alvarez* for discovering resonance states of elementary particles.
1969 (P) *Murray Gell-Mann* for classifying elementary particles (1963).
1970 (P) *Hannes Alfvén* for developing magnetohydrodynamic theory and *Louis Eugène Félix Néel* for discovering antiferromagnetism and ferrimagnetism (1930s).

1971 (P) *Dennis Gabor* for developing holography (1947).
(C) *Gerhard Herzberg* for studying the structure of molecules spectroscopically.

1972 (P) *John Bardeen, Leon N. Cooper,* and *John Robert Schrieffer* for explaining superconductivity (1957).

1973 (P) *Leo Esaki* for discovering tunneling in semiconductors, *Ivar Giaever* for discovering tunneling in superconductors, and *Brian D. Josephson* for predicting the Josephson effect, which involves tunneling of paired electrons (1958–1962).

1974 (P) *Anthony Hewish* for discovering pulsars and *Martin Ryle* for developing radio interferometry.

1975 (P) *Aage N. Bohr, Ben R. Mottelson,* and *James Rainwater* for discovering why some nuclei take asymmetric shapes.

1976 (P) *Burton Richter* and *Samuel C. C. Ting* for discovering the J/psi particle, the first charmed particle (1974).

1977 (P) *John H. Van Vleck, Nevill F. Mott,* and *Philip W. Anderson* for studying solids quantum-mechanically.
(C) *Ilya Prigogine* for extending thermodynamics to show how life could arise in the face of the second law.

1978 (P) *Arno A. Penzias* and *Robert W. Wilson* for discovering the cosmic background radiation (1965) and *Pyotr Kapitsa* for his studies of liquid helium.

1979 (P) *Sheldon L. Glashow, Abdus Salam,* and *Steven Weinberg* for developing the theory that unified the weak and electromagnetic forces (1958–1971).

1980 (P) *Val Fitch* and *James W. Cronin* for discovering CP (charge-parity) violation (1964), which possibly explains the cosmological dominance of matter over antimatter.

1981 (P) *Nicolaas Bloembergen* and *Arthur L. Schawlow* for developing laser spectroscopy and *Kai M. Siegbahn* for developing high-resolution electron spectroscopy (1958).

1982 (P) *Kenneth G. Wilson* for developing a method of constructing theories of phase transitions to analyze critical phenomena.

1983 (P) *William A. Fowler* for theoretical studies of astrophysical nucleosynthesis and *Subramanyan Chandrasekhar* for studying physical processes of importance to stellar structure and evolution, including the prediction of white dwarf stars (1930).

1984 (P) *Carlo Rubbia* for discovering the W and Z particles, verifying the electroweak unification, and *Simon van der Meer,* for developing the method of stochastic cooling of the CERN beam that allowed the discovery (1982–1983).

1985 (P) *Klaus von Klitzing* for the quantized Hall effect, relating to conductivity in the presence of a magnetic field (1980).

1986 (P) *Ernst Ruska* for inventing the electron microscope (1931), and *Gerd Binnig* and *Heinrich Rohrer* for inventing the scanning-tunneling electron microscope (1981).

1987 (P) *J. Georg Bednorz* and *Karl Alex Müller* for the discovery of high temperature superconductivity (1986).

1988 (P) *Leon M. Lederman, Melvin Schwartz,* and *Jack Steinberger* for a collaborative experiment that led to the development of a new tool for studying the weak nuclear force, which affects the radioactive decay of atoms.

1989 (P) *Norman Ramsay* (U.S.) for various techniques in atomic physics; and

Hans Dehmelt (U.S.) and *Wolfgang Paul* (Germany) for the development of techniques for trapping single charge particles.

1990 (P) *Jerome Friedman, Henry Kendall* (both U.S.), and *Richard Taylor* (Canada) for experiments important to the development of the quark model.

1991 (P) *Pierre-Gilles de Gennes* for discovering that methods developed for studying order phenomena in simple systems can be generalized to more complex forms of matter, in particular to liquid crystals and polymers.

1992 (P) *George Charpak* for developing detectors that trace the paths of evanescent subatomic particles produced in particle accelerators.

1993 (P) *Russell Hulse* and *Joseph Taylor* for discovering evidence of gravitational waves.

1994 (P) *Bertram N. Brockhouse* and *Clifford G. Shull* for pioneering work in neutron scattering.

APPENDIX F

Spreadsheet Problems

OVERVIEW

Students come to introductory physics courses with a wide variety of computing experience. Many are already accomplished programmers in one or more programming languages (BASIC, Pascal, FORTRAN, and so forth). Others have never even turned on a computer. To further complicate matters, a wide variety of hardware environments exists, although most can be classified as IBM/compatible (MS-DOS) or Macintosh environments. We have designed the end-of-chapter spreadsheet problems and the text ancillary, *Spreadsheet Investigations in Physics,* to be usable by and useful to students in all these diverse situations. Our goal is to enable students to investigate a range of physical phenomena and obtain a feel for the physics. Merely "getting the right answer" by plugging numbers into a formula and comparing the result to the answer in the back of the book is discouraged.

Spreadsheets are particularly valuable in exploratory investigations. Once you have constructed a spreadsheet, you can simply vary the parameters and see instantly how things change. Even more important is the ease with which you can construct accurate graphs of relations between physical variables. When you change a parameter, you can view the effects of the change upon the graphs simply by pressing a key. "What if" questions can be easily addressed and depicted graphically.

HOW TO USE THE TEMPLATES

The computer spreadsheet problems are arranged by level of difficulty. The least difficult problems are coded in black. For most of these problems, spreadsheets are provided on disk, and only the input parameters need to be changed. Problems of moderate difficulty, coded in blue, require additional analysis, and the provided spreadsheets must be modified to solve them. The most challenging problems are coded in magenta. For most of these, you must develop your own spreadsheets. The emphasis should be on understanding what the results mean rather than just getting an answer. For example, one spreadsheet problem explores how the distance of the horizon varies with height above the ground. You can explore why you can see farther distances when you're on top of a tall building than when you are on the ground. Why were lookouts on sailing ships placed in the crow's nest at the top of the mast?

SOFTWARE REQUIREMENTS

The spreadsheet templates are provided on a high-density (1.44 Megabyte) MS-DOS diskette using the Lotus 1-2-3 WK1 format. This format was introduced with versions 2.x of the Lotus 1-2-3 program and can be read by all subsequent versions. It can also be read directly by all the other major spreadsheet programs, including the latest Windows versions of Lotus 1-2-3, Microsoft Excel, Microsoft Works, and Novell/Wordperfect Quattro Pro as well as Microsoft Excel for the Macintosh. The program f(g) Scholar can import WK1 spreadsheets; however, some minor format changes of the templates are needed.

The Lotus WK1 format was chosen so that the templates will be usable in the widest possible variety of computing environments. Even though most spreadsheet programs operate in basically the same way, many of the latest spreadsheet programs have very powerful formatting and graphing capabilities along with many other useful features. The user of these powerful programs can exploit these capabilities to improve the appearance of their spreadsheets.

HARDWARE REQUIREMENTS

You will need a microcomputer that can run one of the spreadsheet programs in one of their many versions. Your computer should be connected to a printer that can print text and graphics. Older versions of the software will run on a 8086/8088 MS-DOS system with just a single floppy disk drive or on a 512K Macintosh with two floppy drives. Newer versions require a more powerful computer to run effectively. For example, to run Excel for Windows, Version 5.0, you must have a hard disk with about 15 megabytes of disk memory available and four to eight megabytes of RAM. Your software manuals will tell you exactly what you need to run the particular version of your spreadsheet program. However, all problems require only a minimal computer system.

There are many different software versions and many different computer configurations. You might have one floppy drive, two floppy drives, a hard disk, a local area network, and so on. The combinations are almost endless. Our best suggestion is to read your software manual and ask your instructor or computer laboratory personnel how to start your spreadsheet program.

SPREADSHEET TUTORIAL

Some students will have the required computer and spreadsheet skills to start working with the templates immediately. Other students, who have not had experience with a spreadsheet program, will need some instruction. We have written an ancillary entitled *Spreadsheet Investigations in Physics* for both these groups of students. The first part of this ancillary contains a spreadsheet tutorial that the novice student can use *independently* to gain the required spreadsheet skills. A two- or three-hour initial session with the tutorial and the computer is all that most students need to get started. Once the students have mastered the basic operations of the spreadsheet program, they should try one or two of the easier problems.

Because very few introductory physics students have studied numerical methods, we have also included a brief introduction to numerical methods in this ancillary. This section covers numerical interpolation, differentiation, integration, and the solution of simple differential equations. The student should not try to master all of this material at one time; only study the sections that are needed to solve the currently assigned problems.

The templates supplied on the distribution diskettes with *Spreadsheet Investigations in Physics* constitute an outline. You must enter the appropriate data and parameters. The parameters must be adjusted to fit the needs of your problem. Feel free to change any parameters, to expand or decrease the number of rows of output, and to change the size of increments (for example, in time or distance).

Most templates have graphs associated with them. You may have to adjust the ranges of variables plotted and the scales for the axes. See the tutorials for how to adjust the appearance of your graphs.

COPY YOUR DISTRIBUTION DISKETTES

Since you will be modifying the spreadsheets on the distribution diskettes, copy the distribution diskettes and place the originals in a safe place.

Answers to Odd-Numbered Problems

Chapter 1

1. 2.80 g/cm^3
3. 184 g
3A. $m = \frac{4}{3}\pi\rho(r_2^3 - r_1^3)$
5. (a) 4 u = 6.64 × 10^{-27} kg
   (b) 56 u = 9.30 × 10^{-26} kg
   (c) 207 u = 3.44 × 10^{-25} kg
7. (a) 72.58 kg   (b) 7.82 × 10^{26} atoms
9. It is.
11. It is.
13. (b) only
15. L^3/T^3, L^{3}T
17. The units of G are m^3/(kg·s^2).
19. 1.39 × 10^3 m^2
21. 8.32 × 10^{-4} m/s
23. 11.4 × 10^3 kg/m^3
25. (a) 6.31 × 10^4 AU   (b) 1.33 × 10^{11} AU
27. (a) 127 y   (b) 15 500 times
29. (a) 1 mi/h = 1.609 km/h   (b) 88.5 km/h
    (c) 16.1 km/h
31. 1.51 × 10^{-4} m
33. 1.00 × 10^{10} lb
35. 5 m
37. 5.95 × 10^{24} kg
39. 2.86 cm
41. ≈10^6
43. 1.79 × 10^{-9} m
45. 3.84 × 10^8 m
47. 34.1 m
49. ≈10^2
51. (a) (346 ± 13) m^2 (b) (66.0 ± 1.3) m
53. 195.8 cm^2 ± 0.7%
55. 3, 4, 3, 2
57. 5.2 m^3 ± 3%
59. It is not.
61. 0.449%
63. (a) 1000 kg   (b) 5 × 10^{-16} kg, 300 g, 0.01 g
65. (a) 10^6 (b) 10^7 (c) 10^3
S1. (a) Jud's horizon is 5.060 m and Spud's horizon is 4.382 m. It makes no difference whether you use d, s, or l. (b) On the Moon, Jud's horizon is 2.638 m and Spud's horizon is 2.285 m.
S3. (a) A plot of log T versus log L for the given data shows an approximate linear relationship. Applying the least-squares method to the data yields 1.017 647 for the slope and 0.890 686 for the intercept of the straight line that best fits the data. The data points and the best-fit line, log T = 1.017 647 log m + 0.890 686 are shown in the figure. Paying attention to significant figures, a reasonable interpretation of this result is that n = 1.1. In terms of T and m, our best-fit equation is $T = Cm^n$, where C = 10$^{1.1}$ = 12.6.

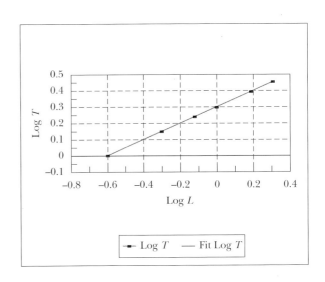

Chapter 2

1. (a) 2.3 m/s   (b) 16.1 m/s   (c) 11.5 m/s
3. (a) 5.0 m/s   (b) 1.2 m/s   (c) −2.5 m/s
   (d) −3.3 m/s   (e) 0
5. (a) 3.75 m/s   (b) 0
7. 50.0 km/h
9. (a) −2.5 m/s   (b) −3.4 m/s   (c) 4.2 s
11. (b) 1.60 m/s
13. (a) 5.0 m/s   (b) −2.5 m/s   (c) 0
    (d) 5.0 m/s
15. −4.00 m/s^2
17. 20.0 m/s; 6.00 m/s^2
19. (a) 20.0 m/s, 5.00 m/s   (b) 262 m
21. (a) 2.00 m   (b) −3.00 m/s   (c) −2.00 m/s^2
23. (a) 1.3 m/s^2 (b) 2.2 m/s^2 at 3.0 s
    (c) At t = 6.2 s and for the interval 10 s < t < 12 s
    (d) −2.0 m/s^2 at 8.2 s

25. -16.0 cm/s^2
27. 160 ft
29. (a) 12.7 m/s   (b) -2.30 m/s
31. (a) 20.0 s   (b) no
33. (a) 8.94 s   (b) 89.4 m/s
35. (a) 0.444 m/s^2 (b) 1.33 m/s   (c) 2.12 s
    (d) 0.943 m/s
37. (a) 45.7 s   (b) 574 m   (c) 12.6 m/s   (d) 765 s
39. (a) $x = 30t - t^2$, $v = 30 - 2t$ (b) 225 m
41. 11.4 s, 212 m
43. (a) -662 ft/s^2 (b) 649 ft
45. (a) 96 ft/s   (b) 3.08×10^3 ft/s^2
    (c) 3.12×10^{-2} s
47. (a) 10.0 m/s   (b) -4.68 m/s
49. (a) 1.53 s   (b) 11.5 m
    (c) -4.60 m/s, -9.80 m/s^2
51. (a) 29.4 m/s   (b) 44.1 m
53. 7.96 s
55. (a) 7.82 m   (b) 0.782 s
57. (a) 7.00 m/s   (b) -5.35 m/s   (c) -9.8 m/s^2
59. $a(t) = a_0 + Jt$, $v(t) = v_0 + a_0 t + \frac{1}{2}Jt^2$,
    $x(t) = x_0 + v_0 t + \frac{1}{2} a_0 t^2 + \frac{1}{6}Jt^3$
61. 0.509 s
63. (a) 3.00 s   (b) -15.3 m/s
    (c) -31.4 m/s, -34.8 m/s
65. (a) -6.26 m/s   (b) 6.02 m/s   (c) 1.25 s
67. 4.63 m
69. (a) 5.45 s   (b) 73.0 m
    (c) $v_{STAN} = 22.6$ m/s, $v_{KATHY} = 26.7$ m/s
71. (a) 12.5 s   (b) -2.29 m/s^2 (c) 13.1 s
73. (a) 5.28 m/s   (b) 4.83×10^{-4} m/s^2
75. $t_{AB} = t_{CD} = 2.00$ min, $t_{BC} = 1.00$ min
77. (a) 5.43 m/s^2, 3.83 m/s^2 (b) 10.9 m/s, 11.5 m/s
    (c) Maggie, 2.62 m
79. 155 s; 129 s
81. $0.577v$
S4. (a) At approximately $t = 37.0$ s after the police car starts, or 42.0 s after the first policeman sees the speeder.   (b) 74.0 m/s   (c) 1370 m

Chapter 3

1. (a) 8.60 m   (b) 4.47 m at 297°; 4.24 m at 135°
3. $(-2.75$ m, -4.76 m)
5. (2.24 m, 26.6°)
7. (2.05 m, 1.43 m)
9. 70.0 m
11. 310 km at 57° south of west
13. (a) 10.0 m   (b) 15.7 m   (c) 0
15. (a) 5.20 m at 60.0°   (b) 3.00 m at 330°
    (c) 3.00 m at 150°   (d) 5.20 m at 300°
17. 421 ft at $-2.63°$
19. 86.6 m, -50.0 m
21. 5.83 m at 59.0° to the right of first direction
23. 47.2 units at 122°
25. (b) $5\mathbf{i} + 4\mathbf{j} = 6.40$ at 38.7°;
    $-\mathbf{i} + 8\mathbf{j} = 8.06$ at 97.2°
27. 7.21 m at 56.3°
29. (a) $2\mathbf{i} - 6\mathbf{j}$ (b) $4\mathbf{i} + 2\mathbf{j}$ (c) 6.32   (d) 4.47
    (e) 288°; 26.6°
31. (a) $(-11.1$ m$)\mathbf{i} + (6.40$ m$)\mathbf{j}$
    (b) $(1.65$ cm$)\mathbf{i} + (2.86$ cm$)\mathbf{j}$
    (c) $(-18.0$ in.$)\mathbf{i} - (12.6$ in.$)\mathbf{j}$
33. 9.48 m at 166°
35. 390 mph at 7.37° north of east
37. $(2.60$ m$)\mathbf{i} + (4.50$ m$)\mathbf{j}$
39. (a) $8\mathbf{i} + 12\mathbf{j} - 4\mathbf{k}$ (b) $2\mathbf{i} + 3\mathbf{j} - \mathbf{k}$
    (c) $-24\mathbf{i} - 36\mathbf{j} + 12\mathbf{k}$
41. (a) 5.92   (b) 19.0
43. (a) $-3\mathbf{i} + 2\mathbf{j}$ (b) 3.61 at 146°   (c) $3\mathbf{i} - 6\mathbf{j}$
47. (a) 49.5, 27.1   (b) 56.4 at 28.7°
49. 240 m at 237°
51. (10.0 m, 20.0 m)

Chapter 4

1. $(8a + 2b)\mathbf{i} + 2c\mathbf{j}$
3. (a) 4.87 km at 209° from east   (b) 23.3 m/s
   (c) 13.5 m/s at 209°
5. (a) $(2\mathbf{i} + 3\mathbf{j})$ m/s^2
   (b) $(3t + t^2)\mathbf{i} + (-2t + 1.5t^2)\mathbf{j}$
7. (a) $(0.8\mathbf{i} - 0.3\mathbf{j})$ m/s^2 (b) at 339°
   (c) $(360\mathbf{i} - 72.8\mathbf{j})$ m; at 345°
9. (a) $\mathbf{v} = (-5\mathbf{i} + 0\mathbf{j})$ m/s, $\mathbf{a} = (0\mathbf{i} - 5\mathbf{j})$ m/s^2
   (b) $\mathbf{r} = -5\mathbf{i} \sin t + 4\mathbf{j} - 5\mathbf{j} \cos t$
   $\mathbf{v} = -5\mathbf{i} \cos t + 5\mathbf{j} \sin t$
   $\mathbf{a} = 5\mathbf{i} \sin t + 5\mathbf{j} \cos t$
   (c) a circle of radius 5 m centered at (0, 4 m)
11. (a) 3.34 m/s at 0°   (b) 309°
13. (a) 2.67 s   (b) 29.9 m/s   (c) $v_x = 29.9$ m/s;
    $v_y = -26.2$ m/s
15. 9.91 m/s
17. (a) clears by 0.89 m   (b) while falling 13.3 m/s
19. (a) 1.69 km/s   (b) 6490 s
21. (a) 277 km   (b) 284 s
23. 53.1°
25. 22.4° or 89.4°
27. (a) 7.90 km/s   (b) 5070 s
29. 377 m/s^2
31. (a) 1.02 km/s   (b) 2.72 mm/s^2
33. 54.4 m/s^2
35. (a) 13.0 m/s^2 (b) 5.70 m/s   (c) 7.50 m/s^2
37. 1.48 m/s^2
39. (a) $-30.8\mathbf{j}$ m/s^2 (b) $70.4\mathbf{j}$ m/s^2
41. (a) 26.9 m/s   (b) 67.3 m   (c) $(2\mathbf{i} - 5\mathbf{j})$ m/s^2
43. 2.02×10^3 s; 21.0% longer
45. 2.50 m/s
47. 153 km/h at 11.3° north of west
49. (a) 57.7 km/h at 210°   (b) 28.9 km/h down
51. (a) 10.1 m/s^2 south at 75.7° below horizontal
    (b) 9.80 m/s^2 down
53. (a) $4.00\mathbf{i}$ m/s   (b) (4.00 m, 6.00 m)

55. (a) 20.0 m/s, 5.00 s   (b) 31.4 m/s at 301°
    (c) 6.53 s   (d) 24.5 m away
57. 3.14 m
59. 20.0 m
61. (a) 0.60 m   (b) 0.40 m
    (c) 1.87 m/s² toward center   (d) 9.80 m/s² down
63. 4.12 m
65. (a) $\sqrt{gR}$ (b) $(\sqrt{2} - 1)R$
67. 10.8 m
69. (a) 6.80 km   (b) directly above explosion
    (c) 66.2°
71. (a) $(-7.05 \text{ cm})\mathbf{j}$ (b) $(7.61 \text{ cm})\mathbf{i} - (6.48 \text{ cm})\mathbf{j}$
    (c) $(10.0 \text{ cm})\mathbf{i} - (7.05 \text{ cm})\mathbf{j}$
73. (a) 22.2° or 67.8°   (b) 235 m   (c) 235 m
75. (a) 407 km/h at 10.6° N of E   (b) 10.8° S of E
77. (a) 5.15 s   (b) 4.85 m/s at 74.5° N of W
    (c) 19.4 m
79. (a) 43.2 m   (b) $(9.66 \text{ m/s})\mathbf{i} - (25.6 \text{ m/s})\mathbf{j}$
81. $(18.8 \text{ m}, -17.4 \text{ m})$
83. (a) 46.5 m/s   (b) $-77.6°$ (c) 6.34 s
85. 2.98 km/s forward and 1.96° inward
S1. There are an infinite number of solutions to the problem; however, there is one practical limitation. We can estimate that the maximum speed that the punter can give the ball is about 20 to 30 m/s. Use a speed in this range.
S3. There are an infinite number of solutions. For example if $v_0 = 22.32$ m/s at 45°, then the ball clears the crossbar in 3.01 s. Or if $v_0 = 24.60$ m/s at 30°, then the ball clears the crossbar in 2.23 s. The time it takes the ball to clear the crossbar is immaterial, since in all likelihood time will run out before the clock is stopped.

Chapter 5

1. (a) 1/3   (b) 0.750 m/s²
3. (a) 3.00 s   (b) 20.1 m   (c) $(18.0\mathbf{i} - 9.0\mathbf{j})$ m
5. 312 N
7. $(6\mathbf{i} + 15\mathbf{j})$ N, 16.2 N
9. (a) 556 N   (b) 56.7 kg
11. (a) 4.47×10^{15} m/s² outward
    (b) 2.09×10^{-10} N inward
13. 1.72 kg, 6.97 m/s²
15. (c) Forces of brick on spring and spring on brick; of spring on table and table on spring; of table on Earth and Earth on table; of brick on Earth and of Earth on brick; of spring on Earth and of Earth on spring.
17. (a) 5.10 kN   (b) 2.65×10^3 kg
19. (a) 2.0 m/s²   (b) 170 N   (c) 2.93 m/s²
21. (a) $(2.50 \text{ N})\mathbf{i} + (5.00 \text{ N})\mathbf{j}$ (b) 5.59 N
23. 1.24 ft/s²
25. 640 N for $0 \leq t \leq 1.0$ s, 627 N at $t = 1.3$ s, 589 N at $t = 2.0$ s
27. $(14.7 \text{ N})\mathbf{i} - (2.5 \text{ N})\mathbf{j}$, 14.9 N
29. 613 N
31. (b) $T_1 = 513.5$ N, $T_2 = 557.4$ N, $T_3 = 325.0$ N
33. (a) 33.9 N   (b) 39.0 N
35. (a) $g \tan \theta$ (b) 4.16 m/s²
37. (a) $F_x > 19.6$ N   (b) $F_x \leq -78.4$ N
39. survival chance better with force on larger mass
41. (a) 4.90 m/s²   (b) 3.13 m/s   (c) 1.35 m
    (d) 1.14 s   (e) no
43. (a) 706 N   (b) 814 N   (c) 706 N   (d) 648 N
45. 21.8 m/s
47. 36.9 N
49. 81.0 m/s
51. $\mu = 0.0773$
53. (a) 0.161   (b) 1.01 m/s²
55. (b) 27.2 N, 1.286 m/s²
57. (a) 1.78 m/s²   (b) 0.368   (c) 9.37 N
    (d) 2.67 m/s
59. (a) $a_1 = 2.31$ m/s² down, $a_2 = 2.31$ m/s² left, $a_3 = 2.31$ m/s² up,
    (b) $T_{\text{Left}} = 30.0$ N, $T_{\text{Right}} = 24.2$ N
61. 0.293
63. 182.5 m
65. (a) $Mg/2, Mg/2, Mg/2, 3Mg/2, Mg$ (b) $Mg/2$
67. (a) $\mu_s = \dfrac{h}{\sqrt{L^2 - h^2}}$ (b) $a = \dfrac{2L}{t^2}$
    (c) $\sin \theta = h/L$ (d) $\mu_k = \dfrac{h - \dfrac{2L^2}{gt^2}}{\sqrt{L^2 - h^2}}$
69. (a) 0.232 m/s²   (b) 9.68 N
71. (a) F forward   (b) $3F/2Mg$
    (c) $F/(M + m_1 + m_2 + m_3)$
    (d) $m_1 F/(M + m_1 + m_2 + m_3)$,
    $(m_1 + m_2)F/(M + m_1 + m_2 + m_3)$,
    $(m_1 + m_2 + m_3)F/(M + m_1 + m_2 + m_3)$
    (e) $m_2 F/(M + m_1 + m_2 + m_3)$
73. (a) friction between the two blocks   (b) 34.7 N
    (c) 0.306
75. (a) 0.408 m/s²   (b) 83.3 N
77. (a) 4.00 m   (b) 3.72 m/s
81. (a) $(-45\mathbf{i} + 15\mathbf{j})$ m/s   (b) at 162°
    (c) $(-225\mathbf{i} + 75\mathbf{j})$ m   (d) $(-227 \text{ m}, 79 \text{ m})$
83. $(M + m_1 + m_2) m_2 g/m_1$
85. $T_1 = 74.5$ N, $T_2 = 34.7$ N, $\mu_k = 0.572$
87. (a) 2.20 m/s²   (b) 27.37 N
89. (a) 30.7°   (b) 0.843 N
91. 6.00 cm

Chapter 6

1. (a) 8.0 m/s   (b) 3.02 N
3. (a) 5.40 kN down   (b) 1.60 kN down
    (c) seatbelt tension plus gravity
5. $0 < v < 8.08$ m/s
7. (a) 1.52 m/s²   (b) 1.66 km/s   (c) 6820 s
9. $v \leq 14.3$ m/s
11. (a) 9.80 N   (b) 9.80 N   (c) 6.26 m/s
13. (a) static friction   (b) 0.085
15. 3.13 m/s

17. (a) 4.81 m/s (b) 700 N up
19. (a) $(-0.163 \text{ m/s}^2)\mathbf{i} + (0.233 \text{ m/s}^2)\mathbf{j}$
    (b) 6.53 m/s (c) $(-0.181 \text{ m/s}^2)\mathbf{i} + (0.181 \text{ m/s}^2)\mathbf{j}$
21. no
23. (a) 0.822 m/s^2 (b) 37.0 N (e) 0.0839
25. (a) 17.0° (b) 5.12 N
27. (a) 491 N (b) 50.1 kg (c) 2.00 m/s
31. (a) 3.47×10^{-2} s^{-1} (b) 2.50 m/s (c) $a = -cv$
33. (a) 1.47 N·s/m (b) 2.04×10^{-3} s
    (c) 2.94×10^{-2} N
35. (a) 8.32×10^{-8} N (b) 9.13×10^{22} m/s^2
    (c) 6.61×10^{15} rev/s
37. (a) $\mathbf{T} = (68.6 \text{ N})\mathbf{i} + (784 \text{ N})\mathbf{j}$ (b) 0.857 m/s^2
39. (a) The true weight is greater than the apparent weight.
    (b) $w = w' = 735$ N at the poles; $w' = 732.4$ N at the equator
41. 780 N
43. 12.8 N
45. (a) 967 lb (b) 647 lb up
47. (a) 6.67 kN (b) 20.3 m/s
47A. (a) $mg - \dfrac{mv^2}{R}$ (b) $\sqrt{gR}$
49. (b) 2.54 s, 23.6 rev/min
51. (a) 1.58 m/s^2 (b) 455 N (c) 329 N
    (d) 397 N upward and 9.15° inward
53. (a) 0.0132 m/s (b) 1.03 m/s (c) 6.87 m/s
S1. Spreadsheet 6.1 typifies the solution of second-order differential equations. In this spreadsheet, we want to solve $a = dv/dt$, where $v = dx/dt$. To find v as a function of t, we assume that a is constant over the time interval dt. Hence, we can use Euler's method to integrate $dv/dt = a$. Therefore, $v_{i+1} = v_i + a_i \Delta t$. In the Lotus 1-2-3 Spreadsheet 6.1, cells F35 and down implement this equation. To find x as a function of t, we know that v varies over the time interval, so we use Euler's modified method to integrate $dx/dt = v$. Or, $x_{i+1} = x_i + 1/2(v_{i+1} + v_i) \Delta t$. Cells F35 and down implement this equation.
S2. All objects fall faster when there is no air resistance than when there is air resistance. As the mass of the objects increases, the difference between their positions at the same time with and without air resistance becomes smaller.
S4. $F_{\text{min}} \cong 252$ N at 31°.
S5. Try large positive speeds, that is $v_0 > 100$ m/s. You may need to increase Δt.
S7. If the terminal speeds are to be equal, then the relationship between b_1 (for $n = 1$) and b_2 (for $n = 2$) is $b_1 = b_2 mg$.

Chapter 7

1. 30.6 m
3. (a) 31.9 J (b) 0 (c) 0 (d) 31.9 J
5. 5.88 kJ
7. (a) 900 J (b) -900 J (c) 0.383
9. (a) 137 W (b) -137 W
11. (a) 79.4 N (b) 1.49 kJ (c) -1.49 kJ
11A. (a) $\mu_k mg/(\cos\theta + \mu_k \sin\theta)$
     (b) $\mu_k mg \, d \cos\theta/(\cos\theta + \mu_k \sin\theta)$
     (c) $-\mu_k mg \, d \cos\theta/(\cos\theta + \mu_k \sin\theta)$
13. 14.0
13A. $r_1 r_2 \cos(\theta_1 - \theta_2)$
17. (a) 16.0 J (b) 36.9°
19. $\mathbf{s} = 2\mathbf{i} + 23.5\mathbf{j}$ or $22\mathbf{i} + 8.5\mathbf{j}$
21. (a) 11.3° (b) 156° (c) 82.3°
23. (a) 7.50 J (b) 15.0 J (c) 7.50 J (d) 30.0 J
25. (a) 575 N/m (b) 46.0 J
25A. (a) F/d (b) $\frac{1}{2}Fd$
27. 0.299 m/s
29. (b) mgR
31. 12.0 J
31A. $3W$
33. (a) 4.10×10^{-18} J
    (b) 1.14×10^{-17} N (c) 1.25×10^{13} m/s^2
    (d) 240 ns
35. (a) 2.00 m/s (b) 200 N
35A. (a) $v = \sqrt{2W/m}$ (b) $\bar{F} = W/d$
37. (a) 650 J (b) -588 J (c) 62.0 J
    (d) 1.76 m/s
39. 6.34 kN
41. (a) $\sqrt{\dfrac{2\,mgh}{m + M/4}}$ (b) $\sqrt{\dfrac{2\,mgh - \mu_k Mgh}{m + M/4}}$
43. 1.25 m/s
45. 2.04 m
47. (a) -168 J (b) 184 J (c) 500 J (d) 148 J
    (e) 5.65 m/s
49. (a) 4.51 m (b) no, since $f > mg \sin\theta$
51. (a) 63.9 J (b) -35.4 J (c) -9.51 J
    (d) 19.0 J
53. 875 W
55. (a) 7.92 hp (b) 14.9 hp
57. (a) 7.5×10^4 J (b) 2.50×10^4 W (33.5 hp)
    (c) 3.33×10^4 W (44.7 hp)
57A. (a) $\dfrac{1}{2}mv^2$ (b) $\dfrac{mv^2}{2t}$ (c) $\dfrac{mv^2 t_1}{t^2}$
59. 220 ft·lb/s
61. 685
63. 80.0 hp
65. (a) 1.35×10^{-2} gal (b) 73.8 (c) 8.08 kW
67. 5.90 km/liter
69. (a) 5.37×10^{-11} J (b) 1.33×10^{-9} J
71. 3.70 m/s
75. (a) $(2 + 24t^2 + 72t^4)$ J (b) $a = 12t$ m/s^2; $F = 48t$ N
    (c) $(48t + 288t^3)$ W (d) 1.25×10^3 J
77. 878 kN
79. (a) 4.12 m (b) 3.35 m
81. (a) -5.60 J (b) 0.152 (c) 2.29 rev
83. (a) $W = mgh$ (b) $\Delta K = mgh$
    (c) $K_f = mgh + mv_0^2/2$

85. 1.94 kJ
87. (b) 8.49×10^5 kg/s (c) 7.34×10^7 m³
    (d) 1.53 km
89. 1.68 m/s
S1. (a) A plot of F as the independent variable and L as the dependent variable shows that the last four points tend to vary the most from a straight line. However, because the first point has the largest percentage deviation from a straight line, we probably are not justified in throwing any of the data points out. The slope of the best straight line obtained from the least-squares fit is 8.654 545 5 mm/N or $k = 0.116$ N/mm = 116 N/m. (b) The least-squares fit we used was of the form $L = aF + b$; solving for F gives $F = L/a - b/a = 0.116L - 0.561$. Hence, for $L = 105$ mm, $F = 12.7$ mm.

Chapter 8

1. (a) -196 J   (b) -196 J   (c) -196 J
   The force is conservative.
3. (b) conservative 62.7 J, nonconservative 20.7 J
   (c) $\mu = 0.330$
5. (a) 125 J   (b) 50.0 J   (c) 66.7 J
   (d) nonconservative, since W is path-dependent
7. (a) 40.0 J   (b) -40.0 J   (c) 62.5 J
9. (a) -9.00 J; No. A constant force is conservative.
   (b) 3.39 m/s   (c) 9.00 J
11. $v_A = \sqrt{3gR}$; 0.098 N downward
13. (a) $v = (gh + v_0^2)^{1/2}$ (b) $v_x = 0.6v_0$;
    $v_y = -(0.64v_0^2 + gh)^{1/2}$
15. (a) 18.5 km, 51.0 km   (b) 10.0 MJ
17. (a) 4.43 m/s   (b) 5.00 m
17A. (a) $\sqrt{2(m_1 - m_2)gh/(m_1 + m_2)}$
    (b) $2m_1h/(m_1 + m_2)$
19. (a) -160 J   (b) 73.5 J   (c) 28.8 N
    (d) 0.679
21. 489 kJ
23. (a) -4.1 MJ   (b) 9.97 m/s   (c) 50.8 m
    (d) It is better to keep the engine with the train.
25. 3.74 m/s
29. 914 N/m
31. (a) -28.0 J   (b) 0.446 m
33. 10.2 m
33A. $(kd^2/2mg) - d$
35. 0.327
37. (a) $F_r = A/r^2$
39. $F = (7 - 9x^2y)i - 3x^3j$
41. (b) $x = 0$ (c) $v = \sqrt{0.80 \text{ J}/m}$
43. (a) $v_B = 5.94$ m/s; $v_C = 7.67$ m/s   (b) 147 J
45. (a) 1.50×10^{-10} J   (b) 1.07×10^{-9} J
    (c) 9.15×10^{-10} J
47. (a) 0.225 J   (b) 0.363 J
    (c) No. The normal force varies with position, and so the frictional force also varies.
49. $\dfrac{h}{5}(4\sin^2\theta + 1)$
51. (a) 349 J, 676 J, 741 J   (b) 174 N, 338 N, 370 N
    (c) yes
53. (a) $\Delta U = -\dfrac{ax^2}{2} - \dfrac{bx^3}{3}$ (b) $\Delta U = \dfrac{A}{\alpha}(1 - e^{\alpha x})$
55. 0.115
59. 1.24 m/s
61. (b) 7.42 m/s
63. (a) 3.19 m   (b) 2.93 m/s
65. (a) 0.400 m   (b) 4.10 m/s
    (c) It reaches the top.
67. 3.92 kJ
67A. $m_1gd(m_2 - \mu_k m_1 \cos\theta - m_1 \sin\theta)/(m_1 + m_2)$
69. (a) 0.378 m   (b) 2.30 m/s   (c) 1.08 m
S1. If the particle has an initial energy less than 2471 J, it will be trapped in the potential well; for example, if $E_T = 1000$ J, it will be confined approximately to -3.15 m $\leq x \leq 4.58$ m.
S2. $x = 0$ is a point of stable equilibrium; $x = 8.33$ m is a point of unstable equilibrium.

Chapter 9

1. $(9.00i - 12.0j)$ kg·m/s, 15.0 kg·m/s
3. 1.60 kN
5. (a) 12.0 kg·m/s   (b) 6.00 m/s   (c) 4.00 m/s
7. (a) 13.5 kg·m/s   (b) 9.00×10^3 N
   (c) 18.0×10^3 N
9. 87.5 N
11. (a) 7.50 kg·m/s   (b) 375 N
13. (a) 13.5 kg·m/s toward the pitcher
    (b) 6.75×10^3 N toward the pitcher
15. 260 N toward the left in the diagram
15A. $\dfrac{-2mv\sin\theta}{t}i$
17. (a) 0.125 m/s   (b) 8 times
19. 120 m
21. (a) 1.15 m/s   (b) -0.346 m/s
23. 4.01×10^{-20} m/s
25. 301 m/s
27. (a) 20.9 m/s east   (b) 8.74 kJ into thermal energy
29. (a) 0.284, or 28.4%   (b) $K_n = 1.15 \times 10^{-13}$ J,
    $K_c = 4.54 \times 10^{-14}$ J
31. 3.75 kN; no
33. (a) 0.571 m/s   (b) 28.6 J   (c) 0.003 97
35. 91.2 m/s
37. 0.556 m
39. 497 m/s
41. $v = (3.00i - 1.20j)$ m/s
43. (a) $v_x = -9.33 \times 10^6$ m/s, $v_y = -8.33 \times 10^6$ m/s
    (b) 4.39×10^{-13} J
45. 3.01 m/s, 3.99 m/s
47. 2.50 m/s at $-60.0°$

51. -0.429 m
53. CM = 454 km, well within the Sun
55. (2.54 m, 4.75 m)
57. 70/6 cm, 80/6 cm
59. (a) $(1.40\mathbf{i} + 2.40\mathbf{j})$ m/s (b) $(7.00\mathbf{i} + 12.0\mathbf{j})$ kg·m/s
61. (a) 2.10 m/s, 0.900 m/s (b) 6.30×10^{-3} kg·m/s, -6.30×10^{-3} kg·m/s
61A. (a) $m_2 v_1/(m_1 + m_2)$ and $m_1 v_1/(m_1 + m_2)$ (b) $m_1 m_2 v_1/(m_1 + m_2)$ toward the CM
63. 200 kN
65. 2150 kg
67. 0.595 m³/s
67A. $F/\rho v$
69. 291 N
71. (a) 1.80 m/s to the left (b) 257 N to the left (c) larger than part b
73. (a) 4160 N (b) 4.17 m/s
75. 32.0 kN; 7.13 MW
77. (a) 6.81 m/s (b) 1.00 m
79. 240 s
81. $(3Mgx/L)\mathbf{j}$
83. (a) As the child walks to the right, the boat moves to the left, but the center of mass remains fixed. (b) 5.55 m from the pier (c) Since the turtle is 7 m from the pier, the boy will not be able to reach the turtle, even with a 1 m reach.
85. (a) 100 m/s (b) 374 J
85A. (a) $v_0 - d\sqrt{\dfrac{kM}{m^2}}$ (b) $v_0 d \sqrt{kM} - \dfrac{1}{2} kd^2 \left(1 + \dfrac{M}{m}\right)$
87. $2v_0$ and 0
89. (a) 3.8 kg·m/s² (b) 3.8 N (c) 3.8 N (d) 2.8 J (e) 1.4 J (f) Friction between sand and belt converts half of the input work into thermal energy.
S1. (a) The maximum acceleration is 100 m/s². It occurs at the end of the burn time of 80 s, when the rocket has its smallest mass. The maximum speed that the rocket reaches is 3.22 km/s. (b) The speed reaches half its maximum after 55.3 s; if the acceleration were constant, it would reach half its maximum speed at 40 s (half the burn time), but the acceleration is always increasing during the burn time.
S3. (b) The disadvantages are that the ship has to withstand twice the acceleration and that it has only traveled half as far when the fuel burns out. The advantage is that it takes half as long to reach its final speed; hence, it travels farther in 100 s.

Chapter 10

1. (a) 4.00 rad/s² (b) 18.0 rad
3. (a) 1.99×10^{-7} rad/s (b) 2.66×10^{-6} rad/s
5. (a) 5.24 s (b) 27.4 rad
7. 13.7 rad/s²
9. (a) 0.18 rad/s (b) 8.10 m/s² toward the center of the track

9A. (a) v/R (b) v^2/R toward center
11. (a) 8.00 rad/s (b) 8.00 m/s, $a_r = -64.0$ m/s², $a_t = 4.00$ m/s² (c) 9.00 rad
13. (a) 126 rad/s (b) 3.77 m/s (c) 1.26 km/s² (d) 20.1 m
15. 29.4 m/s², 9.80 m/s²
15A. $-2g\dfrac{(h-R)}{R}\mathbf{i} - g\mathbf{j}$
17. (a) 143 kg·m² (b) 2.57×10^3 J
19. (a) 92.0 kg·m², 184 J (b) 6.00 m/s, 4.00 m/s, 8.00 m/s, 184 J
23. (a) $(3/2)MR^2$ (b) $(7/5)MR^2$
25. -3.55 N·m
27. 2.79%
29. (a) 0.309 m/s² (b) $T_1 = 7.67$ N, $T_2 = 9.22$ N
29A. (a) $\dfrac{(m_2 \sin\theta - \mu_k)(m_1 + m_2 \cos\theta)}{m_1 + M/2 + m_2} g$ (b) $T_1 = \mu_k m g + m_1 a$, $T_2 = T_1 + \tfrac{1}{2} Ma$
31. (a) 56.3 J (b) 8.38 rad/s (c) 2.35 m/s (d) 0.4% greater
33. (a) $2(Rg/3)^{1/2}$ (b) $4(Rg/3)^{1/2}$ (c) $(Rg)^{1/2}$
35. (a) 11.4 N, 7.57 m/s², 9.53 m/s down (b) 9.53 m/s
37. (a) 1.03 s (b) 10.3 rev
39. 168 N·m (clockwise)
41. (a) 4.00 J (b) 1.60 s (c) yes
43. (a) $\omega = \sqrt{3g/L}$ (b) $\alpha = 3g/2L$ (c) $-\tfrac{3}{2}g\mathbf{i} - \tfrac{3}{4}g\mathbf{j}$ (d) $-\tfrac{3}{2}Mg\mathbf{i} + \tfrac{1}{4}Mg\mathbf{j}$
45. (a) $0.707R$ (b) $0.289L$ (c) $0.632R$
47. 149 rad/s
49. (a) 2.60×10^{29} J (b) -1.65×10^{17} J/day
51. (a) 118 N, 156 N (b) 1.19 kg·m²
51A. (a) $T_1 = m_1(a + g\sin\theta)$, $T_2 = m_2(g - a)$ (b) $m_2 R^2 g/a - m_1 R^2 - m_2 R^2 - m_1 R^2 (g/a)\sin\theta$
53. (a) -0.176 rad/s² (b) 1.29 rev (c) 9.26 rev
S1. The answer is not unique because the torque $\tau = FR$.
S3. Replace X^2 with $(X - H)^2$ in line 110.

Chapter 11

1. (a) 500 J (b) 250 J (c) 750 J
3. (a) $a_{CM} = \tfrac{2}{3}g\sin\theta$ (disk), $a_{CM} = \tfrac{1}{2}g\sin\theta$ (hoop) (b) $\tfrac{1}{3}\tan\theta$
5. 44.8 J
5A. $0.7Mv^2$
7. (a) $-17\mathbf{k}$ (b) 70.5°
9. (a) negative z direction (b) positive z direction
11. 45.0°
13. $|\mathbf{F}_3| = |\mathbf{F}_1| + |\mathbf{F}_2|$, no
15. (17.5 kg·m²/s)$\mathbf{k}$
15A. $\tfrac{1}{2}(m_1 + m_2)vd$
17. $(60$ kg·m²/s$)\mathbf{k}$
19. $mvR\left[\cos\left(\dfrac{vt}{R}\right) + 1\right]\mathbf{k}$
21. $-mg\ell t \cos\theta \, \mathbf{k}$

23. (a) zero (b) $[-mv_0^3 \sin^2 \theta \cos \theta / 2g]\mathbf{k}$
    (c) $[-2mv_0^3 \sin^2 \theta \cos \theta / g]\mathbf{k}$ (d) The downward force of gravity exerts a torque in the $-z$ direction.
25. (a) 0.433 kg·m²/s (b) 1.73 kg·m²/s
27. (a) $\omega = \omega_0 I_1/(I_1 + I_2)$ (b) $I_1/(I_1 + I_2)$
29. (a) 0.360 rad/s in the counterclockwise direction
    (b) 99.9 J
31. (a) 6.05 rad/s (b) 113 J
33. (a) $mv\ell$ down (b) $M/(M + m)$
35. (a) 2.19×10^6 m/s (b) 2.18×10^{-18} J
    (c) 4.13×10^{16} rad/s
37. 0.91 km/s
41. (a) The net torque around this axis is zero.
    (b) Since $\tau = 0$, L = const. But initially, L = 0, hence it remains zero throughout the motion. Consequently, the monkey and bananas move upward with the same speed at any instant. The distance between the monkey and bananas stays constant. Hence, the monkey will not reach the bananas.
47. 30.3 rev/s
49. (a) $v_0 r_0/r$ (b) $T = (mv_0^2 r_0^2) r^{-3}$
    (c) $\frac{1}{2}mv_0^2\left(\frac{r_0^2}{r^2} - 1\right)$ (d) 4.50 m/s, 10.1 N, 0.450 J
51. (a) $F_y = \frac{W}{L}\left(d - \frac{ah}{g}\right)$ (b) 0.306 m
    (c) $(-306\mathbf{i} + 553\mathbf{j})$ N
53. (a) 3.75×10^3 kg·m²/s (b) 1.875 kJ
    (c) 3.75×10^3 kg·m²/s (d) 10.0 m/s
    (e) 7.50 kJ (f) 5.625 kJ
53A. (a) Mvd (b) Mv^2 (c) Mvd (d) $2v$
    (e) $4Mv^2$ (f) $3Mv^2$
55. $\frac{1}{3}L$
57. (c) $(8Fd/3M)^{1/2}$
61. $v_0 = [ag(16/3)(\sqrt{2} - 1)]^{1/2}$
63. F_1 clockwise torque, F_2 zero torque, F_3 and F_4 counterclockwise torque
65. (a) 0.800 m/s², 0.400 m/s²
    (b) 0.600 N (top), 0.200 N (bottom)

Chapter 12

1. 10.0 N up; 6.00 N·m counterclockwise
3. $[(w_1 + w)d + w_1\ell/2]/W_2$
5. $F_W = 480$ N, $F_v = 1200$ N
5A. $\left(\frac{w_1}{2} + \frac{w_2 x}{L}\right)\left(\frac{d}{\sqrt{L^2 - d^2}}\right)$; $w_1 + w_2$
9. -1.50 m, -1.50 m
11. (a) 859 N (b) 1040 N, left and upward at 36.9°
13. 0.789
15. $F_f = 4410$ N, $F_r = 2940$ N
17. $2R/5$
19. $\frac{1}{3}$ by the left string, $\frac{2}{3}$ by the right string
21. $x = \frac{3}{4}L$
23. 4.90 mm
25. (a) 73.6 kN (b) 2.50 mm
27. 29.2 μm
27A. $\dfrac{8m_1 m_2 gL}{\pi d^2 Y(m_1 + m_2)}$
29. (a) 3.14×10^4 N (b) 62.8 kN
31. 1800 atm
33. $N_A = 5.98 \times 10^5$ N, $N_B = 4.80 \times 10^5$ N
35. (b) 69.8 N (c) 0.877ℓ
37. (a) 160 N right (b) 13.2 N right (c) 292 N up
    (d) 192 N
39. (a) $T = w(\ell + d)/\sin \theta(2\ell + d)$ and
    (b) $R_x = w(\ell + d)\cot \theta/(2\ell + d)$; $R_y = w\ell/(2\ell + d)$
41. (a) $F_x = 268$ N, $F_y = 1300$ N (b) 0.324
41A. (a) $\dfrac{m_1 g}{2 \tan \theta} + \dfrac{m_2 gx}{L \tan \theta}$; $(m_1 + m_2)g$
    (b) $\dfrac{m_1/2 + m_2 d/L}{(m_1 + m_2) \tan \theta}$
43. 5.08 kN, $R_x = 4.77$ kN, $R_y = 8.26$ kN
45. $T = 2.71$ kN, $R_x = 2.65$ kN, $R_y = -12.0$ N
47. (a) 20.1 cm to the left of the front edge, $\mu = 0.571$
    (b) 0.501 m
47A. (a) $\dfrac{F \cos \theta}{mg - F \sin \theta}$; $\dfrac{mgw/2 - Fh \cos \theta}{(mg - F \sin \theta)}$ (b) $\dfrac{mgw}{2F \cos \theta}$
49. (a) $W = \dfrac{w}{2}\left(\dfrac{2\mu_s \sin \theta - \cos \theta}{\cos \theta - \mu_s \sin \theta}\right)$
    (b) $R = (w + W)\sqrt{1 + \mu_s^2}$, $F = \sqrt{W^2 + \mu_s^2(w + W)^2}$
51. (a) 133 N (b) $N_A = 429$ N, $N_B = 257$ N
    (c) $R_x = 133$ N, $R_y = -257$ N
53. 66.7 N
55. $F = \frac{3}{8}w$
57. (a) 1.67 N, 3.33 N, 1.67 N (b) 2.36 N
59. (a) 4500 N (b) 4.50×10^6 N/m²
    (c) This is more than sufficient to break the board.
61. $y_{cg} = 16.7$ cm

Chapter 13

1. (a) 1.50 Hz, 0.667 s (b) 4.00 m (c) π rad
   (d) 2.83 m
3. (b) 1.81 s (c) no
5. (a) 13.9 cm/s, 16.0 cm/s² (b) 16.0 cm/s, 0.262 s
   (c) 32.0 cm/s², 1.05 s
7. (b) 6π cm/s, 0.333 s (c) $18\pi^2$ cm/s², 0.500 s
   (d) 12.0 cm
9. (a) 0.542 kg (b) 1.81 s (c) 1.20 m/s²
11. 40.9 N/m
13. (a) 2.40 s (b) 0.417 Hz (c) 2.62 rad/s
15. (a) 0.400 m/s, 1.60 m/s²
    (b) ± 0.320 m/s, -0.960 m/s² (c) 0.232 s
17. (a) 0.750 m (b) $x = -(0.75$ m$)\sin(2.0t)$
17A. (a) v/ω (b) $(v/\omega)\cos(\omega t + \pi/2)$
19. 2.23 m/s
21. (a) quadrupled (b) doubled
    (c) doubled (d) no change
23. ± 2.60 cm

25. (a) 1.55 m   (b) 6.06 s
27. (a) 3.65 s   (b) 6.41 s   (c) 4.24 s
27A. (a) $2\pi\sqrt{L/(g+a)}$ (b) $2\pi\sqrt{L/(g-a)}$
  (c) $2\pi L^{1/2}(g^2+a^2)^{-1/4}$
29. (a) 0.817 m/s   (b) 2.57 rad/s²   (c) 0.634 N
31. increases by 1.78×10^{-3} s
33. 0.944 kg·m²
35. (a) 5.00×10^{-7} kg·m²
  (b) 3.16×10^{-4} N·m/rad
39. 1.00×10^{-3} s^{-1}
41. (a) 1.42 Hz   (b) 0.407 Hz
43. 318 N

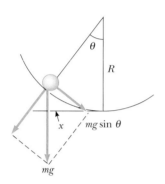

45. 1.57 s
47. (a) $E = \tfrac{1}{2}mv^2 + mgL(1-\cos\theta)$ (b) $U = \tfrac{1}{2}m\omega^2 s^2$
49. Referring to the sketch, we have $F = -mg\sin\theta$ and $\tan\theta = x/R$. For small displacements, $\tan\theta \approx \sin\theta$ and $F = -(mg/R)x = -kx$ and $\omega = (k/m)^{1/2} = (g/R)^{1/2}$.
51. (a) $2Mg$, $T_P = Mg\left(1 + \dfrac{y}{L}\right)$
  (b) $\dfrac{4\pi}{3}\sqrt{\dfrac{2L}{g}} = 2.68$ s
55. $f = \dfrac{1}{2\pi}\sqrt{\dfrac{MgL + kh^2}{ML^2}}$
57. 0.0662 m
57A. $\dfrac{\mu_s g}{4\pi^2 f^2}$
59. (a) 3.00 s   (b) 14.3 J   (c) 0.441 rad
61. 9.19×10^{13} Hz
63. (a) 15.8 rad/s   (b) 5.23 cm   (c) 1.31 cm, π
67. (a) $k = 1.74$ N/m ± 6%   (b) $k = 1.82$ N/m ± 3%; the values of k agree   (c) $m_s = 8$ g ± 12%, in agreement with 7.4 g
S2. Since $E_{tot} = K(t) + U(t) = \tfrac{1}{2}kA^2[\cos^2(\omega t + \delta) + \sin^2(\omega t + \delta)] = \tfrac{1}{2}kA^2$, the total energy is independent of ω and δ. It depends only on k and A.
S3. The periods increase as ϕ_0 increases. These periods are always greater than those calculated using $T_0 = 2\pi\sqrt{L/g}$. For example, if $\phi_0 = 45° = \pi/4$ rad, and $L = 1$ m, then $T_0 = 2.00\ 709$ s, but the actual period is 2.08 s. (*Note:* Due to numerical errors in the integration of the differential equations, the amplitudes may tend to increase. If this occurs, try smaller time steps.)

Chapter 14

1. (a) 3.46×10^8 m   (b) 3.34×10^{-3} m/s²
3. (1.00 m − 61.3 nm)
5. $\dfrac{GM}{\ell^2}\left(\dfrac{2\sqrt{2}+1}{2}\right)$ toward the opposite corner
7. 35.0 N toward the Moon
9. 3.73 m/s²
11. 12.6×10^{31} kg
11A. $\dfrac{2v^3 T}{\pi G}$
13. (a) 4.39×10^{20} N   (b) 1.99×10^{20} N
  (c) 3.55×10^{22} N
15. 1.90×10^{27} kg
17. Y has completed 1.30 revolutions
19. 8.98×10^7 m
21. $2GMr/(r^2 + a^2)^{3/2}$, to the left
23. 3.84×10^4 km from the Moon's center
25. 2.82×10^9 J
27. (a) 1.84×10^9 kg/m³   (b) 3.27×10^6 m/s²
  (c) -2.08×10^{13} J
29. 1.66×10^4 m/s
31. 11.8 km/s
35. 1.58×10^{10} J
35A. $\dfrac{mGM_E(R_E + 2h)}{2R_E(R_E + h)} - \tfrac{1}{2}mv^2$
37. (a) 42.1 km/s relative to the Sun   (b) 2.20×10^{11} m
39. (a) 1.31×10^{14} N/kg   (b) 2.62×10^{12} N/kg
39A. (a) $\dfrac{GM}{(d+\ell/2)^2}$ (b) $\dfrac{GM\ell(2d+\ell)}{d^2(d+\ell)^2}$
41. (a) 7.41×10^{-10} N   (b) 1.04×10^{-8} N
  (c) 5.21×10^{-9} N
43. 2.26×10^{-7}
45. 0.0572 rad/s = 32.7 rev/h
45A. $\omega = \sqrt{2g/d}$
47. 7.41×10^{-10} N
49. (a) $k = \dfrac{GmM_E}{R_E^3}$, $A = \dfrac{L}{2}$
  (b) $\dfrac{L}{2}\left(\dfrac{GM_E}{R_E}\right)^{1/2}$, at the middle of the tunnel
  (c) 1.55×10^3 m/s
51. $\dfrac{2\sqrt{2}Gm}{a^2}(-\mathbf{i})$
53. 2.99×10^3 rev/min
55. (a) $v_1 = m_2\left[\dfrac{2G}{d(m_1 + m_2)}\right]^{1/2}$
  $v_2 = m_1\left[\dfrac{2G}{d(m_1 + m_2)}\right]^{1/2}$
  $v_{rel} = \left[\dfrac{2G(m_1 + m_2)}{d}\right]^{1/2}$
  (b) $K_1 = 1.07 \times 10^{32}$ J, $K_2 = 2.67 \times 10^{31}$ J

57. (a) 7.34×10^{22} kg (b) 1.63×10^3 m/s
    (c) 1.32×10^{10} J
59. 119 km
61. (a) 5300 s (b) 7.79 km/s (c) 6.45×10^9 J
61A. (a) $2\pi(GM_E)^{1/2}(R_E + h)^{3/2}$
    (b) $\sqrt{GM_E/(R_E + h)}$ (c) $\dfrac{mGM_E(R_E + 2h)}{2R_E(R_E + h)}$
63. (a) $M/\pi R^4$ (b) $-GmM/r^2$
    (c) $-(GmM/R^4)r^2$
65. 1.48×10^{22} kg
67. (b) 981 kg/m³
69. (b) $GMm/2R$
S5. The kinetic energy is
$$K = \tfrac{1}{2}mv^2 = \tfrac{1}{2}m(v_x^2 + v_y^2)$$
The potential energy is
$$U = \dfrac{GM_E m}{R_E} - \dfrac{GM_E m}{r}$$
where G is the universal gravitation constant, M_E is the mass of the Earth, R_E is the radius of the Earth, m is the mass of the satellite, and r is the distance from the center of the Earth to the satellite. The total energy, $K + U$, is constant. There may be a small change in the numerical value of the total energy, because of numerical errors during the integration of the differential equations.
S6. According to Kepler's laws of planetary motion the angular momentum, L, is a constant. There may be a small change in the numerical value of L, because of numerical errors during the integration of the differential equations.

Chapter 15

1. 0.111 kg
3. 3.99×10^{17} kg/m³. Matter is mostly free space.
5. 6.24×10^6 Pa
7. 4.77×10^{17} kg/m³
9. $P_{ATM} + \rho\sqrt{g^2 + a^2}\,(L/\sqrt{2})\cos\left(45° - \arctan\dfrac{a}{g}\right)$
11. 1.62 m
13. 77.4 cm²
15. 0.722 mm
17. 9.12 MPa
19. 12.6 cm
21. 10.5 m; no, a little alcohol and water evaporate
23. 1.08 cm
25. 1470 N down
27. (a) 7.00 cm (b) 2.80 kg
29. $\rho_{oil} = 1250$ kg/m³; $\rho_{sphere} = 500$ kg/m³
31. 0.611 kg
33. 1.07 m²
33A. $\dfrac{m}{h(\rho_w - \rho_s)}$

35. 1430 m³
37. (a) 17.7 m/s (b) 1.73 mm
37A. (a) $\sqrt{2gh}$ (b) $(R/\pi)^{1/2}(8/gh)^{1/4}$
39. 0.0128 m³/s
41. 31.6 m/s
43. (a) 28.0 m/s (b) 392 kPa
45. Av/a
47. 103 m/s
49. $2[h(h_0 - h)]^{1/2}$
51. 5 kW at 20°C
53. (b) $\tfrac{1}{2}\rho Av^3$ if the mill could make the air stop; the same
55. 0.258 N
57. 1.91 m
59. 455 kPa
63. 8 cm/s
65. 2.01×10^6 N
69. 90.04% Zn
71. 5.02 GW
73. 4.43 m/s
75. (a) 1.25 cm (b) 13.8 m/s
77. (a) 18.3 mm (b) 14.3 mm (c) 8.56 mm

Chapter 16

1. $y = \dfrac{6}{(x - 4.5t)^2 + 3}$
3. (a) longitudinal (b) 666 s
5.

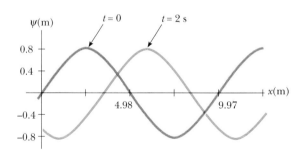

7. (a) 5.00 rad (b) 0.858 cm
9. (a) Wave 1 travels in the $+x$ direction; wave 2 travels in the $-x$ direction. (b) 0.75 s (c) $x = 1.00$ m
11. 520 m/s
13. 13.5 N
15. 586 m/s
17. 0.329 s
19. (b) 0.125 s
21. 0.319 m
23. (b) $k = 18.0$ rad/m, $T = 0.0833$ s, $\omega = 75.4$ rad/s, $v = 4.20$ m/s
    (c) $y(x, t) = (0.20$ m$)\sin(18.0x + 75.4t - 0.151)$
25. $y_1 + y_2 = 11.2 \sin(2.0x - 10t + 63.4°)$
27. (a) $y = (0.0800$ m$)\sin(7.85x + 6\pi t)$
    (b) $y = (0.0800$ m$)\sin(7.85x + 6\pi t - 0.785)$

29. (a) 2.15 cm (b) 0.379 rad (c) 541 cm/s
    (d) $y(x, t) = (2.15 \text{ cm})\cos(80\pi t + 8\pi x/3 + 0.379)$
31. $A = 2.0$ cm; $k = 2.11$ rad/m; $\lambda = 2.98$ m;
    $\omega = 3.62$ rad/s; $f = 0.576$ Hz; $v = 1.72$ m/s
33. (a) $y = (0.200 \text{ mm}) \sin[16.0x - 3140t]$
    (b) $T = 158$ N
35. 1.07 kW
35A. $2\pi^2 M v^3 A^2 / L\lambda^2$
37. (a) remains constant (b) remains constant
    (c) remains constant (d) p is quadrupled
39. (a) 62.5 m/s (b) 7.85 m (c) 7.96 Hz
    (d) 21.1 W
43. (b) $f(x + vt) = \frac{1}{2}(x + vt)^2$ and $g(x - vt) = \frac{1}{2}(x - vt)^2$
45. (a) 3.33 m/s in the positive x direction
    (b) -5.48 cm (c) 0.667 m, 5.00 Hz
    (d) 11.0 m/s
47. (a) 179 m/s (b) 17.7 kW
49. (a) 39.2 N (b) 89.2 cm (c) 83.6 m/s
49A. (a) $2Mg$ (b) $L_0 + 2Mg/k$
    (c) $(2MgL_0/m + 4M^2g^2/km)^{1/2}$
51. (a) 5.00 m/s $+ x$ (b) 5.00 m/s $- x$
    (c) 7.5 m/s $- x$ (d) 24.0 m/s $+ x$
55. 3.86×10^{-4} (at 5.93°C)
57. $(2L/g)^{1/2}$, $L/4$
59. (a) $\dfrac{\mu \omega^3}{2k} A_0^2 e^{-2bx}$ (b) $\dfrac{\mu \omega^3}{2k} A_0^2$ (c) e^{-2bx}

Chapter 17

1. 5.56 km
3. 1430 m/s
5. 332 m/s
7. 1.988 km
9. (a) 27.2 s (b) 25.7 s It is shorter by 5.30%.
11. 1.55×10^{-10} m
13. 5.81 m
15. (a) 2.00 μm, 0.400 m, 54.6 m/s (b) -0.433 μm
    (c) 1.72 mm/s
17. $(0.200 \text{ Pa}) \sin(62.8x - 2.16 \times 10^4 t)$
19. 66.0 dB
23. 100.0 m and 10.0 m
25. (a) 65.0 dB (b) 67.8 dB (c) 69.6 dB
27. 241 W
29. (a) 30.0 m (b) 9.49×10^5 m
31. 50.0 km
33. 46.4°
35. 56.4°
37. 26.4 m/s
39. (a) 338 Hz (b) 483 Hz
41. 2.82×10^8 m/s
43. (a) 56.3 s
    (b) $(56.6 \text{ km})\mathbf{i} + (20.0 \text{ km})\mathbf{j}$ from the observer
45. 130 m/s, 1.73 km
47. 80.0°
49. 1204 Hz
51. (a) 0.948° (b) 4.40°
53. 1.34×10^4 N
55. 95.5 s
55A. $\dfrac{0.3E}{4\pi d^2 I_0} 10^{-\beta/10}$
57. (a) 55.8 m/s (b) 2500 Hz
59. (a) 6.45 (b) 0
61. 1.60
63. The measured wavelengths depend on your monitor. However, the wavelengths should be proportional to those given here. For $u/v = 0.0$, $\lambda_0 = 0.75$ cm. For $u/v = 0.5$, $\lambda_{\text{front}} = 0.37$ cm and $\lambda_{\text{back}} = 1.13$ cm. Therefore,

$$\left| \frac{\Delta \lambda_{\text{front}}}{\lambda_0} \right| = \frac{0.37 \text{ cm} - 0.75 \text{ cm}}{0.75 \text{ cm}} = 0.5$$

and

$$\left| \frac{\Delta \lambda_{\text{back}}}{\lambda_0} \right| = \frac{1.13 \text{ cm} - 0.75 \text{ cm}}{0.75 \text{ cm}} = 0.5$$

We see that $\dfrac{\Delta \lambda}{\lambda_0} = \dfrac{u}{v}$.

Chapter 18

1. (a) 9.24 m (b) 600 Hz
3. 0.500 s
5. (a) The path difference to A is $\lambda/2$.
   (b) $9x^2 - 16y^2 = 144$
7. at 0.0891 m, 0.303 m, 0.518 m, 0.732 m, 0.947 m, and 1.16 m
9. (a) 4.24 cm (b) 6.00 cm (c) 6.00 cm
   (d) $x = 0.5$ cm, 1.5 cm, 2.5 cm
11. 25.1 m, 60.0 Hz
13. (a) 2.00 cm (b) 2.40 cm
15. (a) $0, \pm 2\pi/3k, \pm 4\pi/3k, \ldots$
    (b) $(\pm \pi - 2\omega t)/k$, $(\pm 3\pi - 2\omega t)/k$, $(\pm 5\pi - 2\omega t)/k$
17. (a) 60.0 cm (b) 30.0 Hz
19. $L/4$, $L/2$
21. 0.786 Hz, 1.57 Hz, 2.36 Hz, 3.14 Hz
23. 2.80 g
23A. $m = \dfrac{Mg}{4Lf_1^2 \tan \theta}$
25. (a) $T = 163$ N (b) 660 Hz
27. 19.976 kHz
29. 338 N
31. 20.5 kg
31A. $m_w = \rho_w A \left(L - \dfrac{v_s}{4f} \right)$
33. 50.4 cm, 84.0 cm
35. 35.8 cm, 71.7 cm
37. 349 m/s
39. (a) 531 Hz (b) 4.25 cm
41. 328 m/s
43. (a) 350 m/s (b) 114 cm

45. $n(206\text{ Hz})$ and $n(84.5\text{ Hz})$, where $n = 1, 2, 3, \ldots$
47. 1.88 kHz
49. (a) 1.59 kHz   (b) odd   (c) 1.11 kHz
51. It is.
53. (a) 1.99 Hz   (b) 3.38 m/s
55. (a) 3.33 rad   (b) 283 Hz
57. 85.7 Hz
59. $f = 50.0\text{ Hz}; L = 1.70\text{ m}$
61. (a) 78.9 N   (b) 211 Hz
63. $\lambda = 4.86\text{ m}$
65. 3.87 m/s *away* from the station *or* 3.78 m/s *toward* the station
67. (a) 59.9 Hz   (b) 20 cm
69. (a) 0.5   (b) $\dfrac{n^2 F}{(n+1)^2}$ (c) $\dfrac{F'}{F} = \dfrac{9}{16}$

S4. The following steps can be used to modify the spreadsheet.
  1. MOVE the entire Y_T column — one column to the right.
  2. COPY the Y_2 column — one column to the right. EDIT the heading label to Y_3.
  3. COPY the second wave input data block to the right and EDIT the labels to reflect the third wave.
  4. EDIT the Y_3 column to reflect the third wave data block addresses.
  5. EDIT the Y_T column to include Y_1, Y_2, and Y_3 in the sum.
S6. Plot $Y(t)$ versus ωt. You may want to start with three terms in the series and then add additional terms and watch how $Y(t)$ changes.

Chapter 19

1. (a) $-273.5°C$ (b) 1.27 atm, 1.74 atm
3. (a) 30.4 mm Hg   (b) 18.0 K
5. 139 K, $-134°C$ (b) 6.56 kPa
7. (a) $-320°F$ (b) 77.3 K
9. $-297°F$
11. (a) 810°F   (b) 450 K
13. (a) 90.0°C   (b) 90.0 K
15. $-40.0°C$
17. 3.27 cm
19. 1.32 cm
21. 0.548 gal
23. 217 kN
25. 1.20 cm
27. (a) 437°C   (b) 2100°C. No; they melt first.
29. (a) 99.4 cm³   (b) 0.943 cm
31. 1.08 L
33. (a) 0.176 mm   (b) 8.78 μm   (c) 93.0 mm³
35. 7.95 m³
37. 4.39 kg
39. 472 K
41. 2.28 kg
41A. $\dfrac{MP_0 V}{RT_1}\left(1 - \dfrac{T_1}{T_2}\right)$

43. 1.61 MPa = 16.1 atm
45. 594 kPa
47. 400 kPa, 448 kPa
49. 1.13
51. (a) $A = 1.85 \times 10^{-3}\ (1/°\text{C})$, $R_0 = 50.0\ \Omega$
   (b) 421 °C
53. $\alpha \Delta T \ll 1$
55. 3.55 cm
55A. $\Delta h = \dfrac{V}{A}\beta \Delta T$
57. (a) 94.97 cm   (b) 95.03 cm
59. (b) 1.33 kg/m³
63. 2.74 m
63A. $y = \frac{1}{2}\sqrt{(L+\Delta L)^2 - L^2}$, where $\Delta L = \alpha L\ \Delta T$
65. 30.4°C
67. (a) 18.0 m   (b) 277 kPa
69. (a) 7.06 mm   (b) 297 K
71. (a) 0.0374 mol   (b) 0.732 atm
73. (a) 6.17×10^{-3} kg/m   (b) 632 N
   (c) 580 N; 192 Hz
75. (a) $(\alpha_2 - \alpha_1)\ L\ \Delta T/(r_2 - r_1)$
   (c) It bends the other way.
S1. From the figure below, note that

$$h = \dfrac{L}{20}(1 - \cos\theta)$$

and

$$\dfrac{L}{L_0} = (1 + \alpha\ \Delta T) = \dfrac{\theta}{\sin\theta} = 1.000\ 055$$

Solving this transcendental equation, $\theta = 0.018\ 165$ rad and $h = 4.54$ m.

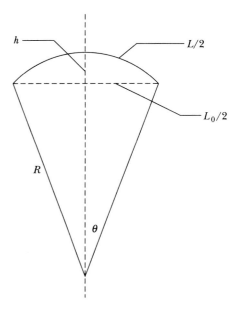

Chapter 20

1. 0.105°C
3. 10.117°C
5. 0.234 kJ/kg·°C
7. 85.3°C
9. 29.6°C
11. 34.7°C
13. 19.5 kJ
15. 50.7 ks
17. 720 cal
19. 47.1°C
21. (a) 0°C   (b) 115 g
23. 2.99 mg
25. (a) 25.760°C   (b) no
27. 0.258 g if the bullet is at 0°C
29. 810 J, 506 J, 203 J
31. 466 J
31A. $nR\,\Delta T$
33. 1.18 MJ
35. (a) 7.65 L   (b) 305 K
37. (a) -567 J   (b) 167 J
37A. (a) $W = P\Delta V$ (b) $\Delta U = Q - W$
39. (a) 12.0 kJ   (b) -12.0 kJ
41. (a) 23.1 kJ   (b) 23.1 kJ
43. 0.0962 g
45. (a) 7.50 kJ   (b) 900 K
47. 2.47 L
47A. $V_i = V_f e^{-W/nRT_0}$, where $W = M_w c_w (T_h - T_c)$
49. (a) 48.6 mJ   (b) 16.2 kJ   (c) 16.2 kJ
51. (a) 10.0 L·atm   (b) 0.0100 atm   (c) 7.00 kJ
53. 1.34 kW
55. 51.0°C
57. 138 MJ
59. 0.0222 W/m·°C
59A. $k = \dfrac{Pt}{A\,\Delta T}$
61. (a) 61.1 kWh   (b) $3.67
63. (a) 0.89 ft²·°F·h/BTU
    (b) 1.85 ft²·°F·h/BTU   (c) 2.08
65. 781 kg
67. 1.87 kJ
69. (a) 16.8 L   (b) 0.351 L/s
71. The bullet will partially melt.
73. 44.5°C
75. 5.26°C
79. 9.32 kW
81. 5.31 h
83. 800 J/kg·°C
S1. By varying h until $T = 45$ °C at $t = 180$ s, one finds $h = 0.010\,855$ cal/s·cm²·°C. Examine the associated graph for each choice of h.
S2. (b) $\Delta U = 4669$ J

Chapter 21

1. 2.43×10^5 m²/s²
3. 2.30 kmol
5. 3.32 mol
7. 8.76×10^{-21} J
9. (a) 40.1 K   (b) 6.01 km/s
11. 477 m/s
13. 109 kPa
15. 75.0 J
17. (a) 316 K   (b) 200 J
17A. (a) $dT = \dfrac{Q}{C_V} - \dfrac{Q}{C_p}$ (b) $\dfrac{QR}{C_p}$
19. (a) 3.46 kJ   (b) 2.45 kJ   (c) 1.01 kJ
21. 24.0 kJ; 68.7 kJ
23. (a) 118 kJ   (b) 6.03×10^3 kg
25. (a) 1.39 atm   (b) 366 K, 254 K
27. (a) 2.06×10^{-4} m³   (b) 560 K   (c) 8.72 K
29. (a) 0.118, so the compression ratio $V_i/V_f = 8.50$
    (b) 2.35
31. 227 K
33. 91.2 J
33A. $9\,P_0 V_0$
35. 1.51×10^{-20} J
37. 2.33×10^{-21} J
39. (a) 7.27×10^{-20} J/molecule   (b) 2.21 km/s
    (c) 3510 K
41. (a) 5.63×10^{18} m, 1.00×10^9 y
    (b) 5.63×10^{12} m, 1.00×10^3 y
43. (a) 1.028   (b) ^{35}Cl
45. (a) 2.01×10^4 K   (b) 902 K
47. (a) 3.21×10^{12} molecules
    (b) 778 km   (c) 6.42×10^{-4} s^{-1}
49. 193
51. 4.65×10^{-8} cm
55. (a) $3.65\,v$ (b) $3.99\,v$ (c) $3.00\,v$
    (d) $106\,mv^2/V$ (e) $7.98\,mv^2$
57. zero, 2.70×10^{20}
59. 0.625
63. (c) 2.0×10^3
65. (b) 5.47 km
67. (a) 10^{82} (b) 10^{12} m   (c) 10^{58} moles
69. (a) 0.510 m/s   (b) 20 ms
S1. (a) At $T = 100$ K, $f(1000)\,dv = 0.0896$ and $f(3000)\,dv = 0.000\,05$.
    (b) At $T = 273$ K, $f(1000)\,dv = 0.0428$ and $f(3000)\,dv = 0.011\,09$.
    (c) At $T = 1000$ K, $f(1000)\,dv = 0.0084$ and $f(3000)\,dv = 0.028\,77$.

Chapter 22

1. (a) 6.94%   (b) 335 J
3. (a) 0.333   (b) 0.667
5. (a) 1.00 kJ   (b) 0
7. (a) 0.375   (b) 600 J   (c) 2.00 kW
9. 0.330
11. (a) 5.12%   (b) 5.27 TJ
13. (a) 0.672   (b) 58.8 kW

15. 0.478°C
17. (a) 0.268   (b) 0.423
19. 453 K
21. 146 kW, 70.8 kW
23. 192 J
25. (a) 24.0 J   (b) 144 J
27. 72.2 J
27A. $W = \dfrac{\Delta T}{T_c} Q$
29. $\Delta S = -90.2$ J/K
31. 195 J/K
33. 717 J/K
35. 3.27 J/K
37. 5.76 J/K, no temperature change
39. 18.4 J/K
41. (a) 154.5 J/K   (b) 54.2 kJ
43. (a) 5.00 kW   (b) 763 W
45. (a) 4.10 kJ   (b) 14.2 kJ
    (c) 10.1 kJ   (d) 28.8%
47. (a) $2nRT_0 \ln 2$ (b) 0.273
49. (a) $10.5nRT_0$ (b) $8.5nRT_0$ (c) 0.190
    (d) 0.833
51. $nC_p \ln 3$
53. 5.97×10^4 kg/s
53A. $\dfrac{dm}{dt} = \dfrac{P}{c_w \Delta T}\left(\dfrac{T_h}{T_h - T_c}\right)$
57. (a) 96.9 W   (b) 1.19°C/h
59. $e = \dfrac{2(T_2 - T_1) \ln (V_2/V_1)}{3(T_2 - T_1) + 2T_2 \ln (V_2/V_1)}$
61. (b) 12.0 kJ   (c) -12.0 kJ
63. -8.26×10^5 J

Chapter 23

1. 5.14×10^3 N
3. (a) 1.59 nN away   (b) 1.24×10^{36} times larger
   (c) 8.61×10^{-11} C/kg
5. (a) 57.1 TC   (b) 3.48×10^6 N/C
7. 0.873 N at 330°
9. 40.9 N at 263°
11. 2.51×10^{-10}
13. 3.60 MN down on the top and up on the bottom of the cloud
15. (a) $(-5.58 \times 10^{-11}$ N/C) j
    (b) $(1.02 \times 10^{-7}$ N/C) j
17. (a) 18.8 nC   (b) 1.17×10^{11} electrons
19. (a) $(1.29 \times 10^4$ N/C) j   (b) $(-3.87 \times 10^{-2}$ N) j
21. (a) $k_e qx (R^2 + x^2)^{-3/2}$
23. (a) at the center   (b) $\left(\dfrac{\sqrt{3}\, k_g}{a^2}\right)$ j
25. (a) $5.91 k_e\, q/a^2$ at 58.8°   (b) $5.91 k_e\, q^2/a^2$ at 58.8°
27. $-\pi^2 k_e\, qi/6a^2$
29. $-\left(\dfrac{k_e \lambda_0}{x_0}\right)$ i

31. (a) $(6.65 \times 10^6$ N/C) i   (b) $(2.42 \times 10^7$ N/C) i
    (c) $(6.40 \times 10^6$ N/C) i   (d) $(6.65 \times 10^5$ N/C) i
33. (a) $\dfrac{k_e Qi}{h} [(d^2 + R^2)^{-1/2} - ((d+h)^2 + R^2)^{-1/2}]$
    (b) $\dfrac{2k_e Qi}{R^2 h} [h + (d^2 + R^2)^{1/2} - ((d+h)^2 + R^2)^{1/2}]$
35. (a) 9.35×10^7 N/C away from the center; 1.039×10^8 N/C is 10.0% larger
    (b) 515.1 kN/C away from the center; 519.3 kN/C is 0.8% larger
37. 7.20×10^7 N/C away from the center; 1.00×10^8 N/C axially away
39. $(-21.6$ MN/C) i
41.

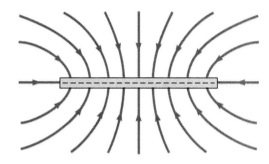

43. (a) $\dfrac{q_1}{q_2} = -1/3$ (b) q_1 is negative and q_2 is positive
45. (a) 6.14×10^{10} m/s²   (b) 19.5 µs
    (c) 11.7 m   (d) 1.20 fJ
47. 1.00×10^3 N/C in the direction of the beam
47A. K/ed parallel to v
49. (a) $(-5.75 \times 10^{13}$ m/s²) i   (b) 2.84×10^6 m/s
    (c) 49.4 ns
51. (a) 111 ns   (b) 5.67 mm
    (c) (450 km/s) i + (102 km/s) j
53. (a) 36.9°, 53.1°   (b) 167 ns, 221 ns
55. (a) 10.9 nC   (b) 5.43×10^{-3} N
55A. (a) $\dfrac{mg}{E_x \cot \theta + E_y}$ (b) $\dfrac{mg\, E_x}{E_x \cos \theta + E_y \sin \theta}$
57. (a) $\theta_1 = \theta_2$
59. 204 nC
63. (a) $-\left(\dfrac{4k_e q}{3a^2}\right)$ j   (b) (0, 2.00 m)
65. (a) 307 ms   (b) Yes; neglecting gravity causes a 2.28% error.
65A. (a) $2\pi \left(\dfrac{L}{g + qE/m}\right)^{1/2}$ (b) Yes. If qE is small compared to mg, then gravity determines its period.
67. 5.27×10^{17} m/s²; 0.854 mm
71. (a) $F = \dfrac{k_e q^2}{s^2} (1.90)(i + j + k)$
    (b) $F = 3.29 \dfrac{k_e q^2}{s^2}$ in a direction away from the vertex diagonally opposite to it

S4. (a) at $d < 0.005$ m ($Y/L = 0.1$); (b) at $d > 0.124$ m ($Y/L = 2.48$); (c) same as parts (a) and (b); (d) at $d < 0.010$ m ($Y/L = 0.1$) and at $d > 0.248$ m ($Y/L = 2.48$); (e) In terms of Y/L, the answers do not change.

S5. The dipole approximation of the electric field $E = 2k_e p/x^3$ is within 20 percent of the actual value when $x > 6.2$ cm. It is within 5 percent when $x > 12.6$ cm.

Chapter 24

1. 0
3. (a) aA   (b) bA   (c) 0
5. 4.14×10^6 N/C
5A. $4\Phi/\pi d^2$
7. EhR
9. 1.87×10^3 Nm²/C
11. (a) $q/2\epsilon_0$ (b) $q/2\epsilon_0$
    (c) Plane and square look the same to the charge.
13. (a) 1.36×10^6 Nm²/C   (b) 6.78×10^5 Nm²/C
    (c) No, the same field lines go through spheres of all sizes.
13A. (a) Q/ϵ_0 (b) $Q/2\epsilon_0$
    (c) No. As the radius increases, the area increases but the field decreases to compensate.
15. -6.89×10^6 Nm²/C. The number of lines entering exceeds the number leaving by 2.91 times or more.
17. 0 if $R < d$; $2\lambda(R^2 - d^2)^{1/2}/\epsilon_0$ if $R > d$
19. 28.3 N·m²/C
21. (a) 761 nC   (b) It may have any distribution. Any and all point and smeared-out charges, positive and negative, must add algebraically to +761 nC.
    (c) Total charge is -761 nC.
23. (a) $\dfrac{Q}{2\epsilon_0}$ (out of the volume enclosed)
    (b) $-\dfrac{Q}{2\epsilon_0}$ (into it)
25. (a) 0   (b) 7.20×10^6 N/C away from the center
27. (a) 0.713 µC   (b) 5.7 µC
29. (a) 0   (b) $(3.66 \times 10^5$ N/C$)\hat{r}$
    (c) $(1.46 \times 10^6$ N/C$)\hat{r}$ (d) $(6.50 \times 10^5$ N/C$)\hat{r}$
31. $E = (a/2\epsilon_0)\hat{r}$
33. (a) 5.14×10^4 N/C outward   (b) 646 Nm²/C
35. $\mathbf{E} = (\rho r/2\epsilon_0)\hat{r}$
37. 5.08×10^5 N/C up.
39. (a) 0   (b) 5.40×10^3 N/C
    (c) 540 N/C, both radially outward
41. (a) 80.0 nC/m² on each face
    (b) $(9.04 \times 10^3$ N/C$)\mathbf{k}$
    (c) $(-9.04 \times 10^3$ N/C$)\mathbf{k}$
43. (a) -99.5 µC/m²   (b) $+382$ µC/m²
43A. (a) $-q/4\pi a^2$ (b) $(Q + q)/4\pi b^2$
45. (a) 0   (b) $(8.00 \times 10^7$ N/C$)\hat{r}$ (c) 0
    (d) $(7.35 \times 10^6$ N/C$)\hat{r}$
47. (a) $-\lambda, +3\lambda$ (b) $\left(\dfrac{3\lambda}{2\pi\epsilon_0 r}\right)\hat{r}$
49. (b) $\dfrac{Q}{2\epsilon_0}$ (c) $\dfrac{Q}{\epsilon_0}$

51. (a) $\mathbf{E} = \left(\dfrac{\rho r}{3\epsilon_0}\right)\hat{r}$ for $r < a$; $\mathbf{E} = \left(\dfrac{k_e Q}{r^2}\right)\hat{r}$ for $a < r < b$;
    $\mathbf{E} = 0$ for $b < r < c$; $\mathbf{E} = \left(\dfrac{k_e Q}{r^2}\right)\hat{r}$ for $r > c$
    (b) $\sigma_1 = -\dfrac{Q}{4\pi b^2}$ inner; $\sigma_2 = +\dfrac{Q}{4\pi c^2}$ outer
53. (c) $f = \dfrac{1}{2\pi}\sqrt{\dfrac{k_e e^2}{mR^3}}$ (d) 102 pm
57. $\mathbf{g} = \left(\dfrac{GM_e r}{R_e^3}\right)\hat{r}$
59. (a) σ/ϵ_0 to the left   (b) zero
    (c) σ/ϵ_e to the right
63. $\mathbf{E} = \dfrac{\rho a}{3\epsilon_0}\mathbf{j}$

Chapter 25

1. 1.80 kV
3. (a) 152 km/s   (b) 6.50×10^6 m/s
5. (a) 2.7 keV   (b) 509 km/s
7. 6.41×10^{-19} C
9. 2.10×10^6 m/s
11. 1.35 MJ
13. 432 V; 432 eV
15. -38.9 V; the origin
17. (a) 20.0 keV, 83.8 Mm/s   (b) 7.64×10^{-23} kg·m/s
19. (a) 0.400 m/s   (b) The same
19A. (a) $\sqrt{2E\lambda d/\mu}$ (b) The same
21. 2.00 m
23. 119 nC, 2.67 m
25. 4.00 nC at $(-1, 0)$ and -5.01 nC at $(0, 2)$
27. -11.0 MV
29. (a) -386 nJ. Positive binding energy would have to be put in to separate them.   (b) 103 V
31. (a) -27.3 eV   (b) -6.81 eV   (c) 0
35. 1.74 m/s
35A. $((1 + \sqrt{2}/4) k_e q^2/Lm)^{1/2}$
37. (a) 1.00 kV $- (1.41$ kV/m$)x + (1.44$ kV$) \ln\left(\dfrac{3\text{ m}}{3\text{ m} - x}\right)$
    (b) $+633$ nJ
37A. (a) $V_0 - \dfrac{\sigma x}{2\epsilon_0} + \dfrac{\lambda}{2\pi\epsilon_0} \ln\left(\dfrac{d}{d-x}\right)$
    (b) $qV_0 - \dfrac{q\sigma d}{8\epsilon_0} + \dfrac{q\lambda}{2\pi\epsilon_0} \ln(4/3)$
39. $E_x = -5 + 6xy$ $E_y = 3x^2 - 2z^2$ $E_z = -4yz$
    7.07 N/C
41. (a) 10.0 V, -11.0 V, -32.0 V   (b) $(7.00$ N/C$)\mathbf{i}$
43. $E_x = \dfrac{3 E_0 a^3 xz}{(x^2 + y^2 + z^2)^{5/2}}$ $E_y = \dfrac{3E_0 a^3 yz}{(x^2 + y^2 + z^2)^{5/2}}$
    $E_z = E_0 + \dfrac{E_0 a^3 (2z^2 - x^2 - y^2)}{(x^2 + y^2 + z^2)^{5/2}}$ outside the sphere, and
    $E = 0$ inside.
45. $-(0.553) k_e Q/R$

47. (a) C/m²   (b) $k_e \alpha \left[L - d \ln \dfrac{d+L}{d} \right]$

49. $(\sigma/2\epsilon_0)(\sqrt{x^2 + b^2} - \sqrt{x^2 + a^2})$

51. 1.56×10^{12} electrons removed

53. (a) 0, 1.67 MV
(b) 5.85×10^6 N/C away, 1.17 MV
(c) 1.19×10^7 N/C away, 1.67 MV

55. (a) 4.50×10^7 N/C outward, 30.0 MN/C outward
(b) 1.80 MV

57. (a) 450 kV   (b) 7.50 μC

59. 5.00 μC

61. (a) 6.00 m   (b) $-2.00\ \mu$C

63. (a) 13.3 μC   (b) 20.0 cm

65. (a) 180 kV   (b) 127 kV

67. (a) $2 k_e Q d^2 (3x^2 - d^2)(x^3 - xd^2)^{-2} \mathbf{i}$
(b) (609 MN/C)$\mathbf{i}$

69.

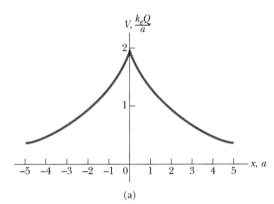

(a)

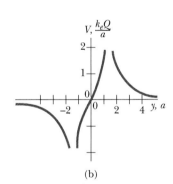

(b)

71. (a) $1.26\ \sigma_0$ (b) $1.26\ E_0$ (c) $1.59\ V_0$

73. $k_e \dfrac{Q^2}{2R}$

75. $V_2 - V_1 = -\dfrac{\lambda}{2\pi\epsilon_0} \ln\left(\dfrac{r_2}{r_1}\right)$

79. $E_y = k_e \dfrac{Q}{\ell y}\left[1 - \dfrac{y^2}{\ell^2 + y^2 + \ell\sqrt{\ell^2 + y^2}}\right]$

81. (a) $E_r = \dfrac{2k_e p \cos\theta}{r^3}$; $E_\theta = \dfrac{k_e p \sin\theta}{r^3}$; yes; no

(b) $\mathbf{E} = \dfrac{3k_e p xy \mathbf{i}}{(x^2 + y^2)^{5/2}} + \dfrac{k_e p (2y^2 - x^2)\mathbf{j}}{(x^2 + y^2)^{5/2}}$

83. $\dfrac{3}{5}\left(\dfrac{k_e Q^2}{R}\right)$

85. $x = 0.57735$ m and $x = 0.24859$ m

Chapter 26

1. 13.3 kV
3. (a) 1.00 μF   (b) 100 V
5. 684 μC
7. (a) 1.33 μC/m²   (b) 13.3 pF
9. (a) 423 fF   (b) 0.652
9A. (a) $4\pi\epsilon_0 (R_1 + R_2)$ (b) R_1/R_2
11. 1.52 mm
13. 4.42 μm
15. (a) 1.11×10^4 N/C toward the negative plate
(b) 98.3 nC/m²   (c) 3.74 pF   (d) 74.8 pC
17. (69.1 pF) $(\pi - \theta)$
17A. $\epsilon_0 NR^2(\pi - \theta)/d$
19. 2.13×10^{16} m³
19A. $7C^3/384\pi^2\epsilon_0^3$
21. (a) 2.68 nF   (b) 3.02 kV
23. (a) 15.6 pF   (b) 256 kV
25. 66.7 nC
27. 18.0 μF
29. (a) 4.00 μF   (b) 8.00 V, 4.00 V, 12.0 V, 24.0 μC, 24.0 μC, 24.0 μC
31. (a) 5.96 μF   (b) 89.2 μC, 63.1 μC, 26.4 μC, 26.4 μC
33. 120 μC; 80.0 μC and 40.0 μC
35. $60R/37k_e$
37. 10
39. 83.6 μC
41. 12.9 μF
43. $\dfrac{\epsilon_0 A}{(s - d)}$
45. 90.0 mJ
47. (a) 55.9 μC   (b) 4.65 V
49. 800 pJ, 5.79 mJ/m³
55. (a) 369 pC   (b) 118 pF, 3.12 V   (c) -45.5 nJ
55A. (a) $\epsilon_0 AV/d$ (b) $C_f = \kappa\epsilon_0 A/d$, $V_f = V/\kappa$
(c) $-\epsilon_0 AV^2(\kappa - 1)/2d\kappa$
57. 16.7 pF, 1.62 kV
59. 1.04 m
61. $\kappa = 8.00$
63. 22.5 V
63A. 1.5 V
65. 416 pF
67. 1.00 μF and 3.00 μF
69. (b) $4\pi\epsilon_0/(a^{-1} + b^{-1})$
71. 2.33
71A. $1 + q/q_0$
73. (a) 243 μJ   (b) 2.30 mJ
75. 4.29 μF
75A. $CV/(V_0 - V)$
77. 480 V
79. 0.188 m²
81. 3.00 μF

83. (b) $Q/Q_0 = \kappa$
85. 2/3
89. 19.0 kV
91. 3.00 μF

Chapter 27

1. (b) 1.05 mA
3. 400 nA
3A. $q\omega/2\pi$
5. 0.265 C
7. (a) 221 nm (b) No
9. (a) 1.50×10^5 A (b) 5.40×10^8 C
11. 13.3 μA/m^2
13. 1.32×10^{11} A/m^2
15. 1.59 Ω
17. 1.98 A
19. 1.33 Ω
19A. $R/9$
21. $1.56R$
23. (a) 1.82 m (b) 280 μm
25. 6.43 A
27. (a) 3.75 kΩ (b) 536 m
29. (a) 3.15×10^{-8} $\Omega \cdot$m (b) 6.35×10^6 A/m^2
    (c) 49.9 mA (d) 6.59×10^{-4} m/s (assume 1 conduction electron per atom) (e) 0.400 V
31. 0.125
33. 20.8 Ω
35. 67.6°C
37. 26.2°C
39. 3.03×10^7 A/m^2
41. 21.2 nm
43. 0.833 W
45. 36.1%
47. (a) 0.660 kW$\cdot$h (b) 3.96¢
49. (a) 133 Ω (b) 9.42 m
51. 28.9 Ω
51A. $V^2 t/mc\,(T_2 - T_1)$
53. 26.9 cents/day
55. (a) 184 W (b) 461°C
57. 2020°C
59. (a) 667 A (b) 50.0 km
63. (a) $R = \dfrac{\rho L}{\pi(r_b^2 - r_a^2)}$ (b) 37.4 MΩ
    (c) $R = \dfrac{\rho}{2\pi L}\ln\left(\dfrac{r_b}{r_a}\right)$ (d) 1.22 MΩ
69. Average $\rho = 1.47$ $\mu\Omega \cdot$m; they agree.
S1. (a) The savings are $254.92.

Chapter 28

1. (a) 7.67 Ω (b) 1.76 W
3. (a) 1.79 A (b) 10.4 V
5. 12.0 Ω
7. (a) 6.73 Ω (b) 1.98 Ω
9. (a) 4.59 Ω (b) 8.16%
11. 0.923 Ω $\leq R \leq$ 9.0 Ω
13. 1.00 kΩ
15. 55.0 Ω
17. 1.41 Ω
17A. $\sqrt{2}\,R$
19. 14.3 W, 28.5 W, 1.33 W, 4.00 W
21. (a) 0.227 A (b) 5.68 V
23. 470 Ω; 220 Ω
23A. $\frac{1}{2}R_s + (R_s^2/4 - R_s R_p)^{1/2}$ and $\frac{1}{2}R_s - (R_s^2/4 - R_s R_p)^{1/2}$
25. (a) -10.4 V (b) 141 mA, 915 mA, 774 mA
27. $\frac{11}{13}$ A, $\frac{6}{13}$ A, $\frac{17}{13}$ A
29. Starter: 171 A   Battery: 0.283 A
31. 3.50 A, 2.50 A, 1.00 A
33. (a) $I_1 = \dfrac{5}{13}$ mA; $I_2 = \dfrac{40}{13}$ mA; $I_3 = \dfrac{35}{13}$ mA
    (b) 69.2 V; c
35. (a) 12.4 V (b) 9.65 V
37. (a) 909 mA (b) -1.82 V
39. 800 W, 450 W, 25.0 W, 25.0 W
41. 3.00 J
41A. $U_0/4$
43. (a) 1.50 s (b) 1.00 s
    (c) 200 μA + (100 μA)$e^{-t/(1.00\text{ s})}$
45. (a) 12.0 s (b) $i(t) = (3.00\ \mu\text{A})e^{-t/12}$
    $q(t) = (36.0\ \mu\text{C})[1 - e^{-t/12}]$
47. (a) 6.00 V (b) 8.29 μs
49. 425 mA
51. 1.60 MΩ
51A. $t/C \ln 2$
53. 16.6 kΩ
55. 0.302 Ω
57. 49.9 kΩ
59. (b) 0.0501 Ω, 0.451 Ω
61. 0.588 A
63. 60.0
65. (a) 12.5 A, 6.25 A, 8.33 A
    (b) 27.1 A; No, it would not be sufficient since the current drawn is greater than 25 A.
67. (a) 0.101 W (b) 10.1 W
69. (a) 16.7 A (b) 33.3 A
    (c) The 120-V heater requires four times as much mass.
71. 6.00 Ω; 3.00 Ω
71A. $P_s/2I^2 + (P_s^2/4I^4 - P_s P_p/I^4)^{1/2}$ and
     $P_s/2I^2 - (P_s^2/4I^4 - P_s P_p/I^4)^{1/2}$
73. (a) 72.0 W (b) 72.0 W
75. (a) 40 W (b) 80 V, 40 V, 40 V
77. (a) $R \leq 1050$ Ω (b) $R \geq 10.0$ Ω
79. (a) $R \to \infty$ (b) $R \to 0$ (c) $R \to r$
81. (a) 9.93 μC (b) 3.37×10^{-8} A
    (c) 3.34×10^{-7} W (d) 3.37×10^{-7} W
83. $T = (R_A + 2R_B)C \ln 2$
85. $R = 0.521$ Ω, 0.260 Ω, 0.260 Ω, assuming resistors are in series with the galvanometer
87. (a) 1/3 mA for R_1, R_2 (b) 50 μC
    (c) $(0.278\text{ mA})e^{-t/0.18\text{s}}$ (d) 0.290 s

89. (a) $1.96\ \mu C$ (b) $53.3\ \Omega$
91. (a) $\ln\dfrac{\mathcal{E}}{V} = (0.0118\ \text{s}^{-1})t + 0.0882$

(b) $85\ \text{s} \pm 6\%$; $8.5\ \mu F \pm 6\%$
93. 48.0 W
93A. $1.50P$

S1. Kirchhoff's equations for this circuit can be written as

$$\mathcal{E}_1 - I_1 R_1 - I_4 R_4 = 0$$
$$\mathcal{E}_2 - I_2 R_2 - I_4 R_4 = 0$$
$$\mathcal{E}_3 - I_3 R_3 - I_4 R_4 = 0$$

and

$$I_1 + I_2 + I_3 = I_4$$

In matrix form, they are

$$\begin{bmatrix} R_1 & 0 & 0 & R_4 \\ 0 & R_2 & 0 & R_4 \\ 0 & 0 & R_3 & R_4 \\ 1 & 1 & 1 & -1 \end{bmatrix} \begin{bmatrix} I_1 \\ I_2 \\ I_3 \\ I_4 \end{bmatrix} = \begin{bmatrix} \mathcal{E}_1 \\ \mathcal{E}_2 \\ \mathcal{E}_3 \\ 0 \end{bmatrix}$$

In Lotus 1-2-3 use the /DataMatrixInvert and /DataMatrixMultiply commands and in Excel use the array formulas =MINVERSE(array) and =MMULTI(array1, array2) to carry out the calculations. Using the data given in the problem, we find

$I_1 = -1.26$ A, $I_2 = 0.87$ A, $I_3 = 1.08$ A, and $I_4 = 0.69$ A.

Chapter 29

1. (a) West
   (b) zero deflection (c) up (d) down
3. B_x is indeterminate; $B_y = -2.62$ mT; $B_z = 0$
5. $48.8°$ or $131°$
7. 26.0 pN west
9. 2.34×10^{-18} N
11. zero
13. $(-2.88\ \text{N})\mathbf{j}$
15. 0.245 T east
17. (a) 4.73 N (b) 5.46 N (c) 4.73 N
19. 196 A east if $\mathbf{B} = 50.0\ \mu\text{T}$ north
21. $F = 2\pi rIB \sin\theta$, up
23. $9.98\ \text{N}\cdot\text{m}$, clockwise as seen looking in the negative y-direction
23A. $NBabI \cos\theta$, clockwise as seen looking down
25. (a) $376\ \mu A$ (b) $1.67\ \mu A$
27. (a) $3.97°$ (b) $3.39\ \text{mN}\cdot\text{m}$
27A. (a) $\tan^{-1}(IBL/2mg)$ (b) $\tfrac{1}{4} IBLd \cos\theta$
29. 1.98 cm
31. 65.6 mT
33. $r_\alpha = r_d = \sqrt{2} r_p$
35. 7.88 pT
37. 2.99 u; $^3_1\text{H}^+$ or $^3_2\text{He}^+$
39. 5.93×10^5 N/C
41. $mg = 8.93 \times 10^{-30}$ N down, $qE = 1.60 \times 10^{-17}$ N up, $qvB = 4.74 \times 10^{-17}$ N down
43. 0.278 m

45. 31.2 cm
47. (a) 4.31×10^7 rad/s (b) 5.17×10^7 m/s
49. 70.1 mT
51. $3.70 \times 10^{-9}\ \text{m}^3/\text{C}$
53. 4.32×10^{-5} T
55. 7.37×10^{28} electrons/m^3
57. 128 mT north at $78.7°$ below the horizon
59. (a) $(3.52\mathbf{i} - 1.60\mathbf{j}) \times 10^{-18}$ N (b) $24.4°$
61. 0.588 T
65. 4.38×10^5 Hz
67. 3.82×10^{-25} kg
69. $3.70 \times 10^{-24}\ \text{N}\cdot\text{m}$
71. (a) $(-8.00 \times 10^{-21}\ \text{kg}\cdot\text{m/s})\mathbf{j}$ (b) $8.91°$

S1. The figure below shows a plot of the data and the best straight-line fit to the data. The slope is $1.58\ \mu\text{A/rad}$. The torsion constant κ is given by $\kappa = NAB \times$ slope $= 4.74 \times 10^{-10}\ \text{N}\cdot\text{m/rad} = 8.28 \times 10^{-12}\ \text{N}\cdot\text{m/deg}$.

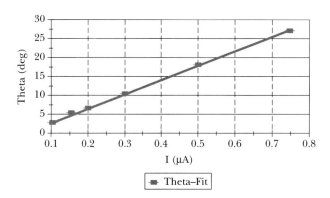

S2. This problem is somewhat similar to Problem 14.4, which dealt with gravitation and orbits, except that here the acceleration is velocity dependent. The equation of motion is

$$\mathbf{F} = q\mathbf{E} + q(\mathbf{v} \times \mathbf{B}) = m\mathbf{a} = q(E_x - v_z B_y)\mathbf{i} + qv_x B_y \mathbf{k}$$

The acceleration in the y direction is zero, so the y component of the velocity of the electron is constant.

Chapter 30

1. 200 nT
3. 31.4 cm
5. (a) $28.3\ \mu T$ (b) $24.7\ \mu T$ into the page
7. 54.0 mm
9. $\dfrac{\mu_0 I}{4\pi x}$ into the plane of the page
11. $26.2\ \mu T$ into the plane of the page
11A. $\mu_0 I/8R$ into the plane of the page
13. $\dfrac{\mu_0 I (a^2 + d^2 - d\sqrt{a^2 + d^2})}{2\pi a d \sqrt{a^2 + d^2}}$ into the page
15. $80.0\ \mu\text{N/m}$
17. 2000 N/m; attractive

19. $(-27.0\ \mu\text{N})\mathbf{i}$
21. 13.0 μT directed downward
21A. $\dfrac{\mu_0}{2\pi m}(400\ I_1^2 + 69.4\ I_2^2)^{1/2}$ at $\tan^{-1}\left(\dfrac{18.5\ I_1 + 3.21\ I_2}{7.69\ I_2 - 7.69\ I_1}\right)$ below the x-axis
23. (a) 3.98 kA   (b) ≈ 0
25. 5.03 T
27. (a) 3.60 T   (b) 1.94 T
29. Assuming constant current, make the wire long and thin. The solenoid radius does not affect the magnetic field.
31. (a) 6.34×10^{-3} N/m, inward
    (b) **F** is greatest at the outer surface.
33. 464 mT
33A. $\mu_0 \mathcal{E}\pi r/2\rho L$
35. 4.74 mT
37. (a) 3.13 mWb   (b) zero
39. (a) $BL^2/2$ (b) $BL^2/2$
41. (a) $(8.00\ \mu\text{A})e^{-t/4}$ (b) 2.94 μA
43. (a) 11.3×10^9 V·m/s   (b) 100 mA
45. 1.0001
47. 191 mT
49. 277 mA
51. 150 μWb
53. M/H
55. 1.27×10^3 turns
57. 2.02
59. (a) 9.39×10^{45} (b) 4.36×10^{20} kg
61. 675 A down
63. 81.7 A
63A. $2\pi d\mu g/\mu_0 I_A$
65. $\mathbf{B} = \dfrac{\mu_0 I}{2\pi \omega}\ln\left(\dfrac{b+w}{b}\right)\mathbf{k}$
67. 594 A east
69. 1.43×10^{-10} T directed away from the center
69A. $\dfrac{\mu_0 \omega q}{8\sqrt{2}\pi R}$
73. (a) $B = \tfrac{1}{3}\mu_0 b r_1^2$ (b) $B = \dfrac{\mu_0 b R^3}{3 r_2}$
75. (a) 2.46 N   (b) 107.3 m/s^2
77. (a) 12.0 kA-turns/m   (b) 2.07×10^{-5} T·m/A
79. 933 μN to the right perpendicular to the straight segments
79A. $\dfrac{\mu_0 I_1 I_2 L}{\pi R}$ to the right perpendicular to the straight segments
83. $\dfrac{\mu_0 I}{4\pi}(1 - e^{-2\pi})$ perpendicularly out of the paper
85. 20.0 μT toward the top of the page
87. $\dfrac{4}{3}\rho\mu_0\omega R^2$
89. $\dfrac{4}{15}\pi\rho\omega R^5$
S3. When integrating $F(r/R)$ when $r = R$, there is a singularity in the integrand at $\theta = \pi/2$ rad. The magnetic field at the wire is infinite, which we would expect even for a straight wire. We suggest you calculate the integral at $r/R = 0.0, 0.2, 0.4, 0.6, 0.8$, and 0.9.

Chapter 31

1. 500 mV
3. 160 A
5. + 121 mA
7. 61.8 mV
9. $(200\ \mu\text{V})e^{-t/7}$
11. $Nn\pi R^2 \mu_0 I_0 \alpha e^{-\alpha t} = (68.2\ \text{mV})e^{-1.6t}$ counterclockwise
11A. $Nn\pi R^2 \mu_0 I_0 \alpha e^{-\alpha t}$ counterclockwise
13. 272 m
15. -6.28 V
15A. $-2\pi R^2 B/t$
17. 763 mV, with the left-hand wingtip positive
19. (a) 3.00 N to the right   (b) 6.00 W
21. 2.00 mV; the west end is positive in the northern hemisphere
23. 2.83 mV
25. 145 μA
25A. $Bd = \left(\dfrac{v_2 R_3 - v_3 R_2}{R_1 R_2 + R_1 R_3 + R_2 R_3}\right)$
27. (a) to the right   (b) to the right   (c) to the right
    (d) into the plane of the paper
29. 0.742 T
31. 114 μV clockwise
31A. $N\mu_0 (I_1 - I_2)\pi r^2/\ell\ \Delta t$ clockwise
33. 1.80×10^{-3} N/C counterclockwise
35. (a) $(9.87 \times 10^{-3}\ \text{V/m})\cos(100\pi t)$ (b) clockwise
37. (a) 1.60 A counterclockwise
    (b) 20.1 μT   (c) up
37A. (a) $\dfrac{n\mu_0\pi r_2^2}{2R}\dfrac{\Delta I}{\Delta t}$ counterclockwise   (b) $\dfrac{n\mu_0^2 \pi r_2^2}{4R r_1}\dfrac{\Delta I}{\Delta t}$
    (c) up
39. 12.6 mV
41. (a) 7.54 kV   (b) **B** is parallel to the plane of the loop.
43. $(28.6\ \text{mV})\sin(4\pi t)$
45. (a) 0.640 N·m   (b) 241 W
47. 0.513 T
49. (a) $F = \dfrac{N^2 B^2 w^2 v}{R}$ to the left   (b) 0
    (c) $F = \dfrac{N^2 B^2 w^2 v}{R}$ to the left
51. $(-2.87\mathbf{j} + 5.75\mathbf{k}) \times 10^9$ m/s^2
53. It is, with the top end in the picture positive.
57. (a) 36.0 V   (b) 600 mWb/s   (c) 35.9 V
    (d) 4.32 N·m
59. (a) 97.4 nV   (b) clockwise
61. Moving east; 458 μV
63. It is, with the left end in the picture positive
65. 1.20 μC
65A. Ba^2/R
67. 6.00 A
67A. b^2/aR if $t \geq b/3a$
69. (a) 900 mA   (b) 108 mN   (c) b   (d) No

71. (a) Counterclockwise (b) $\dfrac{K\pi r^2}{R}$

73. (a) $\dfrac{\mu_0 IL}{2\pi}\ln\left(\dfrac{h+w}{h}\right)$ (b) $-4.80\ \mu V$

Chapter 32
1. 100 V
3. 1.36 μH
5. $\mathcal{E}_0/k^2L$
7. (a) 188 μT (b) 33.3 nWb (c) 375 μH
   (d) field and flux
9. 21.0 μWb
11. $(18.8\ V)\cos(377t)$
13. $\tfrac{1}{2}$
15. (a) 15.8 μH (b) 12.6 mH
17. (500 mA) $(1-e^{-10t/s})$, $1.50\ A - (0.25\ A)e^{-10t/s}$
19. 1.92 Ω
21. (a) 1.00 kΩ (b) 3.00 ms
21A. (a) $\sqrt{L/C}$ (b) $\sqrt{LC}$
23. (a) 139 ms (b) 461 ms
25. (a) 5.66 ms (b) 1.22 A (c) 58.1 ms
27. (a) 113 mA (b) 600 mA
29. (a) 0.800 (b) 0
31. (a) 1.00 A (b) 12.0 V, 1.20 kV, 1.21 kV
    (c) 7.62 ms
33. 2.44 μJ
35. (a) 20.0 W (b) 20.0 W (c) 0 (d) 20.0 J
37. (a) 8.06×10^6 J/m^3 (b) 6.32 kJ
39. 44.3 nJ/m^3 in the **E** field; 995 μJ/m^3 in the **B** field
41. 30.7 μJ and 72.2 μJ
43. (a) 668 mW (b) 34.3 mW (c) 702 mW
45. 1.73 mH
47. 80.0 mH
49. 553 μH
51. (a) 18.0 mH (b) 34.3 mH (c) -9.00 mV
53. 400 mA
55. (a) -20.0 mV (b) $-(10.0\ MV/s^2)t^2$
    (c) 63.2 μs
55A. (a) $-LK$ (b) $-Kt^2/2C$ (c) $2\sqrt{LC}$
57. 608 pF
59. (a) 36.0 μF (b) 8.00 ms
61. (a) 6.03 J (b) 0.529 J (c) 6.56 J
63. (a) 2.51 kHz (b) 69.9 Ω
65. (a) 4.47 krad/s (b) 4.36 krad/s (c) 2.53%
69. 979 mH
71. (a) 20.0 ms (b) -37.9 V (c) 3.03 mV
    (d) 104 mA
73. 95.6 mH
79. (a) 72.0 V; b (c) 75.2 μs
81. $\dfrac{\mu_0}{\pi}\ln\left(\dfrac{d-a}{a}\right)$
83. 300 Ω
85. 45.6 mH
85A. $9t^2/\pi^2 C$
S1. (a) $\tau = 0.35$ ms, (b) $t_{50\%} = 0.24$ ms $= 0.686\tau$, and
    (c) $t_{90\%} = 0.80$ ms $= 2.29\tau$; $t_{99\%} = 1.6$ ms $= 4.57\tau$.

Chapter 33
3. 2.95 A, 70.7 V
5. 3.38 W
7. 14.6 Hz
9. (a) 42.4 mH (b) 942 rad/s
11. 7.03 H
13. 5.58 A
15. 3.80 J
17. (a) $f > 41.3$ Hz (b) $X_C < 87.5\ \Omega$
19. 2.77 nC
19A. $\sqrt{2}\ CV$
21. 100 mA
23. 0.427 A
25. (a) 78.5 Ω (b) 1.59 kΩ (c) 1.52 kΩ
    (d) 138 mA (e) $-84.3°$
27. (a) $17.4°$ (b) voltage leads the current
29. 1.88 V
29A. $R_b V_2/\sqrt{R_b^2 + 1/4\ \pi^2 f^2 C_s^2}$ where V_2 is the transformer secondary voltage
31. (a) 146 V (b) 213 V (c) 179 V (d) 33.4 V
33. (a) 194 V (b) current leads by $49.9°$
35. 132 mm
35A. $\sqrt{800\ Ppd/\pi V^2}$ where ρ is the material's resistivity.
37. 8.00 W
39. $v(t) = (283\ V)\sin(628t)$
41. (a) 16.0 Ω (b) $-12.0\ \Omega$
43. 159 Hz
45. 1.82 pF
47. (a) 124 nF (b) 51.5 kV
49. 242 mJ
49A. $\dfrac{4\pi RCV^2\sqrt{LC}}{9L + 4CR^2}$
51. (a) 613 μF (b) 0.756
53. 8.42 Hz
55. (a) True, if $\omega_0 = (LC)^{-1/2}$ (b) 0.107, 0.999, 0.137
57. 0.317
59. 687 V
61. (a) 9.23 V (b) 4.55 A (c) 42.0 W
63. 87.5 Ω
63A. $R_s = \dfrac{R_L N_1 (N_2 V_s - N_1 V_2)}{V_2 N_2^2}$
65. 56.7 W
67. (a) 1.89 A
    (b) $V_R = 39.7$ V, $V_L = 30.1$ V, $V_C = 175$ V
    (c) $\cos\phi = 0.265$
    (d)

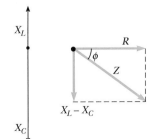

69. 99.6 mH
71. (a) 2.90 kW (b) 5.80×10^{-3}
    (c) If the generator is limited to 4500 V, no more than 17.5 kW could be delivered to the load, never 5000 kW.
73. (a) Circuit (a) is a high-pass filter, (b) is a low-pass filter.
    (b) $\dfrac{V_{out}}{V_{in}} = \dfrac{\sqrt{R_L^2 + X_L^2}}{\sqrt{R_L^2 + (X_L - X_C)^2}}$ for circuit (a);
    $\dfrac{V_{out}}{V_{in}} = \dfrac{X_C}{\sqrt{R_L^2 + (X_L - X_C)^2}}$ for circuit (b)
75. (a) 172 MW (b) 17.2 kW (c) 172 W
77. (a) 200 mA; voltage leads by 36.8° (b) 40.0 V; $\phi = 0°$ (c) 20.0 V; $\phi = -90°$
    (d) 50.0 V; $\phi = +90°$
83. (a) 173 Ω (b) 8.66 V
85. (a) 1.838 kHz
87.

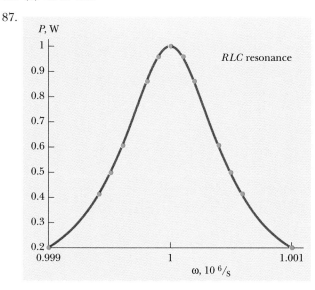

S1. (a)

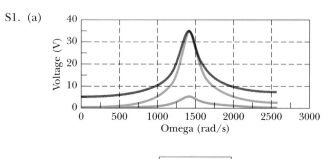

$-V_R - V_C - V_L$

(b) $|V_L| = |V_C|$; (c) for $\omega < \omega_0$, $V_C > V_L$; (d) for $\omega > \omega_0$, $V_L > V_C$

Chapter 34

1. 2.68×10^3 AD
5. (a) 162 N/C (b) 130 N/C

7. 38.0 pT
9. (a) 6.00 MHz (b) $(73.3 \text{ nT})(-\mathbf{k})$
   (c) $\mathbf{B} = (73.3 \text{ nT}) \cos(0.126x - 3.77 \times 10^7 t)(-\mathbf{k})$
11. (a) $B = 333$ nT (b) $\lambda = 628$ nm
    (c) $f = 4.77 \times 10^{14}$ Hz
13. 5.16 m
15. 307 μW/m²
17. 66.7 kW
19. (a) 332 kW/m² (b) 1.88 kV/m, 222 μT
21. (a) 971 V/m (b) 16.7 pJ
23. (a) 540 N/C (b) 2.58 μJ/m³ (c) 774 W/m²
    (d) This is 77.4% of the flux mentioned in Example 34.3 and 57.8% of the 1340 W/m² intensity above the atmosphere. It may be cloudy at this location, or the sun may be setting.
25. 83.3 nPa
27. (a) 1.90 kV/m (b) 50.0 pJ
    (c) 1.67×10^{-19} kg·m/s
29. (a) 11.3 kJ (b) 5.65×10^{-5} kg·m/s
31. (a) 2.26 kW (b) 4.71 kW/m²
33. 7.50 m
35. (a) $\mathcal{E}_m = 2\pi^2 r^2 f B_m \cos \theta$, where θ is the angle between the magnetic field and the normal to the loop. (b) Vertical, with its plane pointing toward the broadcast antenna
37. 56.2 m
37A. $2\pi cm/qB$
39. (a) 6.00 pm (b) 7.50 cm
41. The radio audience hears it 8.41 ms sooner.
43. (a) 4.17 m to 4.55 m (b) 3.41 to 3.66 m
    (c) 1.61 m to 1.67 m
45. 3.33×10^3 m²
47. (a) 6.67×10^{-16} T (b) 5.31×10^{-17} W/m²
    (c) 1.67×10^{-14} W (d) 5.56×10^{-23} N
49. 6.37×10^{-7} Pa
51. 95.1 mV/m
53. 7.50×10^{10} s = 2370 y; out the back
55. 3.00×10^{-2} degrees
57. (a) $B_0 = 583$ nT, $k = 419$ rad/m, $\omega = 1.26 \times 10^{11}$ rad/s; xz
    (b) 40.6 W/m² in average value (c) 271 nPa
    (d) 406 nm/s²
59. (a) 22.6 h (b) 30.5 s
61. (b) 1.00 MV/s
63. (a) 3.33 m; 11.1 ns; 6.67 pT
    (b) $\mathbf{E} = (2.00 \text{ mV/m}) \cos 2\pi \left(\dfrac{x}{3.33 \text{ m}} - \dfrac{t}{11.1 \text{ ns}} \right) \mathbf{j}$
    $\mathbf{B} = (6.67 \text{ pT}) \cos 2\pi \left(\dfrac{x}{3.33 \text{ m}} - \dfrac{t}{11.1 \text{ ns}} \right) \mathbf{k}$
    (c) 5.31×10^{-9} W/m² (d) 1.77×10^{-14} J/m³
    (e) 3.54×10^{-14} Pa

Chapter 35

1. 299.5 Mm/s
3. (b) 294 Mm/s
5. 114 rad/s

7. 198 Gm
9. (a) 4.74×10^{14} Hz (b) 422 nm
   (c) 2.00×10^8 m/s
11. 70.5° from the vertical
13. 61.3°
15. 19.5°, 19.5°, 30.0°.
17. (a) 327 nm (b) 287 nm
19. 59.83°, 59.78°, 0.0422°
21. 30.0°, 19.5° at entry; 40.5°, 77.1° at exit
23. 0.171°
23A. $\sin^{-1}\left(\dfrac{\sin\theta}{n_{700}}\right) - \sin^{-1}\left(\dfrac{\sin\theta}{n_{400}}\right)$
27. 18.4°
29. 86.8°
31. 4.61°
33. 62.4°
35. (a) 24.4° (b) 37.0° (c) 49.8°
37. 1.00008
39. $\theta < 48.2°$
41. 53.6°
43. 2.27 m
45. 2.37 cm
49. 90°, 30°, No
51. 62.2%
53. $\sin^{-1}[(n^2 - 1)^{1/2} \sin\phi - \cos\phi]$ If $n \sin\phi \leq 1$, $\theta = 0$
55. 82
57. 27.5°
59. 7.91°
59A. $\phi = \sin^{-1}\left[n \sin\left(90° - 2\theta + \sin^{-1}\left(\dfrac{\sin\theta}{n}\right)\right)\right]$
61. (a) 1.20 (b) 3.40 ns
S1. Using Snell's Law and plane geometry, the following relationships can be derived:
$$\sin\theta_1 = n \sin\rho_1$$
$$\rho_2 = A - \rho_1$$
$$n \sin\rho_2 = \sin\theta_2$$
$$\theta = \theta_1 + \theta_2 - A$$

These equations are used to find ρ_1, ρ_2, θ_2, and θ in the spreadsheet. IF statements are used in the calculation of θ_2 and θ so that error messages are not printed when ρ_2 exceeds the critical angle for total internal reflection.

Chapter 36

3. 2′11″
3A. $h/2$
5. 30 cm
7. (a) $q = 45.0$ cm, $M = -\frac{1}{2}$
   (b) $q = -60.0$ cm, $M = 3.00$
   (c) Similar to Figures 36.8 and 36.12b
9. (a) 15.0 cm (b) 60.0 cm
9A. (a) $d/2$ (b) $2d$
11. concave with radius 40.0 cm
13. (a) 2.08 m (concave)
    (b) 1.25 m in front of the object
15. (a) $q = -12.0$ cm, $M = 0.400$
    (b) $q = -15.0$ cm, $M = 0.250$
    (c) The images are erect
17. 11.8 cm above the floor
19. 1.50 cm/s
19A. v/n
21. 8.57 cm
23. 3.88 mm
25. 2
27. (a) 16.4 cm (b) 16.4 cm
29. 25.0 cm; -0.250
31. (a) 13.3 cm
    (b) A trapezoid

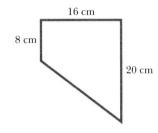

    (c) 224 cm²
33. (a) -12.3 cm, to left of lens (b) $+0.615$
    (c)

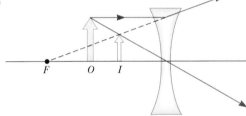

35. (a) 39.0 mm (b) 39.5 mm
37. 2.84 cm
39. $\dfrac{f}{1.41}$
41. -4 diopters, a diverging lens
43. 0.558 cm
45. 20.0 m
45A. $vf\Delta t/\Delta f$
47. 3.5
49. -800, image is inverted
51. -18.8
53. 2.14 cm
55. (a) 1.99 (b) 10.0 cm to the left of the lens, -2.50
    (c) inverted since the overall magnification is negative.
57. (a) 20.0 cm to the right of the second lens, -6.00
    (b) inverted (c) 6.67 cm to the right of the second lens, -2.00, inverted
59. 10.7 cm beyond the curved surface
61. 20.0 cm
61A. f
63. (a) 44.6 diopters (b) 3.03 diopters
65. $d = 8$ cm

67. 21.3 cm

67A. $h_1 \cot\left(\sin^{-1}\dfrac{nh_1}{R} - \sin^{-1}\dfrac{h_1}{R}\right) + \sqrt{R^2 - h_1^2}$
$- h_2 \cot\left(\sin^{-1}\dfrac{nh_2}{R} - \sin^{-1}\dfrac{h_2}{R}\right) - \sqrt{R^2 - h_2^2}$

69. (a) 52.5 cm   (b) 1.50 cm
71. $q' = 1.5$ m, $h' = -13.1$ mm
73. (a) -0.400 cm   (b) $q' = -3.94$ mm, $h' = 535$ μm
75. (a) 30.0 cm and 120 cm   (b) 24.0 cm
   (c) real, inverted, diminished
77. real, inverted, actual size
S1. See Figure 36.1. Applying the law of sines to triangle PAC, one finds $r \sin\theta_1 = (r + p)\sin\alpha$. Snell's Law gives $n_1 \sin\theta_1 = n_2 \sin\theta_2$. Using plane geometry gives $\gamma = \theta_1 - \alpha - \theta_2$. Applying the law of sines to triangle ACP' gives $q' - r = r\sin\theta_2/\sin\gamma$. These equations are used to calculate θ_1, θ_2, γ, and q in the spreadsheet. (a) The maximum angle α for which the error in the image distance is 3 percent or less is 2.35°. (b) 2.25° (d) No. (For a plot of the percent error versus α, just call up the associated graph.)

Chapter 37

1. (a) 2.62×10^{-3} m   (b) 2.62×10^{-3} m
3. 515 nm
5. (a) 36.2°   (b) 5.08 cm   (c) 508 THz
7. 11.3 m
9. 2.50 m
9A. $c/2f$
13. 4.80×10^{-5} m
15. 423.5 nm
17. 343 nm
17A. $\lambda/2(n - 1)$
19. 641
21. (a) 2.63 rad   (b) 246 nm
23. (a) 7.95 rad   (b) 0.453
27. $10 \sin(100\pi t + 0.93)$
29. $\pi/2$
31. $360°/N$
33. No reflection maxima in the visible spectrum
35. 512 nm
37. 85.4 nm, or 256 nm, or 427 nm . . .
39. (a) Green   (b) Purple
41. 167 nm
43. 4.35 μm
45. 3.96×10^{-5} m
47. 654 dark fringes
49. (a) 5.99 m   (b) 2.99 m
49A. (a) $2\sqrt{4h^2 + d^2} - 2d$ (b) $\sqrt{4h^2 + d^2} - d$
51. 3.58°
53. 421 nm
55. 2.52 cm
57. 1.54 mm
59. 3.6×10^{-5} m

61. $x_{\text{bright}} = \dfrac{\lambda \ell (m + \frac{1}{2})}{2hn}$, $x_{\text{dark}} = \dfrac{\lambda \ell m}{2hn}$
63. (b) 115 nm
65. 1.73 cm
69. (b) 266 nm
S1. (a) 9:1   (b) 1:9

Chapter 38

1. 632.8 nm
3. 560 nm
5. $\cong 10^{-3}$ rad
7. 0.230 mm
9. $\phi = \beta/2 = 1.392$ rad

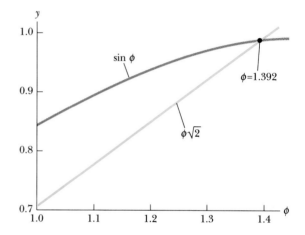

11. 1.62×10^{-2}
13. 3.09 m
15. 51.2 cm
17. 20 cm
17A. (0.8 μm) (vt/nd)
19. 15.7 km
21. 13 m
23. 105 m
27. (a) 5.7; 5 orders is the maximum
    (b) 10; 10 orders to be in the short-wavelength region.
29. 0.0683°
31. (a) 2800   (b) 4.72 μm
33. 2
35. 7.35°
37. 14.7°
39. 31.9°
41. (a) 0°, 19.9°, 42.9°   (b) 8.33 cm, 135 cm
43. (a) 0.281 nm   (b) 0.18%
45. 14.4°
47. 1.11
49. 31.2°
51. (a) 18.4°   (b) 45.0°   (c) 71.6°
53. (a) 54.7°   (b) 63.4°   (c) 71.6°
55. 60.5°

57. 2.2
59. 5.9°
61. 0.244 rad ≈ 14.0°
63. 545 nm
65. (a) 12 000, 24 000, 36 000   (b) 0.0111 nm
67. (a) 3.53×10^3 lines/cm
    (b) Eleven maxima can be observed.
69. $\frac{1}{8}$
71. (a) 6   (b) 7.50°
73. (a) 41.8°   (b) 0.593   (c) 0.262 m
S1. (a) 9:1   (b) 1:9

Chapter 39

5. (a) 60 m/s   (b) 20 m/s   (c) 44.7 m/s
5A. (a) $v_b + v$ (b) $v - v_b$ (c) $\sqrt{v^2 + v_b^2}$
7. (a) 39.2 μs   (b) Accurate to one digit. More precisely, he aged 1.78 μs less on each orbit.
9. $0.800c$
9A. $v = cL_p / \sqrt{L_p^2 + c^2 t^2}$
11. $0.436 L_0$
13. $0.789c$
15. $0.696c$
17. $0.960c$
19. 42 g/cm³
19A. $\dfrac{m}{V(1 - v^2/c^2)}$
21. 1625 MeV/c
25. (a) 939.4 MeV   (b) 3.008×10^3 MeV
    (c) 2.069×10^3 MeV
27. $0.864c$
29. (a) 0.582 MeV   (b) 2.45 MeV
31. (a) 3.91×10^4 (b) $0.9999999997c$ (c) 7.66 cm
33. 4 MeV and 29 MeV
35. (a) 3.29 MeV   (b) 2.77 MeV
37. 4.2×10^9 kg/s
39. (a) $0.800c$ (b) $0.929c$
39A. (a) $\dfrac{2v}{1 + v^2/c^2}$ (b) $\dfrac{3v + v^3/c^2}{1 + 3v^2/c^2}$

41. 0.7%
43. (a) 6.67 ks   (b) 4.00 ks
43A. (a) $\dfrac{2d}{v + c}$ (b) $\dfrac{2d\sqrt{1 - v^2/c^2}}{v + c}$
47. (a) 76.0 min   (b) 52.1 min
47A. (a) $\dfrac{t_{II}}{\sqrt{1 - \left(\dfrac{(v_I + v_{II})\,c}{c^2 + v_I v_{II}}\right)^2}}$ (b) $\dfrac{t_{II}}{\sqrt{1 - v_{II}^2/c^2}}$

49. 15.0 h in the spacecraft frame, 10.0 h less than in the Earth frame.
51. (a) $0.554c, 0.866c$ (b) 0.833 m
51A. (a) $u = \dfrac{v - u'}{1 - vu'/c^2}$ where $u' = c\sqrt{1 - L'^2/L_p^2}$
    (b) $L_p \sqrt{1 - u^2/c^2}$
53. (a) 83.3 m and 62.5 m   (b) 27.0 m and 20.3 m
    (c) 6.00 ks   (d) 5.33 ks   (e) 7.10 ks
    (f) Rocket one is destroyed with crew aboard – they were going too fast.
55. (b) For $v \ll c$, $a = qE/m$ as in the classical description. As $v \to c$, $a \to 0$, describing how the particle can never reach the speed of light.
    (c) Perform $\displaystyle\int_0^v \left(1 - \dfrac{v^2}{c^2}\right)^{-3/2} dv = \int_0^t \dfrac{qE}{m}\, dt$ to obtain $v = \dfrac{qEct}{\sqrt{m^2 c^2 + q^2 E^2 t^2}}$ and then $\displaystyle\int_0^x dx = \int_0^t \dfrac{qEtc}{\sqrt{m^2 c^2 + q^2 E^2 t^2}}\, dt$

S1. (a)

Z	0.2	0.5	1.0	2.0
v/c	0.180	0.385	0.600	0.800

    (b) For $Z = 3.8$, $v/c = 0.870$.
S3. For $Z = 0.2$, $r = 2710$ Mly; $Z = 0.5$, $r = 5770$ Mly; $Z = 1.2$, $r = 9860$ Mly; and $Z = 3.8$, $r = 13\,800$ Mly.

Index

Page numbers in *italics* indicate illustrations; page numbers followed by an "n" indicate footnotes, page numbers followed by "t" indicate tables.

Aberrations, chromatic, *1071*, 1071–1072
  of lenses, *1070*, 1070–1072, *1071*
  spherical, 1056, *1056*, *1070*, 1070–1071
Absolute pressure, 427
Absolute temperature scale, 534, 624
Absolute zero, 535, 624, 624n
Absorber, ideal, 574, 576
Absorption, selective, polarization of light waves by, *1133*, 1133–1134, *1134*
ac circuits. *See* Circuit(s), alternating current.
Accelerated electron, 669
Accelerated frames, motion in, 151–154
Acceleration, 28–32, *29*
  angular, and angular velocity, 277–278
    average, 278
    instantaneous, 278, 278n, 296
    of wheel, *291*, 291–292
    torque and, *289*, 289–292
  average, 29, 44, 73, *73*
    and instantaneous, 31
  centripetal, 85, 93–94, 145, 162
  constant, 35–36
    two-dimensional motion with, 74–77, *75*
  constant angular, rotational motion with, 278–279
  dimensions of, 11t
  free-fall, 37, 39, *39*, 44, 394–395, 411
    and projectile motion, 77, 77n
    measurement of, 120–121, *121*
    variation of, with altitude, 395
  in simple harmonic motion, 362–363, *367*
  instantaneous, 29, 44, 73–74
  inverse-square law and, 396
  linear and angular, 280, *280*
  magnitude of, of connected objects with friction, *128*, 128
  mass and, 111
  of automobile, 35
    up hill, 190–191, *191*
  of center of mass, 256
  of mass, on frictionless incline, *119*, 119–120
  of moon, *396*, 396–397

Acceleration *(Continued)*
  of two blocks of masses, magnitude of, 122, *122*
  of two unequal connected masses, *121*, 121–122
  on frictionless surface, resultant force and, 113, *113*
  positive, of point charge, 668, *668*
  radial, *87*, 87–89, 93–94
  relative, 89–92, *90*
  tangential, *87*, 87–89, 280
  total, 87
  transverse, 468–469
  units of, 112t
  velocity, and position, graphical relations of, 30
Acceleration-time graph, *30*, 30–31, 32, *33*
Acceleration vector, displacement and velocity vectors, 71–74
Accommodation in eye, 1074
Action forces, 114–115, *115*
Adiabatic expansion, reversible, *PV* diagram for, 594–595, *595*, 608
Adiabatic free expansion, *563*, 565, 575
Adiabatic process, 565, 575
  first-law equation for, 565
  for ideal gas, 594–596
  reversible, definition of, 594
Air drag, 154
  at high speeds, 156–157, *157*
Air friction, resistive forces and, 189
Air resistance, and projectile motion, 77, 77n
Algebra, A.15–A.20
Ammeter, *813*, 813–814
Ampère, Andre, 833, 871
Ampère-Maxwell law, 880, *880*, 880n, 924, 925, 994
Ampère-meter 2, 842
Ampère (unit), 654, 773, 776, 789
  definition of, 870
Ampère's law, 864, 870–874, 875, 876, 890
  generalized, 924

Ampère's law *(Continued)*
  displacement current and, 879–881, *880*, 891
Amplitude, displacement, of sound waves, 481–482, *482*
  in simple harmonic motion, 361
  of wave, 457, *457*
  pressure, of sound waves, 481–482, *482*, 491
Analyzer, 1133
Angle, Brewster's, 1135
  phase, 971, 983
  polarizing, 1134, 1139
  solid, 696
Angular frequency, 466
  of ac generator, 965
Angular moment, 882
Angular speed, 277, 278n, 296
Angular wave number, 466
Antenna, production of electromagnetic wave by, 1008–1011, *1010*, *1011*
Antiderivative, finding of, 41
Antilogarithm, A.20
Antinodes, 503, *504*, 519
Archimedes, 428
Archimedes' principle, 427
  buoyant forces and, 427–431, *428*, *429*, 441
Area, dimensions of, 11t
  of rectangle, 16
  of room, 16
Associative law of addition, 58, *58*
Astigmatism, 1076
Atmosphere(s), law of, 600
  molecules in, 600
Atomic clock(s), 6
  time dilation and, 1166n, 1166–1167
Atomic energy levels, *601*, 601–602
Atomic mass(es), density and, 9–10
  table of, A.4–A.13
Atomizer, 437, *437*
Atom(s), 8, 10
  ground state of, 221
  hydrogen, 657
  in cube, 10

I.1

Atom(s) *(Continued)*
 in solid, 14
 magnetic moments of, 881–883, *882*, 883t
 metal, 655n
 "seeing of, 1126
Atwood's machine, 120–121, *121*, 292, *292*
Aurora Australis, 847
Aurora Borealis, 832, 846–847
Automobile(s), acceleration of, 35
  up hill, 190–191, *191*
 car lift for, 425–426
 collision into wall, average force and, 241, *241*
 collision of, at intersection, 250, *250*
 compact, gas consumed by, 190
 energy and, 188–191
 in U.S., gas used by, 14
 speed of, coefficient of static friction and, 147
Avogadro's number, 541
 definition of, 10, 17

Bainbridge mass spectrometer, 848, *848*
Barometer, 427, *427*
Baseball player, displacement and, 24–25, *25*
Battery(ies), conducting rods connected to, 1009, *1010*
 positive charge flows in, 786, *786*, 786n
 power supplied to resistor by, 786
 terminal voltage of, 799
Beam, uniform, horizontal, *343*, 343–344
Beats, 513–515, *515*
Bednorz, George, 782
Bell, Alexander Graham, 484n
Bernoulli, Daniel, 434
Bernoulli principle, 422, 434
Bernoulli's equation, 433–435, *434*
 applications of, *436*, 436–437, *437*
Bicycle wheel, spinning, 321, *321*
Big Bang theory, 161
Billiard balls, collision of, 251, *251*
Bimetallic strip, 539, *539*
Biot, Jean Baptiste, 865
Biot-Savart law, 864, 865–869, *866*, 890
Black body, 574, 576
Black hole, formation of, 1182
Body, rigid. *See* Rigid body(ies).
Bohr, Niels, 92, 324–325
Bohr magneton, 883
Boltzmann, Ludwig, 599
Boltzmann distribution law, 599–602
Boltzmann's constant, 542, 590, 607
Boyle's law, 541
Bragg, W.L., 1132
Bragg's law, 1132
Brahe, Tycho, 395
Brewster, Sir David, 1135
Brewster's angle, 1135
Brewster's law, 1135, 1139

Bridge(s), suspension, collapse of, 380
 Tacoma Narrows, collapse of, 380
British engineering system, 7
British thermal unit, definition of, 553
Brown, Robert, 529
Brownian motion, 529
Bubble chamber, 846
Building materials, R values for, 571, 571t
Bulk modulus, 346, 347–348, *348*, 349
Buoyant force(s), 155n, 427–428, 441
 and Archimedes' principle, 427–431, *428*, *429*, 441

Calcite, *1136*, 1136–1137
Calculus, differential, A.24–A.27
 integral, A.27–A.30
 kinematic equations derived from, 41–43
Calorie, 553
calorie (unit of heat), 553, 553n
Calorimeters, 556
Calorimetry, 556, 560
Camera, *1072*, 1072–1073
 box moving past, *1164*, *1164*
 calculation of exposure time and, 1073
Capacitance, and geometry, 760t
 calculation of, 742–745
 definition of, 741–742, 760, 969
 equivalent, 748–749, *749*
 of filled capacitors, 752, 752n
 SI unit of, 742, 760
Capacitive reactance, 970, 983
Capacitor(s), 741–742, *742*, 742n, 760, 761–762
 charged, energy stored in, 749n, 749–751, 754–755, *755*
 charging, charge versus time for, 810, *810*
  current versus time for, 810, *810*
 charging of, 808–810, *809*
 combinations of, *745*, 745–749, *746*, *747*
 cylindrical, *744*, 744–745
 discharging, charge versus time for, 811
  current versus time for, 811
 discharging of, 811, *811*
 displacement current in, 881
 electrolytic, 753–754, *754*
 filled, capacitance of, 752, 752n
 high voltage, 753, *754*
 in ac circuit, *969*, 969–971, 975
 in RC circuit, charging of, 812, *812*
  discharging of, 812–813
 instantaneous voltage drop across, 970
 maximum charge on, 809
 paper-filled, 754
 parallel-plate, *743*, 743–744, *744*, 758, *758*
  and nonuniform electric field, *755*, 755–756
  effect on metal slab, 759–760, *760*
  energy stored in, 750
 partially filled, 759, *759*
 phase relationships in, 971–972, *972*
 spherical, 745, *745*

Capacitor(s) *(Continued)*
 tubular, 753, *754*
 two charged, rewiring of, 750–751, *751*
 types of, 753–756, *754*, *755*
 with dielectrics, *751*, 751–756
Car lift, 425–426
Carburetor, 437
Carlson, Chester, 726
Carnot, Sadi, 620
Carnot cycle, 620, 621, *621*, 638
 PV diagram for, 621, *621*
 ratio of heats for, 622–623
Carnot-cycle heat engine, 627–628
Carnot engine, 620–624, 630
 efficiency of, 622, 623, 624
Carnot's theorem, 620, 638
Cartesian coordinate system, points in, 54, *54*
Cathode ray tube (CRT), 669–670, *670*
Cavendish, Henry, 393
Cavendish apparatus, 393, *393*
Celsius, Anders, 536n
Celsius temperature scale, 534, 536
Centripetal acceleration, 85, 93–94, 145, 162
cgs system of units, 7
Charge carrier, 773
 in conductor, motion of, 783, *783*
 in electric field, motion of, 783–784, *784*
 motion of, in conductor, 774, *774*
Charge density, 777
Charge(s), conservation of, 797, 819
 electric, conserved, 650
 continuous charge distribution, 661–662
 continuous distribution, of electric field, 661–665, *662*, 672
 cylindrically symmetric distribution of, 692, *692*
 induced, 652
 linear density, 662
 negative, 650, *650*, 650–651
 nonconducting plane sheet of, 693, *693*
 on spheres, *657*, 657–658
 positive, 650
 properties of, 649–651, *651*, 671
 quantized, 651
 spherically symmetric distribution of, *690*, 690–691, *691*
 surface density, 662
 two, electric field due to, *660*, 660–661
 volume density, 662
 finite line of, of electric potential, 719, *719*
 maximum, 809
 point. *See* Point charge(s).
 uniform ring of, of electric field, 663–664, *664*
Charles and Gay-Lussac's law, 541
Circle, equation of, A.21
Circuit(s), alternating current, 964–993, 983
 capacitors in, *969*, 969–971
 inductors in, *967*, 967–969

Circuit(s) *(Continued)*
  power in, 975–976
  purely capacitive, 971
  purely inductive, 969
  resistors in, *965*, 965–967, *966*, 966n, 967t, 983
  sources and phasors, 965
 direct current, 797–831
 elements of, impedance values and phase angles for, 973, *973*
 filter, *979*, 979–980, *980*
 household, wiring diagram for, 817, *817*
 LC, 948
  charge versus time for, 950, *951*, 954
  current versus time for, 950–951, *951*, 954
  energy transfer in, 948, *949*
  oscillations in, 948–952
   angular frequency of, 950
  oscillatory, 952, *952*
  total energy in, 950, 951, *951*, 955
 multiloop, 807–808, *808*
 RC, 808–813
 RL, *941*, 941–944, *942*
  current versus time for, 944, *944*
  initial current in, 944, 954
  time constant of, *942*, 942–943, 944, *944*, 954
 RLC, *952*, 952–953, 955
  charge versus time for, 953, *953*
  overdamped, 953, *953*
 RLC series, 971–975, *972*
  analysis of, 974
  average power in, 976
  average power versus frequency for, 978, *978*
  current versus frequency for, 977, 977–978
  impedance of, 973
  instantaneous power in, 975–976
  resonance in, 977–980
  resonating, 979
 single-loop, *806*, 806–807
 steady state, 797, 808
 time constant of, 819
Circular motion, 314, *314*
 nonuniform, *150*, 150, *151*
  of mass suspended by string, 88–89, *89*, 150–151, *151*
 uniform, *85*, 85–87, 144–170
  and simple harmonic motion, compared, *375*, 375–377, *376*
  Newton's second law of motion and, 144–149, *145*, *146*, 162
  with constant speed, 377
Classical mechanics, 1–450
Clausius, Rudolf, 628, 995, 996
Clausius statement, 618, 618n
Clocks, atomic, 6
 time dilation and, 1166n, 1166–1167
 synchronization of, 1157–1158
Coaxial cable, 946, *946*
 resistance of, 780, *780*

Coefficient(s), average, of linear expansion, 537–538, 538t, 543
 of volume expansion, 538, 543
 drag, 158–160
 Hall, 852
 of friction, 124–125, 126t
  measurement of, 127, *127*
 of kinetic friction, 125–126
  determination of, *127*, 127–128
 of performance, 627, 628
 of viscosity, *440*, 440t, 440–441
 temperature, of resistivity, 780–781
  for various materials, 777t
Coil(s), inductance of, 940, 954
 induction, 996
 induction of emf in, 910
 primary and secondary, 907
Collision frequency, 605
Collision(s), *242*, 242–243, 249
 as independent event, 784n
 average time between, 784–785
 elastic, *244*, 244–246, 262
  and inelastic, in one dimension, 243–248
 electron, in copper, 785
 inelastic, 244, 261
 linear momentum and, 235–275
 mass-spring, 185–186, 214–215, *215*
 of automobile(s), at intersection, 250, *250*
  into wall, average force and, 241, *241*
 of billiard balls, 251, *251*
 perfectly inelastic, 244, *244*
 projectile-cylinder, 319, *319*
 proton-proton, 250–251
 slowing down of neutrons by, 247–248
 total momentum of system and, 243
 two-body, with spring, 247, *247*
 two-dimensional, *248*, 248–251
Colors, in peacock's feathers, 1103
Commutative law of addition, *57*, *57*
Compact disc, diffraction grating and, 1129, *1129*
Compound microscope, 1078–1080, *1079*
Compressibility, 348
Compression ratio, 626
Computers, hardware requirements and spreadsheets, A.40
 software requirements and spreadsheets, A.39–A.40
 spreadsheet problems, A.39–A.41
 spreadsheet tutorial for, A.40–A.41
Concrete, prestressed, 348, *348*
Condensation(s), heat of, 558n
 of sound waves, 479, 481
Conducting loop, 908, *908*
Conduction, charging object by, 652
 electrical, model for, 783–785
 heat, 568–571, *569*, 576, 632
 law of, 569
Conduction electrons, 783
Conductivity(ies), 775n, 784, 789
 and resistivity, 784, 790
 thermal, 569, 570t, 576

Conductor(s), 569
 cavity within, *722*, 722–723
 charge carrier in, motion of, 774, *774*, 783, *783*
 charged, potential of, 720–723, *721*, *722*
 current-carrying, field created by, 866, 870, *870*
 in uniform magnetic field, 838, *838*, 839, *839*, 854
 magnetic forces on, 837–841, *838*
 current in, 774, *774*, 789
 definition of, 651, 671
 grounded, 652
 in electrostatic equilibrium, 693–695, *694*, 698
 magnetic field surrounding, 866–867, *867*
 metals as, 652
 parallel, magnetic force between, *869*, 869–870
 power loss in, 786–787, 790
 resistance of, 776, 779, 790
 semicircular, force on, *840*, 840–841
 uniform, resistance of, 777
Constant-volume gas thermometer, 534, *534*
Constant-volume process, first-law equation for, 566
Contact lenses, 1077n
Continuity, equations of, 433, *433*
Convection, 572–573, *573*, 576
 forced, 572
 natural, 572
Conversion factors, A.1–A.2
Coordinate systems, and frames of reference, 54–55
Coordinates, Galilean transformation of, *1150*, 1150–1151
Copernicus, Nicolaus, 395
Copper, Hall effect for, 853
Copper wire, drift speed of, 774–775
Cornea of eye, 1073
Correspondence principle, 92
Cosmic rays, 845–847
Coulomb, Charles, 647, 651, 653, 654
Coulomb constant, 654, 671
Coulomb (unit), definition of, 870
Coulomb's law, 649, 653–658, 655t, 662, 666, 671, 687–688, 866
 experimental proof of, 696, *696*
Coulomb's torsion balance, 651, *651*
Couple, 339, *339*
Critical angle, for total internal reflection, 1040
 index of refraction and, 1041, *1041*
Cross product. *See* Vector product(s).
Crystal ball, refraction and, 1063, *1063*
Crystals, birefringent, *1136*, 1136–1137
 diffraction of x-rays by, 1130n, 1130–1132, *1131*
 double-refracting, indices of refraction for, 1136t
Curie, Pierre, 887

Curie temperature, 887, *887*, 887t
Curie's constant, 887
Curie's law, 887
Current, and motion of charged particles, 773, *773*
 and resistance, 772–796
 average, 773
 definition of, 773
 density, 775, 784, 789
 direction of, 773, *773*
 displacement, 880, 891, 994
  capacitors in, 881
 in conductor, 774, *774*, 789
 in electron beam, *788*, 788–789
 induced, 906, 906n, *907*, 912, *912*, *915*, 915–916
  in motor, 922
 instantaneous, 773
 maximum, 809
 root mean square (rms), 966, *966*, 966n, 967, 983, 984
 SI unit of, 773, 776, 789
 sinusoidal, infinite current sheet carrying, 1008
 versus time, for *RL* circuit, 944, *944*
Current loop, circular, axis of, magnetic field on, *868*, 868–869, *869*
 magnetic moment of, 842, 854
 torque on, in uniform magnetic field, *841*, 841–843, *842*
Current sheet, infinite, carrying sinusoidal current, 1008
Curvature of space-time, 1181–1182
Cyclic process, 565, 575
 irreversible, 638
 PV diagram for, 617, *617*
 reversible, 638
Cyclotron, 849–850, *850*
Cyclotron frequency, 844
Cylinder, net torque on, 288–289, *289*
 raising of, 345
 rolling, total energy of, 307–308
 rolling motion of, 307, *307*
 solid, uniform, moment of inertia of, 285, *285*

Dam, force on, 426–427, *427*
D'Arsonval galvanometer, 814, *814*
Decibel(s), 484, 484n
 sound levels in, 484, 484t
Dees, 849–850
Demagnetization, nuclear, 624n
Democritus, 8
Density(ies), and atomic mass, 9–10
 charge, 777
 current, 775, 784, 789
 definition of, 17
 energy, average, 1003
  in electric field, 750
  magnetic, 945
  total instantaneous, 1003

Density(ies) *(Continued)*
 linear charge, 662
 mass, 777
 of cube, 13
 of various substances, 9t, 422, 422t
 surface charge, 662, 775n
 volume charge, 662
Depth, apparent, and true depth, 1064, *1064*
 variation of pressure with, *424*, 424–427
Derivative(s), partial, 716
Descartes, Rene, 1029n, 1037
Descartes' law, 1029n
Deuteron, binding energy of, 1178
Deviation, angle of, 1033–1035, *1034*
Dewar, Sir James, 574n
Dewar flask, 574, 574n
Diamagnetism, 881, 884, 887–888
Diatomic molecule, rigid-rotor model of, 324
Dielectric constant(s), 751, 752t, 761
Dielectric strength(s), 752t, 753
Dielectric(s), atomic description of, 757–760, *758*
 capacitors with, *751*, 751–756
 definition of, 751
Differential, perfect, A.29
Differential calculus, A.24–A.27
Differential equations, 155
Diffraction, definition of, 1117, 1118, 1138
 Fraunhofer, 1118, *1119*, *1120*, *1121*, 1123, *1123*, 1138
 Fresnel, *1118*, 1119, *1119*
 introduction of, 1117–1119, *1118*
 of x-rays, by crystals, 1130n, 1130–1132, *1131*
 pattern, circular aperture, resolution of, *1125*, 1125–1126
 single-slit, resolution of, 1124
 single-slit, *1119*, 1119–1124, *1120*
  intensity of, *1121*, 1121–1123
 types of, 1118
Diffraction grating, 1127–1130, *1128*, 1139
 compact disc and, 1129, *1129*
 intensity distribution for, 1128, *1128*
 orders of, 1129
 resolving power of, 1129–1130
Diffraction grating spectrometer, 1128, *1128*
Dimensional analysis, 10–12, 17
Diopters, 1076
Dipole, electric, 761
 electric field lines for, 666, *666*
 in electric field, potential energy of, *756*, 757
 in external electric field, 756, 756–757
 potential energy of, 756–757, 761
 torque on, 756, *756*, 761
 electric field of, 661, *661*
 electric potential of, *716*, 716–717
 oscillating electric, 1010–1011, *1011*
Dipole moment, electric, 756, 761

Disk, uniformly charged, electric field of, *664*, 664–665
Dispersion, 456n
Displacement, 43
 amplitude, of sound waves, 481–482, *482*
 angular, 277, *277*
 as function of time, 38, *38*
 of particle, as vector quantity, 56, *56*
 resultant, 59, *59*, 63
 velocity and speed, 24–26, *25*
 velocity as function of, 33–34
 versus time for simple harmonic motion, 361, *361*, 367
Displacement current, 880, 891, 994
 in capacitors, 881
Displacement vector, 71–72, *72*
 of projectile, 78, *79*
 velocity and acceleration vectors, 71–74
Domain walls, 885, *885*
Doppler, Christian Johann, 487n
Doppler effect, *487*, 487–491, *488*, 492
Dot product(s), 174, 192
 of unit vectors, 175
Double-slit experiment, Young's, 1092–1096, *1093*, 1108
Double-slit interference patterns, average intensity of, 1108
 intensity distribution of, *1096*, 1096–1098, *1098*
 phasor diagrams for, 1099, *1100*
Drag, 436
Drag coefficients, 158–160
Drift speed, 774, 784, 784n, 790
 in copper wire, 774–775
Dulong-Petit law, 598
Dyne, 112
Dyne·cm, 173

Earth, magnetic field(s) of, 888–890, *889*
Eddy currents, *922*, 922–923, *923*
Efflux, speed of, 435, *435*
Einstein, Albert, 92, 454n, 454–455, 529–530, 923, 1022, 1024, 1147, 1156, 1157, 1161, 1167
 and energy and mass, 220, 1174, 1176
 time interval measurement and, 1158
Einstein's general relativity, 1181
Einstein's theory of gravitation, *1180*, 1180–1181
Einstein's theory of relativity, 4, 92–93, 1149, 1149n, 1156n, 1156–1157
Elastic, definition of, 338
Elastic limit, 347, *347*
Elastic modulus, 346, 346t, 349
Elastic potential energy, 204–205
Elasticity, in length, *346*, 346–347
 of shape, 347, *347*
 of solids, 345–349
 static equilibrium and, 337–359
 volume, 347–348, *348*
Electric, "root" of word, 647

Electric charges. *See* Charge(s), electric.
Electric current. *See* Current.
Electric field(s). *See* Field(s), electric.
Electric flux, 684–687, *685, 686,* 686n, 697
  net, through closed surface, 688, *689*
  through cube, 686–687, *687*
Electric force(s). *See* Force(s), electric.
Electric guitar, *909,* 909–910
Electric potential. *See* Potential, electric.
Electrical heater, power in, 787
Electrical instruments, 813–817
Electrical safety, 818
Electrical shock, 818
Electrically charged, 650
Electricity, and magnetism, 647–1019
  Franklin's model of, 650
  magnetism and, 833
Electromagnet, 833
  eddy current and, 922–923
Electromagnetic forces, 161
Electromagnetic induction, 907
  law of. *See* Faraday's law.
Electromagnetic waves. *See* Wave(s), electromagnetic.
Electromagnetism, 648
  laws of, 905, 906
Electromotive force, 787
Electron beam, bending of, 845, *845*
  current in, *788,* 788–789
Electron gas, 783
Electron gun, 670, *670*
Electron volt, 709
Electron(s), accelerated, 669
  acceleration of, 35
  charge on, 655, 671
    and mass of, 655t
  collision of, in copper, 785
  conduction, 783
  free, 655, 655n
  magnetic moments of, *882,* 882–883
  measurement of ratio e/m for, 849, *849*
  momentum of, 1172, 1182
  projection of, into uniform electric field, *668,* 668–669
  speed of, 1148–1149
  speedy, energy of, 1175
  spinning, *882,* 882–883
  wave-like properties of, 785
Electrophotography, 753
Electrostatic equilibrium, conductors in, 693–695, *694,* 698
Electrostatic precipitator, 726, *726*
Electrostatics, applications of, *725,* 725–727, *726, 727, 728*
  Coulomb's law of. *See* Coulomb's law.
Electroweak forces, 161
Elements, periodic table of, A.32–A.33
Elevator motor, power delivered by, 187–188, *188*
Ellipse, equation of, A.21
Elliptical orbits, 398

emf, 787, 797–799, 818
  back, 921
  induced, 905, 908, 909
    and electric fields, *917,* 917–919
    in generator, 920
    in rotating bar, *913,* 913–914
  induction of, in coil, 910
  motional, *911,* 911–914, 925
  self-induced, *940, 940,* 947, 954
Emissivity, 573
Energy, and automomobile, 188–191
  and mass, conservation of, 1177
    equivalence of, 1176–1178, 1183
  and work, 171–201
  binding, 403
    of deuteron, 1178
  carried by electromagnetic waves, 1002–1004
  conservation of, 207–210, 211, 219, 556, 797, 819
  conversion factors for, A.2
  electrical, and power, 786–789
    supplied to resistor, 786, 790
  equipartition of, 596–598, 607
    theorem of, 590
  forms of, 171
  from wind, 437–439, *438*
  in inductor, 946
  in mass-spring collision, 214–215, *215*
  in planetary and satellite motion, 404–407
  in rotational motion, 293–295, 297
  internal, 552, 564, 565, 575
    change in, 564
    of ideal gas, *592, 592*
    total, of solid, 598
  ionization, 221
  kinetic, 171–172, 181, 192, 850, 948
    and work-energy theorem, *180,* 180–186, *182,* 192
    at high speeds, 191–192
    average, per molecule, 590, 607
      temperature and, 590
    change in, work and, 403, 403n
    collisions and, 243–244
    for various objects, 181t
    in nuclear reactor, 1177
    loss, due to friction, 183
    mass and, 404, 404n
    of simple harmonic oscillator, 368, *369,* 381
    relativistic, 191–192, 1173, 1174, *1174,* 1182
    rotational, 281–282, *282,* 308
    total, 590
      of rigid body, 307–308, 325
    loss of, in resistor, 813
  mass in, 220
  mechanical, 207
    changes in, in nonconservative forces, 210–216
    conservation of, 207
    for freely falling body, 208

Energy *(Continued)*
    total, 207, 222
  motion of standing wave and, 503–504, *504*
  of photons, 1024, 1025
  of satellite, 221, *221*
  of simple harmonic oscillator, 368–370, *369*
  of speedy electrons, 1175
  of speedy protons, 1175–1176
  potential, 202–204, 206, 222, 708, 948
    change in, 404, 708, 728
    conservative forces and, 202, 206–207, 217
    due to point charges, electric potential and, *712,* 712–714, *713*
    elastic, 204–205, 222
    function, 206
    function for mass-spring system, 218, *218*
    gravitational, 203, *203,* 204, 222, 401–404, *402, 403,* 411
    of electric dipole, 756–757, 761
      in electric field, *756,* 757
    of simple harmonic oscillator, 368, *369,* 381
    of two point charges, 713, 729
    terms, addition of, 403, 403n
  quantization of, 220–221
  quantized, 1024
  relativistic, 1173–1176
  rest, 1174, 1175, 1182
  rotational, 281–283, *282,* 296
  solar, 1005
  thermal, 616, 618
    definition of, 552
    heat and, 552–554
    temperature and, 558–559, 562–563, *563*
    transfer of, 552, 563, 564
  total, definition of, 1174
    for circular orbits, 405–406, 411
    of simple harmonic oscillator, 368, 381
    of Universe, 219
  transfer of, in LC circuit, 948, *949*
  transmitted by sinusoidal waves on strings, *469,* 469–470
  zero point, 535
Energy density, average, 1003
  in electric field, 750
  total instantaneous, 1003
Energy diagrams, and equilibrium of system, 218–219
Energy-level diagrams, 221, *221*
Energy-momentum relationship, 1174–1175, 1183
Engine(s), Carnot, 620–624, 630
  efficiency of, 622
  efficiency of, 619
  gasoline, *625,* 625–626
  Hero's, 562
  internal combustion, 616
  steam, 623
  Stirling, 625

Entropy, 628–631, 638
and probability, 636
as measure of microscopic disorder, 636, 638
changes in, 628, 629–630, 631, 634–635, 638
in irreversible processes, 631–635
of Universe, 629, 632, 638
on microscopic scale, 635–637, 638
Entropy function, 628
Equation(s), analysis of, 12
Bernoulli's, 433–435, *434*
applications of, *436*, 436–437, *437*
differential, 155
factoring of, A.17
first-law, 564, 564n
for adiabatic process, 565
for constant-volume process, 566
for angular momentum, 294t
for electromagnetic waves, 995–996, 998
for thin lenses, 1065, 1082
for translational motion, 294t
Galilean transformation, 1168
kinematic. *See* Kinematic equations.
Kirchhoff's loop, 942, 965
lens makers', 1065
linear, A.18–A.19
linear wave, 470–472, *471*
mass-energy equivalence, 1174
Maxwell's, 906, 923–925, 994, 997, 1001, 1013
mirror, 1057, 1082
net-force, 210
of circle, A.21
of continuity, 433, *433*
of ellipse, A.21
of motion for simple pendulum, 372, 372n
of parabola, A.22
of rectangular hyperbola, A.22
of state, 541
for ideal gas, 541–542
van der Waals', *606*, 606–607, *607*
of straight line, A.21
quadratic, A.17–A.18
transformation, Galilean, 1168
Lorentz, 1167–1171, 1168n
wave, for electromagnetic waves in free space, 998
general, 998
Equilibrium, 105, 115, 337, 338, 349
electrostatic, conductors in, 693–695, *694*, 698
neutral, 219
objects in, 341
of rigid object, conditions of, *338*, 338–340, *339*
of system, energy diagrams and, 218–219
rotational, 339, 349
stable, 218
static, 337, 339
and elasticity, 337–359
rigid objects in, 341–345

Equilibrium *(Continued)*
thermal, 532, 543
translational, 339, 349
unstable, 219, *219*
Equilibrium position, 364
Equipartition of energy, 596–598, 607
Equipotential surface(s), 710–711, 715, *715*, 728
charged conductor in equilibrium and, 721
Equivalence principle, of Einstein, 220, 1174, 1176, 1181
Erg, 173
Escape speed(s), 405–407, *406*, 407t
Ether, and speed of light, 1151–1152, *1152*, 1154
Euler method, 159–160, 160t
Evaporation process, 603
Exit ramp, curved, design of, 147–148, *148*
Expansion, adiabatic, reversible, *PV* diagram for, 594–595, *595*, 608
adiabatic free, *563*, 565
free, of gas, 563, 619, *619*, 633, *633, 635*, 635–636
of ideal gas, 623
isothermal, of ideal gas, *566*, 566–567
work done during, 567
linear, average coefficient of, 537–538, 538t, 543
of railroad track, 540
thermal, of solids and liquids, 536–541, *537*, 537n
volume, average coefficient of, 538, 543
Exponents, rules of, A.17
Eyeglasses, 1077n
Eye(s), accommodation in, 1074
conditions of, 1074
focusing of, 1074
parts of, 1073–1074, *1074*
resolution of, 1127, *1127*

Fahrenheit, Gabriel, 536n
Fahrenheit temperature scale, 536
Farad, 742, 760
Faraday, Michael, 106, 647, 742, 833, 905, 910
Faraday's experiment, 906–908, *907*
Faraday's ice-pail experiment, 696n
Faraday's law, 905–938, 940, 947, 954, 965, 981, 994
applications of, 909–910
Farsightedness, 1074–1075, *1075*
Fermat's principle, 1042, *1042*, 1042n, 1043
Ferromagnetism, 881, 884–887
Fiber optics, 1041, *1041*
Fictitious forces, 151–152, *152*, 162
Field-ion microscope, 727, *727*, 728
Field(s), electric, 649–683, *658*, 658–661
amplitude, and single-slit diffraction pattern, 1121
between parallel plates of opposite charge, 711, *711*

Field(s) *(Continued)*
calculations, using Gauss's law, 697t
charge carrier in, motion of, 783–784, *784*
definition of, 658
due to charged rod, 663, *663*
due to group of charges, 659, 672
due to point charge, 659, *659*, 672
due to point source, 1004
due to solenoid, *918*, 918–919
due to two charges, *660*, 660–661
energy density in, 750
external, electric dipole in, *756*, 756–757
finding of, 672–673
induced, 918
induced emf and, *917*, 917–919
lines, *665*, 665–667, 672, 715, *715*
around two spherical conductors, 722, *722*
for point charge(s), *665*, 665–667, *666*, *667*
rules for drawing, 666
nonuniform, and parallel-plate capacitor, *755*, 755–756
obtaining of, from electric potential, *715*, 715–717
of continuous charge distribution, 661–665, *662*, 672
of dipole, 661, *661*
of uniform ring of charge, 663–664, *664*
of uniformly charged disk, *664*, 664–665
potential energy of, electric dipole in, *756*, 757
projection of electron into, *668*, 668–669
radiated, 1007, *1007*
SI units of, 709
sinusoidal, 999, 1014
total, 659
uniform, motion of proton, 711–712, *712*
potential differences in, 709–712, *710*, 728
gravitational, 203, 400–401
magnetic, 832–863, 890
along axis of solenoid, *876*, 876–877, *877*
created by current-carrying conductor, 866, 870, *870*
created by current-carrying wire, *872*, 872–873, *873*
created by infinite current sheet, 873–874, *874*
created by toroid, 873, *873*, 890
direction of, 834, *834*
due to point source, 1004
due to wire segment, 867–868, *868*
energy in, 944–946, 954
exponential decrease of, 911, *911*
loop moving through, 916–917, *917*
motion of charged particles in, 834–835, *836*, 843–847, *844*, 854
applications of, 847–850
of Earth, 888–890, *889*
of solenoid, *875*, 875–876, 890, 945

Field(s) *(Continued)*
  on axis of circular current loop, *868*, 868–869, *869*
  patterns, 834, *834*
  properties of, created by electric current, 865
  proton moving in, 837, *837*
  radiated, 1006–1007
  SI units of, 837
  sinusoidal, 999, 1014
  sources of, 864–904
  steady, 836
  strength of, 883
    magnetization and, 883–884
  surrounding conductor, 866–867, *867*
  uniform, current-carrying conductor in, *838*, 838, 839, *839*, 854
    force on, 838, 854
    proton moving perpendicular to, 844
    torque on current loop in, *841*, 841–843, *842*
Films, thin, interference of light waves in, *1103*, 1103–1107
  wedge-shaped, interference in, 1106–1107, *1107*
Fitzgerald, G.F., 1156
Fixed point temperatures, 535, 535t
Fizeau, Armand H.L., 1026
Fizeau's technique for measuring speed of light, 1026, *1026*
Floating ice cube, 430, *430*
Flow, characteristics of, *431*, 431–432, 441
  irrotational, 432
  nonsteady(turbulent), 431, *432*
  steady(laminar), 431, *431*, 432
  tube of, 432, *432*
Flow rate, 433
Fluid dynamics, 422, 431–432, 441
Fluid mechanics, 421–451
Fluid(s), definition of, 421
  force exerted by, 422, *422*
  ideal, 432
  incompressible, 432
  nonviscous, 432
  pressure in, *422*, 423
    depth and, 424, *424*
  device for measuring, 422, *422*
Flux, electric, 684–687, *685*, *686*, 686n, 697
  net, through closed surface, 688, *689*
  through cube, 686–687, *687*
  expulsion, 888
  magnetic, *877*, 877–878, *878*, 891
  unit of, 877
  volume, 433
Focal length, 1057
  of two thin lenses in contact, 1069
Focal point, 1057
  image, 1066
  object, 1066
Foot, as unit of length, 5
Force law(s), 104–105
  Coulomb's. *See* Coulomb's law.

Force(s), action, 114–115, *115*
  applied, power delivered by, 913
  as vector quantity, 56
  average, in collision of automobile into wall, 241, *241*
  between molecules, 588
  buoyant, 155n, 427–428, 441
    and Archimedes' principle, 427–431, *428*, *429*, 441
  central, 145, 162
    speed of mass and, 146
    work done by, 402
  concept of, 105–107
  concurrent, 340, *340*
  conservative, 204, 222
    and potential energy, 202, 206–207, 217
    work done by, 204
  constant, horizontal, block pulled on frictionless surface by, 183–184, *184*
    measurement of, 180, *180*
    of spring, 179
    work done by, *172*, 172–174, 192
  contact, 105, *106*
  conversion factors for, A.1
  electric, 649, 651, 653, *655*, 655–656, 671–672
    and magnetic, 836
    on proton, 660
  electromagnetic, 161
  electromotive, 787
  electroweak, 161
  equivalent, 338
  external, 107–108
  fictitious, 151–152, *152*, 162
    in linear motion, 152–153, *153*
  field, 105–107, *106*
  frictional, and power requirements for car, 189t
    block pulled on rough surface by, 184
    total, 189
  gravitational, 203
    acting on planets, 399, *399*
    and motion of particle, 410, *410*
    between extended object and particle, *407*, 407–408
    between mass and bar, 407–408, *408*
    between particles, 401–402, *402*
    between spherical mass and particle, *408*, 408–410, *409*
    between two particles, 392, *392*
    pendulum and, 371
    properties of, 393
    weight and, 394–395
  gravitational weight and, *123*, 123–124
  horizontal, work done using, 174, *174*
  impulse of, 239, *239*, 261
  impulsive, 240, 261
  in rotating system, 154, *154*
  inertial, 151
  Lorentz, 847, 924
  magnetic, and electric, 836

Force(s) *(Continued)*
  between two parallel conductors, *869*, 869–870
  direction of, 835, *835*, 836, *836*, 853
  magnitude of, 836, 854
  on current-carrying conductor, 837–841, *838*
  on current segment, 874, *874*
  on sliding bar, 914, *914*
  perpendicular, 836
  measurement of, 107, *707*
  net, 105
  nonconservative, 205–206, 222
  normal, 115, 115n
  nuclear, 161
  of dam, 426–427, *427*
  of friction, 124–128, *125*
  of gravity, 203, 203n, 204
  of nature, 106–107, 160–161, 162
  on semicircular conductor, *840*, 840–841
  on uniform magnetic field, 838, 854
  reaction, 114–116, *115*
  relativistic, 1172, 1173
  resistive, 162
    and air friction, 189
    motion in, 154–158
    proportional to speed, 155, *155*
  restoring, 179
  resultant, 105
    acceleration of frictionless surface and, 113, *113*
  location of, 656, *656*
  on zero, *656*, 656–657
  unbalanced, 105
  units of, 112, 112t
  varying, work done by, *176*, 176–180, *178*, 192
Foucault, Jean, 1024
Fourier series, 516
Fourier synthesis, 517, 518, *518*
Fourier's theorem, 516, 517
Franklin, Benjamin, 650
  model of electricity of, 650
Fraunhofer diffraction, 1118, *1119*, *1120*, *1121*, 1123, *1123*, 1138
Free-body diagram(s), 116–117, *117*, 118, *118*, *119*, 129, *130*
Free expansion, adiabatic, *563*, 565
  of gas, 563, 619, *619*, 633, *633*, 635, 635–636
Free-fall, 37
  definition of, 37
  speed of ball in, 209, *209*
Free-fall acceleration, 37, 39, *39*, 44, 394–395, 411
  and projectile motion, 77, 77n
  variation of, with altitude, 395, 395t
Free space, electromagnetic waves in, wave equations for, 998
  impedance of, 1003
  permeability of, 865, 890
  permittivity of, 654–655

Freely falling body(ies), 36–40
  mechanical energy for, 208
  position and velocity versus time for, 40, *40*
Freezing of water, 541
Frequency(ies), angular, 361, 466
  of ac generator, 965
  of motion for simple pendulum, 372
  of rotating charged particle, 844, 854
  average, 516
  beat, 515
  collision, 605
  cyclotron, 844
  for mass attached to spring, 365–366
  fundamental, 506
    of vibrating strings, 507–508
  moving train whistle, 489–490
  observed, 489, 492
  of electromagnetic waves, 995, 996
  of normal modes, 506
  of oscillator, 502, *502*
  of periodic waves, 455
  of simple harmonic motion, 361
  of sinusoidal waves, 466
  of tuning fork, 512, *512*
  of vibration, 367
  precessional, 324
  resonance, 379, 508, *508*, 973, 977, 984
  siren, 490
  with observer in motion, 488, 492
Fresnel, Augustin, 1024, 1118
Fresnel diffraction, *1118*, 1119, *1119*
Friction, as nonconservative force, 206
  coefficients of, 124–125, 126t
    measurement of, 127, *127*
  connected masses with, magnitude of acceleration of, 128, *128*
  forces of, 124–128, *125*
  kinetic, coefficients of, 125–126
    determination of, *127*, 127–128
    force of, 124
  situations involving, *182*, 182–183, *183*
  loss of kinetic energy due to, 183
  static, coefficient of, 124–125
    speed of car and, 147
    force of, 124, 129
Frictional forces, and power requirements for car, 189t
  total, 189
Fringes of light waves, 1095, 1108
  bright, 1120
  dark, 1120–1121
  distance between, 1096
  order number of, 1095, 1108
Ft·lb, 173
Fusion, latent heat of, 557–558, 558t, 559
Fusion reaction, 1177–1178

Galilean transformation(s), 90–91, 92, 94, 1150, 1167
  of coordinates, *1150*, 1150–1151
  velocity, 1151, 1169

Galileo Galilei, 1, 4, 36, 108, 1025, 1152
Galvanometer, 814, *814*, *815*, 906, *907*, 907–908
Gamma rays, 1013
Gas constant, universal, 542
Gas thermometers, 533, 534, *534*, 535
Gas(es), 421
  density of, 422
  electron, 783
  expansion of, quasi-static, *561*, 561n, 561–562, *562*, 566
  free expansion of, 563, 619, *619*, 633, *633*, *635*, 635–636
    adiabatic, 575
  ideal, adiabatic processes for, 594–596
    definition of, 541, 541n, 542
    equation of state for, 541–542
    free expansion of, 637
    internal energy of, 592, *592*
    isothermal expansion of, *566*, 566–567
    law of, 541–542
    macroscopic description of, 541–543, 544
    molar specific heats for, 593, 607
    molecular model of, 587–591, *588*
    quasi-static, reversible process for, 630
    specific heat of, *591*, 591–594, *592*
    temperature scale of, 624
    thermal energy and, 591
  in equilibrium, atmospheric layer of, 599
  kinetic theory of, 586–614
  molar specific heats of, 593, 593t
  molecules in, 542–543, 629, 636, *636*
    average values of, computation of, 600–601
    mean free path for, *604*, 604–605, *605*
    speed distribution of, 603, *603*
  moles of, 542
  work done by, 575
Gasoline, consumed by compact car, 190
  used by cars in U.S., 14
Gasoline engine(s), *625*, 625–626
Gauge pressure, 427
Gauss, Karl Friedrich, 687
Gauss (unit), 837
Gaussian surface, *687*, 694–695, *695*
Gauss's law, *687*, 687–689, 694, 698, 924
  application of, 698–699
  to charged insulators, 690–693
  derivation of, 696–697, *697*
  electric field calculations using, 697t
  experimental proof of, 696, *696*
  in magnetism, 878–879, *879*, 891, 924
Gauss's probability integral, A.31
Generator(s), alternating current, 919, *919*
  angular frequency of, 965
  delivered to circuit, 975
  voltage amplitude of, 965
  and motors, 919–922
  direct current, 920–921, *921*
  emf induced in, 920
  Van de Graaff, 707, *725*, 725–726
  water-driven, 939

Geometry, A.20–A.22
  capacitance and, 760t
Gilbert, William, 647, 833
Golf ball, teeing off, *240*, 240–241
Grain, waves, 455
Gravitation, Einstein's theory of, *1180*, 1180–1181
Gravitational constant, universal, 392, 410
  measurement of, 393–394
Gravitational field(s), 203, 400–401
Gravitational force. *See* Force(s), gravitational.
Gravitational potential energy, 203, *203*, 204, 222, 401–404, *402*, *403*, 411
Gravitational redshift, 1181
Gravity, center of, 254, 340–341, *341*, 349
  force of, 203, 203n, 204
  laws of, 391–420
  and motion of planets, *396*, 396–400
  Newton's laws of, 392–393, 396, 410
Grimaldi, Francesco, 1024
Ground-fault interrupter(s), 818, 909, *909*
Ground state, of atoms, 221
Gyroscope, motion of, 322, *323*

Hall, Edwin, 851
Hall coefficient, 852
Hall effect, *851*, 851–853
  for copper, 853
  quantum, 853
Hall voltage, 852
Hand, weighted, *342*, 342–343
Harmonic motion, simple. *See* Motion, simple harmonic.
Harmonic series, 506–507, *507*, 519
Harmonics, of waveform, *518*
Heat, and thermal energy, 552–554, 574
  and work, 615
  conduction of, 568–571, 632
    law of, 569
  conversion factors for, A.2
  definition of, 552
  flow of, direction of, 632–633
  in thermodynamic processes, 561–563
  latent, 557n, 557–560, *558*, 575
    of fusion, 557–558, 558t, 559
    of vaporization, 558, 558n, 558t
  mechanical equivalent of, *553*, 553–554, 575
  of condensation, 558n
  of solidification, 558n
  specific, 554–557, 555n, 555t, 575
    measurement of, 556, 556n
    molar, 555, 555t
    at constant pressure, 591
    at constant volume, 591, 607
    of gas, 593, 593t, 607
    of hydrogen, 597, *597*
    of ideal gas, *591*, 591–594, *592*
    of solids, 598, *598*
  transfer of, 560, 568–574, *569*

Heat *(Continued)*
  irreversible, 633–634
  through slabs, 570
  to solid object, 632–633
  units of, 553
Heat capacity, 554, 575
Heat engine(s), 616, *616*, 617, *617*, 637
  and second law of thermodynamics, 616–619
  Carnot-cycle, 627–628
  refrigerators as, 617–618, *618*, 628
  thermal efficiency of, 617
Heat pumps, 617, *618*, *627*, 627–628
Height, measurement of, 373
Helium, liquid, boiling of, 560
Helix, 844
Henry, Joseph, 647, 833, 905, 906
Henry (unit), 940, 947
Hero's engine, 562
Hertz, Heinrich, 648, 924–925, 994–995, 996–997, 1024
Home insulation, 571–572
Hooke's law, 107, 179, 364
Hoop, uniform, moment of inertia of, 284, *284*
Horsepower, 187
Household appliances, power connections for, 817, *817*
Household wiring, 817, *817*
Huygens, Christian, 1022, 1024, 1025–1026, 1036, 1037
Huygens' principle, 1036–1039, *1037*, 1043, 1119
  applied to reflection and refraction, *1038*, 1038–1039, *1039*
Hydraulic press, 425, *425*
Hydrogen, molar specific heat of, 597, *597*
Hydrogen atom, 657
Hyperbola, rectangular, equation of, A.22
Hyperopia, 1074–1075, *1075*
Hysteresis, magnetic, 886
Hysteresis loops, *886*, 886–887, *887*

Ice cube, floating, 430, *430*
Ice-pail experiment, Faraday's, 696n
Ideal absorber, 574, 576
Ideal gas. *See* Gas(es), ideal.
Ideal gas law, 541–542
Image distance, 1053
Image(s), focal point, 1066
  formation of, 1053
  formed by converging lenses, 1068
  formed by diverging lenses, 1068
  formed by flat mirrors, 1052–1055, *1053*
  formed by refraction, *1061*, 1061–1064
  formed by spherical mirrors, *1055*, 1055–1061, *1056*
  length contraction, of box, 1164, *1164*
  of lens, location of, 1064, *1064*
  real, 1053
  virtual, 1053

Impedance, 977, 983
  in *RLC* series circuit, 973
  of free space, 1003
Impedance triangle, 973, *973*
Impulse, and linear momentum, 239–242
Impulse approximation, 240
Impulse-momentum theorem, 239, *239*, 261
Inductance, 939–963
  and emf, calculation of, 941
  definition of, 940
  mutual, 947, 947–948
    definition of, 947
    of two solenoids, 947–948, *948*
  of coils, 940, 954
  of solenoid, 941, 954
Induction, charging metalic object by, 652, *652*
  electromagnetic, 907
  Faraday's law of, 905, 906–911, 918, 924, *925*
  mutual, 939, 947
Inductive reactance, 968–969, 983
Inductor(s), 939, 941
  current and voltage across, 968, *968*
  energy in, 946
  in ac circuits, *967*, 967–969
  instantaneous voltage drop across, 969
  maximum current in, 968
  phase relationships in, 971–972, *972*
Inertia, law of, 109–110
  moment(s) of. *See* Moment(s) of inertia.
Inertial forces, 151
Inertial frames, 109–110, 129
  different, laws of mechanics in, 1149–1150, *1150*
Inertial mass, 110–111
Inertial system, definition of, 1149
Infeld, L., 454n, 454–455
Infinite current sheet, carrying sinusoidal current, 1008
  radiation from, *1006*, 1006–1008, *1007*
Infrared waves, 1012
Insulation, home, 571–572
Insulator(s), 652–653, *653*
  application of Gauss's law to, 690–693
  definition of, 651
Integral calculus, A.27–A.30
Integral(s), definite, A.27–A.28, 42
  Gauss's probability, A.31
  indefinite, A.27, A.29–A.30
  line, 708
  path, 708
Integration, 27, 41
  partial, A.28
Intensity(ies), average, of double-slit interference patterns, 1108
  distribution, for diffraction grating, 1128, *1128*
  of double-slit interference patterns, *1096*, 1096–1098, *1098*
  of single-slit diffraction, *1121*, 1121–1123
  of sound waves, 482–484, *483*, 486–487

Intensity(ies) *(Continued)*
  reference, 484
  relative, of maxima, 1123–1124
  wave, 1002
Interference, 460, 499
  and light waves, 1117–1118, *1118*
  conditions for, 1091–1092
  constructive, 460, *461*, 472, 500, *501*, 518
    condition for, 1095, 1097, 1103, 1104, 1108, 1109
    of light waves, 1095
  destructive, 461, *462*, 472, 500, 501, *501*, 518–519
    condition for, 1095, 1103–1104, 1108, 1109
  in soap film, 1106
  in thin films, *1103*, 1103–1107
    problems in, 1105
  in time, 513–515
  in wedge-shaped film, 1106–1107, *1107*
  of light waves, 1091–1116
  patterns of, 461, 1104
    double-slit, average intensity of, 1108
      intensity distribution of, *1096*, 1096–1098, *1098*
      phasor diagrams for, 1099, *1100*
    in water waves, *1093*
    multiple-slit, *1100*, 1101
    three-slit, *1100*, 1100–1101
  spatial, 513
  temporal, 513–515
Interferometer, Michelson, *1107*, 1107–1108, 1154, *1154*
Interrupters, ground-fault, 818, 909, *909*
Inverse-square law, 392, 393
  and acceleration, 396
Ionization energy, 221
Ions, magnetic moments of, 883t
Irreversible process(es), 616, 616n, 619–620
  changes in entropy in, 631–635
Isentropic process, 630
Isobaric process, 566, 575
Isothermal expansion, of ideal gas, *566*, 566–567
  work done during, 567
Isothermal process, 566, 575
  PV diagram for, 606, *606*
  work done in, 566–567
Isovolumetric process, 566, 575

Jewett, Frank Baldwin, *849*
Joule, James, 551, 552, 553
Joule heating, 787, 787n, 913
Joule (unit), 173, 553
Junction, in circuit, 800–801
Junction rule, Kirchhoff's, 804, 805, *805*

Kamerlingh-Onnes, Heike, 782
Kaon, decay of, at rest, 238, *238*
Kelvin, definition of, 535, 543

Kelvin, Lord (William Thomson), 5n, 536n, 618
Kelvin-Planck form of second law of thermodynamics, 617, 618
Kelvin temperature scale, 534, 535, *535*, 536, 624
Kepler, Johannes, 4, 392, 395–396
Kepler's law(s), 395–396, 411
 second, 397, 411
  and conservation of angular momentum, 399–400
 third, 397–398, 411
Kilocalorie, 553
Kilogram(s), 5, 110, 112n
 definition of, 5
Kilowatt hour, 187
Kinematic equations, 44
 choice of, 34
 derived from calculus, 41–43
 for motion in straight line under constant acceleration, 34t
 for rotational and linear motion, 279t
 rotational, 279, 296
Kinematics, 23
 rotational, 278–279
Kinetic energy. *See* Energy, kinetic.
Kinetic friction, situations involving, *182*, 182–183, *183*
Kinetic theory, of gases, 586–614
Kirchhoff's circuit law, 943–944
Kirchhoff's loop equation, 942, 965
Kirchhoff's loop rule, 967, 969
Kirchhoff's rule(s), 797, 804–808, *805*, 819
 application of, 807, *807*
 first, 816, 819
 second, 808, 811, 816, 819

Ladder, uniform, 344, *344*
Laminar flow, 431, *431*, 432
Land, E.H., 1133
Latent heat, 575
Laue, Max von, 1130
Laue pattern, 1131
Lawrence, E.O., 849
Law(s), Ampère-Maxwell, 880, *880*, 880n, 924, 925
 Ampère's, 864, 870–874, 875, 876, 890
  generalized form of, 924
 Biot-Savart, 864, 865–869, *866*, 890
 Boltzmann distribution, 599–602
 Boyle's, 541
 Bragg's, 1132
 Brewster's, 1135, 1139
 Coulomb's. *See* Coulomb's law.
 Curie's, 887
 Descartes', 1029n
 distributive, of multiplication, 175
 Dulong-Petit, 598
 Faraday's. *See* Faraday's law.
 force, 104–105
 Galilean addition, for velocity, 1151

Law(s) *(Continued)*
 Gauss's. *See* Gauss's law.
 Hooke's, 107, 179, 364
 ideal gas, 541–542
 inverse-square, 392, 393
 Kepler's, 395–396, 397–398, 411
 Kirchhoff's circuit, 943–944
 Lenz's, 908, 914n, 914–917, *915*, 925, 940
  application of, 916, *916*
 Lorentz force, 925
 Malus's, 1134, 1134n
 of atmospheres, 600
 of Charles and Gay-Lussac, 541
 of conservation of angular momentum, 318, 325
 of conservation of linear momentum, 237
 of conservation of momentum, 1171
 of electromagnetic induction. *See* Faraday's law.
 of electromagnetism, 905, 906
 of gravity, 391–420
  and motion of planets, *396*, 396–400
  Newton's, 392–393, 396, 410
 of heat conduction, 569
 of ideal gas, 541–542
 of inertia, 109–110
 of motion, 104–143. *See also* Newton's laws of motion.
 of reflection, 1027, *1028*, 1043
 of thermodynamics. *See* Thermodynamics.
 of vector addition, 840
 Ohm's, 775–776, 776n, 784, 789, 801, 973
  resistance and, 775–780
 Pascal's, 424
 power, analysis of, 12
 Snell's. *See* Snell's law.
 Stefan's, 573
 Torricelli's, 435
LC circuits. *See* Circuit(s), *LC*.
Length contraction image, of box, 1164, *1164*
Length(s), 5, 10, 17
 contraction, *1163*, 1163–1165
 conversion factors for, A.1
 dimensional analysis of, 11
 elasticity in, 346, 346–347
 focal, of mirror, 1057, 1082
  of two thin lenses in contact, 1069
 measured, approximate values of, 7t
 proper, 1163
 standards of, 4–8
Lens maker's equation, 1065
Lens(es), aberrations, *1070*, 1070–1072, *1071*
 biconcave, 1066, *1066*
 biconvex, 1065, *1065*, *1066*
 converging, 1082
  final image of, 1070, *1070*
  image formed by, 1068
 diverging, 1082
  image formed by, 1068
 f-number of, 1072–1073

Lens(es) *(Continued)*
 gravitational, 1181, *1181*
 image of, location of, 1064, *1064*
 of eye, 1073
 optical, testing of, 1105
 power of, 1076
 thin, 1064–1070, *1066*
  combination of, 1069
  equations for, 1065, 1082
  ray diagrams for, *1067*, 1067–1068
  shapes of, 1066, *1067*
  sign convention for, 1066, 1066t
  two in contact, focal length of, 1069
  under water, 1069
Lenz, Heinrich, 914n
Lenz's law, 908, 914n, 914–917, *915*, 925, 940
 application of, 916, *916*
Leucippus, 8
Lift, 436–437
Light. *See also* Wave(s), light.
 and optics, 1021–1145
 dispersion of, and prisms, 1033–1036
 dual nature of, 1025
 nature and properties of, 1021–1022
 nature of, 1023–1025
 passing through slab, 1033, *1033*
 photons in, 1092n
 reflection of, 1027–1028
 refraction of, *1029*, 1029–1033, *1030*
 source(s) of, coherent, 1092
  monochromatic, 1092
  wavelength of, 1096
 spectrum of, 1034–1035, *1035*
 speed of, 1151–1154, *1152*
  constancy of, 1156, 1157
  in silica, 1033
  in vacuum, 1030n, 1030–1031, 1043
  measurements of, 1025–1026, 1154
 ultraviolet, 1012–1013
 visible, 835, 1012
Light ray, double-reflected, 1028, *1028*
Lightbulb, cost of operating, 788
 electrical rating of, 788
 three-way, operation of, 803–804, *804*
Lightning bolts, time interval measurement of, 1158–1159, *1159*
Limiting process, 29
Line, straight, equation of, A.21
Line integral, 708
Linear equations, A.18–A.19
Linear expansion, average coefficient of, 537–538, 538t, 543
Linear momentum. *See* Momentum, linear.
Linear motion, 314, *314*
 fictitious forces in, 152–153, *153*
Linear wave equation, 470–472, *471*
Linear waves, 459, 472
Liquids, 422
 sound waves in, 480t, 481
 speed distribution of molecules in, 603
 thermal expansion of, *540*, 540–541

Live wire, 817, 817n
Livingston, M.S., 849
Lloyd's mirror, *1101*, 1101–1102
Load resistance, 798, 982
  power and, 799, *799*
Logarithm(s), A.20
  natural, 810
Loop rule, 805
"Loop-the-loop" maneuver by plane, in vertical circle at constant speed, 149, *149*
Lorentz, Hendrik A., 1156, 1167
Lorentz force, 847, 924
Lorentz force law, 925
Lorentz transformation, 1182
  velocity, 1169–1170, 1172, 1182
Lorentz transformation equations, 1167–1171, 1168n

Mach number, 491
Magnet, permanent, and superconductor, 888
Magnetic bottle, 845, *845*
Magnetic energy density, 945
Magnetic fields. *See* Field(s), magnetic.
Magnetic flux, 877, 877–878, *878*, 891
  through rectangular loop, 878, *878*
Magnetic force. *See* Force(s), magnetic.
Magnetic hysteresis, 886
Magnetic moment(s), of atoms, 881–883, *882*, 883t
  of coil, 843
  of current loop, 842, 854
  of electrons, *882*, 882–883
  of ions, 883t
  orbital, 882, *882*
  SI units of, 842
Magnetic permeability, 884
Magnetic poles, 832, 833
Magnetic resonance imaging, 876
Magnetic substances, classification of, 884–885
Magnetic susceptibility(ies), 884, 884t
Magnetism, electricity and, 647–1019
  Gauss's law in, 878–879, *879*, 891, 924
  in matter, 881–888
  induction of, 833
  knowlege of, history of, 647
Magnetite, 647, 832
Magnetization, and magnetic field strength, 883–884
  saturation, 888
Magnetization curve, 886, *886*, 887
Magnetization vector, 883
Magneton, Bohr, 883
Magnification, angular, 1077, *1078*, 1079, 1081
  lateral, 1053, 1079, 1082
  maximum, 1078
Magnifier, simple, *1077*, 1077–1078
Malus, E.L., 1134

Malus's law, 1134, 1134n
Manometer, open-tube, 427, *427*
Maricourt, Pierre de, 832
Mass density, 777
Mass-energy, conservation of, 1174
Mass-energy equivalence, 220, 1149, 1174
Mass spectrometer, *848*, 848–849
Mass-spring collision, energy of, 214–215, *215*
  speed of mass and, 185–186
Mass-spring system, 368
  mechanical energy for, 208
  potential energy function for, 218, *218*
  simple harmonic motion for, *370*, 381
Mass(es), 17, 104
  acceleration and, 111
  acceleration of, on frictionless incline, *119*, 119–120
  and energy, conservation of, 1177
    equivalence of, 1176–1178, 1183
  and kinetic energy, 404, 404n
  and weight, 111
  as inherent property, 111
  as scalar quantity, 111
  atomic, selected, table of, A.4–A.13
  attached to spring, *364*, 364–368, *366*, *367*
    period and frequency for, 365–366
  center of, 251–255, *252*, *253*
    acceleration of, 256
    location of, 252–253, 254, *254*
    of pair of particles, 252, *252*
    of right triangle, 255, *255*
    of rigid body, 252, *252*, 262
    of rod, *254*, 254–255
    of three particles, 254, *254*
    vector position of, 253, *253*, 262
    velocity of, 256, 262
  connected, 295, *295*, 317, *317*
  in motion, *215*, 215–216
  with friction, magnitude of acceleration of, *128*, 128
  conservation of, 220
  conversion factors for, A.1
  gravitational attraction of, 1179–1180
  gravitational force on, 394
  in free-fall, speed of, 209, *209*
  inertial, 110–111
  inertial property of, 1179–1180
  invariant, 1175
  method for lifting, 216, *216*
  molar, 541
  moving in circular orbit, 404–405, *405*
  Newton's second law of motion applied to, 404
  of spring, 1178
  of Sun, 399
  of various bodies, 7t
  on curved track, *212*, 212–213
  pulled on frictionless surface, speed of, 183–184, *184*
  pulled on rough surface by frictional force, 184

Mass(es) *(Continued)*
  rest, 1177
  rotating, 283, *283*
  SI unit of, 5–6, 110
  sliding down ramp, speed of, 212, *212*
  speed of, central force and, 146
    mass-spring system and, 185–186
  standards of, 4–8
  struck by club, range of launch of, *240*, 240–241
  two blocks of, magnitude of acceleration of, 122, *122*
  two unequal connected, acceleration of, *121*, 121–122
  units of, 112, 112n, 112t
Mathematical notation, 16–17
Mathematics review, A.14–A.30
Matter, building blocks of, 8–9
  magnetism in, 881–888
Maxwell, James Clerk, 602, 647–648, 833, 923, 995, 1022, 1147, 1151
Maxwell-Boltzmann distribution function, 602
Maxwell's equations, 906, 923–925, 994, 997, 1001, 1013
Mean free path, 604–605
Mean free time, 605
Mean solar day, 6
Mean solar second, 6
Measurement(s), expression of, in numbers, 5n
  physics and, 3–22
Mechanical energy. *See* Energy, mechanical.
Mechanical waves, 453–527
  requirements for, 455
Mechanics, 1–450, 1148
  classical, 104
  newtonian, 1158
  statistical, 596
Meissner effect, 888
Mercury barometer, 427, *427*
Mercury thermometers, 533, *533*
Metals, as conductors, 652
  atoms of, 655n
  resistivity versus temperature for, 781, *781*
Meter, 5
  definitions of, 5
Michell, John, 833
Michelson, Albert A., 1107, 1154, 1155
Michelson interferometer, *1107*, 1107–1108, 1154, *1154*
Michelson-Morley experiment, *1153*, 1153n, 1153–1156, 1157
Microscope, compound, 1078–1080, *1079*
  field-ion, 727, *727*, *728*
  limiting resolution of, 1126
Microwaves, 1011, 1012, 1012n
Millikan, Robert A., 651, 724
Millikan oil-drop experiment, *724*, 724n, 724–725
Mirror equation, 1057, 1082

Mirror(s), concave, *1055*, 1055–1057, *1056*, *1057*, 1060–1061
convex, 1058, *1058*, 1061
diverging, 1058
flat, images formed by, 1052–1055
Lloyd's, *1101*, 1101–1102
ray diagrams for, 1058–1060, *1059*
rearview, tilting, 1054–1055, *1055*
sign convention for, 1058t
spherical, images formed by, *1055*, 1055–1061, *1056*
two, multiple images formed by, 1054, *1054*
Modes, normal, frequencies of, 506
 in standing waves, *505*, 505–506
 wavelengths of, 506
Modulus, bulk, 346, 347–348, *348*, 349
 elastic, 346, 346t, 349
 shear, 346, 347, *347*, 349
 Young's, 346, 349
Molar mass, 541
Molar specific heat, 555, 555t
Molecule(s), average kinetic energy per, 590, 607
 collision frequency for, 605
 diatomic, possible motions of, 596, *596*
 forces between, 588
 in atmosphere, 600
 in gas, 629, 636, *636*
 Newton's laws of motion and, 587
 nonpolar, 758
 oxygen, 282
 polar, 757–758
 rms speeds for, 590–591
 speeds of, distribution of, *602*, 602–604, *603*
 symmetrical, polarization and, 757, *757*
 water, *756*, 756–757
Mole(s), 10, 541
 of gas, 542
Moment arm, 287, *288*, 288
Moment(s), angular, as quantized, 882
 magnetic. *See* Magnetic moment(s).
Moment(s) of inertia, calculation of, 283–287
 concept of, use of, 285n, 296
 of rigid objects, 286t, 296
 of uniform hoop, 284, *284*
 of uniform rigid rod, 285, *285*
 of uniform solid cylinder, 285, *285*
Momentum, and radiation pressure, 1004–1006, 1005n, *1105*
 angular, 236n
  as fundamental quantity, 324–325
  conservation of, 317–321
   Kepler's second law and, 399–400
   law of, 318, 325
  equations for, 294t
  net torque and, 322
  of particle, *312*, 312–315, 325
  of rigid body, 315, *315*, 325
  spin, 882

Momentum *(Continued)*
 torque and, 313
 total, system of particles of, 313
 concept of, 235, 236n
 conservation of, for two-particle system, 236, 236–237, 261
 laws of, 1171
 linear, and collisions, 235–275
  and rocket propulsion, 260
  conservation of, 236–238
  definition of, 236, 261
  impulse and, 239–242
  laws of conservation of, 237
  of electron, 1172, 1182
  of recoiling pitching machine, 238, *238*
  relativistic, definition of, 1172
  Newton's laws of motion and, 1171–1173
  rotational, 236n
  total, 237, 261
  of system, in collision of automobiles, 243
  of system of particles, 256
 total linear, of system, 256
Moon, acceleration of, *396*, 396–397
 motion of, 391–392
Morley, Edward W., 1154
Motion, and scalars, 24n
 circular. *See* Circular motion.
 connected blocks in, *215*, 215–216
 horizontal, 81
 in accelerated frames, 151–154
 in elliptical orbit, 400, *400*
 in one direction, 23–52
 in plane, 76–77
 in resistive forces, 154–158
 laws of, 104–143. *See also* Newton's laws of motion.
 linear, 314, *314*
 fictitious forces in, 152–153, *153*
 of charge carrier, in conductor, 774, *774*, 783, *783*
 in electric field, 783–784, *784*
 of charged particles, electric current and, 773, *773*
 in magnetic field, 834–835, 836, 843–847, *844*, 854
  applications of, 847–850
 in uniform electric field, 667–669, 672
 of disturbance, 455
 of gyroscopes and tops, 321–324, *322, 323*
 of longitudinal pulse, 480, *480*
 of moon, 391–392
 of particle, gravitational force and, 410, *410*
 of proton in uniform electric field, 711–712, *712*
 of system of particles, 255–259
 one-dimensional, kinematic equations for, 34t
  with constant acceleration, 32–36
 oscillatory, 360–390
 periodic, 360–390
 planetary, energy in, 404–407

Motion *(Continued)*
 Kepler's laws of, 395–396, 397–398, 411
 laws of gravity and, *396*, 396–400
 precessional, 321–322
 projectile, 77–85, 93
 pure rotational, 276
 reference circle for, 376, *376*, 377
 relative, at high speeds, 92–93
 rolling, angular momentum, and torque, 306–336
 of rigid body, *307*, 307–309, *308*
 pure, 307, *307*
 rotational, equations for, 294t
 with constant angular acceleration, 278–279
 work, power, and energy in, *293*, 293–295, 297
 work and energy in, 293
 work-energy relation for, 294
 satellite, energy in, 404–407
 simple harmonic, 361–364, *362*, 365
 acceleration in, 362–363, *367*
 amplitude in, 361
 and uniform circular motion, compared, *375*, 375–377, *376*
 displacement versus time for, 361, *361, 367*
 for mass-spring system, *370*, 381
 frequency of, 361
 pendulum and, 371, *371*
 period of, 361, 380
 phase constant in, 361
 properties of, 363
 speed in, 361, 362–363, 367
 velocity as function of position for, 369
 three types of, 23
 translational, 23
  equations for, 294t
 two-dimensional, 71–103
  with constant acceleration, 74–77, *75*
 uniform, 109
 uniform circular, *85*, 85–87
  and simple harmonic motion, compared, *375*, 375–377, *376*
  Newton's second law of motion and, 144–149, *145, 146*, 162
 wave, 454–478, *456*
 and vibration, 453
 simulator, 460
Motorcycle, pack leaders, relativistic, 1171, *1171*
 speeding, 1170, *1170*
Motor(s), and generators, 919–922
 definition of, 921
 induced current in, 922
Mueller, E.W., 727
Müller, K. Alex, 782
Multiplication, distributive law of, 175
Muon(s), decay curves for, 1162, *1162*
 time dilation and, 1162, *1162*, 1164
Myopia, 1075, *1076*

National Standard Kilogram No. 20, 6
Nature, fundamental forces of, 160–161, 162
Near point of eye, 1074
Nearsightedness, 1075, *1076*, 1076–1077
Net-force equation, 210
Neutrino, energy and momentum for, 1175
Neutron(s), 8, 9, 10
  charge and mass of, 655t
  slowing down of, by collisions, 247–248
Newton, Isaac, 4, 105, 106, 108, 236, 391, 1021, 1023, 1037, 1082, 1104, 1147
Newtonian mechanics, 1–450, 1148, 1158
Newtonian relativity, principle of, 1149–1154
Newton·meter, 173
Newton's laws of gravity, 392–393, 396, 410
Newton's laws of motion, 3, 4
  applications of, 116–118, 144–170
  first, 108–109, 129, 152
  molecules and, 587
  relativistic momentum and relativistic form of, 1171–1173
  second, 111–113, 112n, 129, 463, 471, 472, 844
    applied to mass, 404
    damped oscillations and, 377–378
    for rotational motion, 375
    for system of particles, 256–257, *257*, 262
    in component form, 117
    linear momentum and, 236, 239, 256
    of mass attached to spring, 364
    torque and, 289–290
    uniform circular motion and, 144–149, *145*, *146*, 162
  third, *114*, 114–116, *115*, 129, 237, 242, 392, 436, 869, 869n
Newton's rings, 1104–1105, *1105*
Newton's theory of light, 1023–1024
Newton(unit), 112
Nichrome wire, resistance of, 779
Nobel prizes, A.35–A.39
Nodes, 503, *504*, 519
Nonohmic materials, 776, 778, *778*
Northern Lights, 832, 846–847
Notation, scientific, A.14–A.15
Nuclear demagnetization, 624n
Nuclear force, strong, 161
  weak, 161
Nuclear reactor, kinetic energy in, 1177
Nucleus, 8, 10
Numbers, expression of measurements in, 5n
Numerical modeling, in particle dynamics, 158–160

Object distance, 1052–1053, *1053*
Ocean, pressure of, 426
Oersted, Hans Christian, 647, 833, 865, 870
Ohm, Georg Simon, 775, 804

Ohm (unit), 776, 969
Ohmic materials, 775–776, 778, *778*
Ohm's law, 775–776, 776n, 784, 789, 801, 973
  resistance and, 775–780
Open-circuit voltage, 798, *798*
Optic axis, 1136
Optical illusion, of levitated professor, 1054, *1054*
Optical lenses, testing of, 1105
Optical stress analysis, 1138, *1138*
Optically active materials, 1137–1138
Optics, fiber, 1041, *1041*
  geometric, 1052–1090
    ray approximation in, *1026*, 1026–1027, *1027*, 1042
  light and, 1021–1145
Orbits, circular, around Earth, satellite moving in, *148*, 148–149
  mass moving in, 404–405, *405*
  total energy for, 405–406, 411
  elliptical, 398
  motion in, 400, *400*
  of planets, 397, *397*, 397n, *399*
Order number, for fringes, 1095, 1108
Order-of-magnitude calculations, 13–14
Oscillating body, 363–364
Oscillation(s), damped, 377–378
  forced, *379*, 379–380
  in *LC* circuit, 948–952
  on horizontal surface, 369–370
Oscillator, critically damped, 378, *378*
  damped, 378, *378*
  frequency of, 502, *502*
  overdamped, 378
  simple harmonic, 467–468, *468*, 468n
    energy of, 368–370, *369*
    position of, 361, 380
  two speakers driven by, 502, *502*
  underdamped, *378*
  velocity and acceleration of, 362, 381
Oscillatory motion, 360–390
Oscilloscope, 669–671, *671*
  cathode ray tube of, 669–670, *670*
Oscilloscope simulator, 974
Otto cycle, efficiency of, 626
  PV diagram for, 625, *626*
Overtones, 506
Oxygen, magnetic field and, 864
Oxygen molecule, 282
Ozone layer, 1013

Parabola, equation of, A.22
Parallel-axis theorem, 285–286, *287*
  application of, 287
Parallel-plate capacitors. *See* Capacitor(s), parallel-plate.
Parallelogram rule of addition, 57, *57*
Paramagnetism, 881, 884, 887

Particle dynamics, numerical modeling in, 158–160
Particle(s), and extended object, gravitational force between, *407*, 407–408
  and solid sphere, gravitational force between, 408–409, *409*
  and spherical shell, gravitational force between, 408, *408*
  angular momentum of, *312*, 312–315, 325
  charged, electromagnetic waves and, 1008–1009, *1009*
  motion of, electric current and, 773, *773*
    in magnetic field, 834–835, 836, 843–847, *844*, 854
    applications of, 847–850
    in uniform electric field, 667–669, 672
  rotating, angular frequency of, 844, 854
  distance traveled by, 56
  distribution of, in space, 599
  falling in vacuum, 158, *158*
  gravitational force between, 401–402, *402*
  moving in simple harmonic motion, properties of, 363
  pair of, center of mass of, 252, *252*
  system(s) of, 587
    conservation of momentum for, *236*, 236–237, 261
    motion of, 255–259
    Newton's second law of motion for, 256–257, *257*, 262
    speed of, 604
    total angular momentum of, 313
    total momentum of, 256
  three, center of mass of, 254, *254*
  three interacting, 403, *403*
  total energy and momentum of, 1175, 1182
  two, gravitational forces between, 392, *392*
  velocity versus time plot for, *41*, 41–42, *42*
Pascal, 441
Pascal, Blaise, 424
Pascal's law, 424
Path, mean free, 604–605
Path difference, and phase angle, 502
  and phase difference, 1121
  of light waves, 1095
Path integral, 708
Path length, 501
Peacock, feathers of, colors in, 1103
Pendulum, 210, *210*, *371*, 371–375
  ballistic, 246, *246*
  conical, free-body diagram for, *146*, 146–147
  in oscillation, 380, *380*
  period of, 1163
  physical, 373–374, *374*, 381
  simple, 372, *373*, 381
    period of motion for, 372–373
    period of oscillation for, 373
  torsional, 374–375, *375*
Performance, coefficients of, 627, 628

Period, for mass attached to spring, 365–366
  of motion, for simple pendulum, 372
  of simple harmonic motion, 361
Periodic motion, 360–390
Periodic table of elements, A.31–A.32
Periodic waves, frequency of, 455
Permeability, magnetic, 884
  of free space, 865, 890
Permittivity of free space, 654–655
Phase angle, 971, 983
  path difference and, 502
Phase change, 557
  due to reflection, 1101–1103, *1102*
Phase constant, 467
  in simple harmonic motion, 361
Phasor diagram(s), 965, 966, 968, 972, 975, *975*, 1097, *1098*, 1099, *1100*, 1109, 1121, *1121*, *1122*
Phasors, 965
Photoelectric effect, 1024, 1025
Photon(s), 1024
  energy and momentum for, 1175
  energy of, 1024, 1025
  in light, 1092n
Physical quantities, symbols, dimensions, and units of, A.3–A.4
Physics, and measurement, 3–22
  areas of, 1
  classical, 4
  modern, 4, 1147–1188
    "architects" of, 1153
Pierce, John R., 924, 924n
Pitch, 516
Pitching machine, recoiling, 238, *238*
Planck, Max, 1022, 1024, 1147
Planck's constant, 324, 882, 1024
Plane electromagnetic waves, 997, 997–1002, 998n, 999, *1001*, 1001n
Plane polar coordinates, 54, *54*
Planets, data on, 397n, 397–398, 398t
  gravitational force acting on, 399, *399*
  motion of, energy in, 404–407
    Kepler's laws of, 395–396, 397–398, 411
    laws of gravity and, 396, 396–400
  orbits of, 397, *397*, 397n, *399*
Plates, in standing waves, 512–513, *513*
Platinum resistance thermometers, 781
Point charge(s), closed surface surrounding, 688, *688*
  electric field due to, 659, *659*, 691, 691–692
  electric field lines for, 665, 665–667, *666*, *667*
  electric potential of, 713
  group of, electric field due to, 659, 672
  outside of closed surface, 688, *688*
  positive, acceleration of, 668, *668*
  potential due to, 714, *714*
  potential energy due to, electric potential and, 712, 712–714, *713*
  several, electric potential of, 713, *713*

Point charge(s) *(Continued)*
  two, electric potential of, *713*, 713n, 713–714
  potential energy of, 713, 729
Poisson, Simeon, 1118
Polar coordinates, 55
Polar molecules, 757–758
Polarization, 652
  induced, 757
  of light waves, *1132*, 1132–1138, *1133*, *1134*, *1135*, *1137*
Polarizer, 1133, *1133*
Polarizing angle, 1134, 1139
Polaroid, 1133
Poles, magnetic, 832, 833
Position, velocity, and acceleration, graphical relations of, 30
  velocity vector as function of, 75
Position-time graph, 24, *24*, 26, *26*, 33
Position vector, 72, *72*, 74, 93
  as function of time, 75
  magnitude and direction of, 64, *64*
Potential, due to two point charges, 714, *714*
  electric, 707–740
    and potential energy due to point charges, 712, 712–714, *713*
    calculation of, 730
    due to continuous charge distributions, 717–720, *718*, 729, 729t
    due to uniformly charged ring, 718, *718*
    obtaining electric fields from, 715, 715–717
    of dipole, *716*, 716–717
    of finite line of charge, 719, *719*
    of point charge, 713
    of several point charges, 713, *713*
    of two point charges, *713*, 713n, 713–714
    of uniformly charged disk, 719, *719*
    potential difference and, 708–709
    SI units of, 709
Potential difference, 776
  and electric potential, 708–709
  in uniform electric field, 709–712, *710*, 728
Potential energy. *See* Energy, potential.
Potentiometer, 816, *816*
Pound, 112
Power, 186–188
  and load resistance, 799, *799*
  average, 186, 976, 978, 983–984
  definition of, 186
  delivered by applied force, 913
  delivered by elevator motor, 187–188, *188*
  delivered to wheels, 190
  electrical energy and, 786–789, 790
  in ac current circuit, 975–976
  in algebra, A.16–A.17
  in electrical heater, 787
  in *RLC* circuit, 978
  in rotational motion, 293–295, 297
  instantaneous, 186–187, 192, 975
  loss in conductor, 786–787, 790
  of lenses, 1076

Power *(Continued)*
  requirements for car, frictional forces and, 189t
  resolving, diffraction grating and, 1129–1130
  transmission of, 981–983
  transmitted by sinusoidal wave, 470, 473
Power connections, for household appliances, 817, *817*
Power cord, three-pronged, 818
Power law, analysis of, 12
Powers of ten, prefixes for, 8, 8t
Poynting vector, 1002, *1002*, 1006, *1006*, 1007, 1010, 1014
Precipitator, electrostatic, 726, *726*
Pressure, 347, 422–423
  absolute, 427
  amplitude, of sound waves, 481–482, *482*, 491
  constant, molar specific heats at, 591
  conversion factors for, A.2
  critical, 607
  definition of, 423, 441
  gauge, 427
  in fluids, *422*, 423
  measurement of, 427, *427*
  of ocean, 426
  variation of, with depth, *424*, 424–427
Prince Rupert drops, 1117
Prism spectroscope, 1035, *1035*
Prism(s), and total internal reflection, *1040*, 1040–1041
  dispersion of light and, 1033–1036
  measurement of index of refraction using, 1036, *1036*
Probability, 637, 637t
  entropy and, 636
Probability integral, Gauss's, A.31
Projectile, and target, *81*, 81–82
  angle of projection and, 82–83, *83*
  displacement vector of, 78, *79*
  from spring-loaded popgun, speed of, 213–214, *214*
  horizontal range of, and duration of flight of, 82, *82*
  and maximum height of, *79*, 79–80
  parabolic trajectory of, 77, *77*
  range of, 80, *80*
  speed of, 82
  struck by club, range of, *240*, 240–241
Projectile-cylinder collision, 319, *319*
Projectile motion, 77–85, 93
Propulsion, rocket, *259*, 259–261, *260*
Proton(s), 8–9, 10
  charge on, 655, 671
  and mass of, 655t
  electric forces on, 660
  motion of, in uniform electric field, 711–712, *712*
  moving in magnetic field, 837, *837*
  moving perpendicular to uniform magnetic field, 844

Proton(s) *(Continued)*
  -proton collision, 250–251
  speedy, energy of, 1175–1176
Ptolemy, Claudius, 395
Pythagorean theorem, A.22, 589, 1160

Quadratic equations, A.17–A.18
Quality factor, *977, 978,* 978n, 978–979
Quantization, of electric charge, 651
Quantum Hall effect, 853
Quantum mechanics, 4
Quantum states, 221, *221*
Quarks, 9, *9,* 655n
  free, 655n

Radial acceleration, *87,* 87–89, 93–94
Radian, 277
Radian measure, A.21
Radiation, 573–574
Radiation belts, Van Allen, 845–847, *846*
Radiation pressure, momentum and, 1004–1006, 1005n, *1105*
Radio waves, 1011
Rarefactions, of sound waves, 479, 481
Ray approximation, in geometric optics, *1026,* 1026–1027, *1027,* 1042
Ray diagram(s), for mirrors, 1058–1060, *1059*
Rayleigh's criterion, 1124–1125, *1125,* 1138–1139
Rays, 486
  extraordinary, 1136, *1136*
  gamma, 1013
  ordinary, 1136, *1136*
  paraxial, *1055,* 1056, 1082
RC circuit, 808–813
  capacitor in, charging of, *812, 812*
    discharging of, 812–813
Reactance, capacitive, 970, 983
  inductive, 968–969, 983
Reaction forces, 114–116, *115*
Rectangle, area of, 16
Rectangular coordinate system, 54, *54*
Reference circle, for motion, 376, *376,* 377
Reflection, 1023
  and refraction, 1027–1033
  change in phase due to, 1101–1103, *1102*
  diffuse, 1027, *1028*
  Huygens' principle applied to, *1038,* 1038–1039
  law of, 1027, *1028,* 1043
  of light, 1027–1028
  polarization of light waves by, 1134–1135, *1135*
  specular, 1027, *1028*
  total internal, *1039,* 1039–1041, *1040, 1041,* 1043
  transmission of waves and, *464,* 464–465, *465*
Refracting surfaces, flat, *1063, 1063*

Refracting surfaces *(Continued)*
  sign convention for, *1061,* 1062t, 1062–1063
Refraction, angle of, for glass, 1032, *1032*
  double, polarization of light waves by, 1135–1137
  Huygens' principle applied to, 1038–1039, *1039*
  images formed by, *1061,* 1061–1064
  index of, 1030–1032, 1031t, *1032,* 1033, *1034,* 1043
  and critical angle, 1041, *1041*
  measurement of, 1032, 1036, *1036*
  reflection and, 1027–1033
  Snell's law of. *See* Snell's law.
Refrigerators, as heat engines, 617–618, *618,* 628
Relativity, 1148–1188
  and electromagnetism, 1178–1179, *1179*
  and simultaneity, of time, 1158–1159
  Einstein's theory of, 4, 92–93, 1149, 1149n, 1156n, 1156–1157
  general, 1179–1182
    Einstein's, 1181
  Newtonian, principle of, 1149–1154
  principle of, 1156, 1157
  special, consequences of, 1157–1167
  special theory of, consequences of, 1182
    postulates of, 1182
Resistance, and Ohm's law, 775–780
  and temperature, 780–782
  current and, 772–796
  definition of, 776, 789
  equivalent, 800, 802, *802,* 818
  finding of, by symmetry arguments, 803, *803*
  load, 798, 982
  power and, 799, *799*
  of coaxial cable, 780, *780*
  of conductors, 776, 779, 790
  of nichrome wire, 779
  of uniform conductors, 777
  SI unit of, 790
Resistive forces, 162
  motion in, 154–158
  proportional to speed, 155, *155*
Resistivity(ies), 776, 777, 784, 789
  conductivity and, 784, 790
  of various materials, 777t
  temperature coefficient of, 780–781
  temperature versus, 781, *781*
Resistor(s), 778
  color code for, 778, *778,* 778t
  current and voltage across, 966
  electrical energy supplied to, 786, 790
  energy loss in, 813
  in ac circuits, *965,* 965–967, *966,* 966n, 967t, 983
  in parallel, 800–804, *801,* 818
  in series, 799–800, *800,* 818
  phase relationships in, 971–972, *972*
  shunt, 814, *815*

Resistor(s) *(Continued)*
  three, in parallel, *802,* 802–803
Resonance, 379, 380, *508,* 508–510, *509,* 519
  in mechanical system, 978
  in *RLC* series circuit, 978, *978*
  vibrations, 380
Resonance frequency, 379, 508, *508,* 973, 977, 984
Rest energy, 1174, 1175, 1182
Rest mass, 1177
Restoring force, 179
Retina of eye, 1074
Reversible process(es), *619,* 619–620
  for quasi-static ideal gases, 630
Right-hand rule, 278, *278,* 842, *842*
  vector product and, 310, *310,* 325
Right triangle, center of mass of, 255, *255*
Rigid body(ies), angular momentum of, 315, *315,* 325
  center of mass of, 252, *252,* 262
  conditions of equilibrium of, *338,* 338–340, *339*
  definition of, 276
  in static equilibrium, 341–345
  moments of inertia of, 286t, 296
  rolling motion of, *307,* 307–309, *308*
  rotating, angular velocity and acceleration of, 279–281, *280,* 281–282, *282*
  work and, 293, *293*
  rotation of, around fixed axis, 276–305, *277, 315,* 315–317, 316n
  torque and, 288, *288,* 289, 289–290
  total kinetic energy of, 307–308, 325
Rings, Newton's, 1104–1105, *1105*
RL circuits. *See* Circuit(s), *RL*.
RLC circuits. *See* Circuit(s), *RLC*.
RLC series circuits. *See* Circuit(s), *RLC* series.
rms current, 966, *966,* 966n, 967, 983, 984
rms speed, 590, 591, 602
rms voltage, 967, 983
Rocket, escape speed of, 406
Rocket propulsion, *259,* 259–261, *260*
Rod(s), center of mass of, *254,* 254–255
  charged, electric field due to, 663, *663*
  conducting, connected to battery, 1009, *1010*
  in standing waves, 512–513, *513*
  rigid, uniform, moment of inertia of, 285, *285*
  rotating. *See* Rotating rod.
  swinging, 374, *374*
Roemer, Ole, 1025
Roemer's method for measuring speed of light, *1025,* 1025–1026
Roentgen, W, 1130
Romognosi, Gian Dominico, 833n
Root mean square(rms) current, 966, *966,* 966n, 967, 983, 984
Root mean square(rms) speed, 590, 602
  for molecules, 591

Root mean square (rms) voltage, 967, 983
Rotating masses, 283, *283*
Rotating platform, angular speed of, 320, *320*
Rotating rigid body. *See* Rigid body(ies), rotating.
Rotating rod, *290*, 290–291, 295, *295*, *316*, 316–317
Rotating system, force in, 154, *154*
Rotating turntable, 281
Rotating wheel, 279
Rotation of rigid body, around fixed axis, 276–305, *277*, *315*, 315–317, 316n
Rotational kinematics, 278–279
Rotational motion. *See* Motion, rotational.
Rowland ring, 885–886, *886*

Satellite receiver-transmitter dish, 994
Satellite(s), energy of, 221, *221*
  in circular orbit around Earth, *148*, 148–149
  motion of, energy in, 404–407
  orbit of, changing of, 405
Savart, Felix, 865
Scalar product, as commutative, 175
  definition of, 192
  of two vectors, 174–176, *175*, 175n
Scalar quantities, vector quantities and, 55–56, 64
Scalars, motion and, 24n
Scattering, polarization of light waves by, *1137*, 1137
Scientific notation, A.14–A.15
Scott, David, 37
Scuba diving, refraction and, 1063
Second, definition of, 6
Seesaw, 342, *342*
Self-inductance, 940–941
Self-induction, 939
Semiconductor(s), 652
  resistivity versus temperature for, *781*, 781–782
Series expansions, A.24
Shape, elasticity of, 347, *347*
Shear modulus, 346, 347, *347*, 349
Shear stress, 347
Shock, electrical, 818
Shock waves, *490*, 491, *491*
Shunt resistor, 814, *815*
SI unit(s), 7
  base, A.34
  derived, A.34
  of average speed, 26
  of capacitance, 742, 760
  of current, 773, 776, 789
  of electric field, 709
  of force, 112
  of inductance, 940
  of magnetic field, 837
  of magnetic moment, 842
  of mass, 5–6, 110

SI unit(s) *(Continued)*
  of potential, 709
  of resistance, 790
  of time, 6
  prefixes for, 8t
Significant figures, 15–16
Simple harmonic oscillator, 467–468, *468*, 468n
Simultaneity, and relativity, of time, 1158–1159
Sinusoidal waves. *See* Wave(s), sinusoidal.
Siren, sound of, frequency of, 490
Ski jumper, vertical component of velocity of, *83*, 83–84
Slide-wire potentiometer, *816*, 816–817
Slug, 112n
Snell, Willebrord, 1029
Snell's law, 1029, 1029n, 1032, 1033, 1039, 1040, 1042, 1043, 1062, 1135
Soap film, interference in, 1106
Sodium chloride, structure, 1131, *1131*, 1135–1136
Software requirements, for spreadsheets, A.40–A.41
Solar cells, nonreflecting coatings for, 1106, *1106*
Solar day, 6n
  mean, 6
Solar energy, 1005
Solar second, mean, 6
Solenoid(s), axis of, magnetic field along, *876*, 876–877, *877*
  electric fields due to, *918*, 918–919
  ideal, 875, *875*
  inductance of, 941, 954
  magnetic field of, *875*, 875–876, 890, 945
  two, mutual inductance of, 947–948, *948*
Solidification, heat of, 558n
Solid(s), 421
  amorphous, 1136
  birefringent, 1136
  crystalline structure of, 1131, *1131*, 1135–1136
  deformation of, 345
  double-refracting, 1136
  elastic properties of, 345–349
  elasticity of, 345–349
  heat transferred to, 567–568
  sound waves in, 480, 480t
  specific heat of, 598, *598*
  thermal expansion of, 536–540, *537*, 537n
  total internal energy of, 598
Sonic boom, 491
Sound levels, in decibels, 484, 484t
Sound waves. *See* Wave(s), sound.
Space-time, curvature of, 1181–1182
Spaceship(s), contraction of, 1163, *1163*, 1164
  relative velocity of, 1170, *1170*
  triangular, *1164*, 1164–1165
  voyage to Sirius, 1165
Spatial interference, 513

Spatial order, 629
Spectrometer, diffraction grating, 1128, *1128*
  mass, *848*, 848–849
Spectroscope, prism, 1035, *1035*
Spectrum(a), of electromagnetic waves, 1011–1013, *1012*, 1014
  of light, 1034–1035, *1035*
  of sodium, lines of, resolving of, 1130
Speed(s), 27, 44, 73
  angular, average, 277
    instantaneous, 277, 278n, 296
  average, 26, 44, 602
  constant, circular motion with, 377
  conversion factors for, A.1
  dimensions of, 11t
  drift, 774, 784, 784n, 790
    in copper wire, 774–775
  escape, 405–407, *406*, 407t
  high, air drag at, 156–157, *157*
  kinetic energy at, 191–192
  relative motion at, 92–93
  in simple harmonic motion, 361, 362–363, *367*
  linear and angular, 280
  molecular, distribution of, *602*, 602–604, *603*
  most probable, 602
  of boat, relative to Earth, *91*, 91–92
    relative to water, 92, *92*
  of car, coefficient of static friction and, 147
  of efflux, 435, *435*
  of electromagnetic waves, 999
  of electron, 1148–1149
  of light. *See* Light, speed of.
  of mass, central force and, 146
    in free-fall, 209, *209*
    sliding down ramp, 212, *212*
  of projectile, 82
  of skier, on incline, 213, *213*
  of sound waves, 480–481, 491, 1152–1154
  of sphere, falling in oil, 156
  of wave pulse, 458
  of waves on strings, 462–464, *463*, 472
  resistive force proportional to, 155, *155*
  root mean square (rms), 590, 591, 602
  terminal, 155, *155*, 162
    for various objects falling through air, 157t
  transverse, 468–469
Sphere(s), charge on, 657, 657–658
  inside spherical shell, 695, *695*
  lead, volume of, 349
  rotating, 316, *316*
    forces acting on, 150–151, *151*
  solid, gravitational force between particle and, 408–409, *409*
  rolling, free-body diagram for, 309, *309*
  speed of, falling in oil, 156
  speed of center of mass of, 308–309
  two, connected charged, 723, *723*

Sphere(s) *(Continued)*
  uniformly charged, of electric potential, 720, *720*
Spherical aberration(s), 1056, *1056, 1070,* 1070–1071
Spherical shell, gravitational force between particle and, 408, *408*
  sphere inside, 695, *695*
  thin, electric field due to, *691,* 691–692
Spin, 865, 882
  angular momentum, 882
Spreadsheet problems, A.40–A.42
Spring, collision of two-body system with, 247, *247*
  deformation of, to measure force, 107, *107*
  elastic potential energy of, 205, *205*
  force constant of, 179
    measurement of, 180, *180*
  mass attached to, *364,* 364–368, *366, 367*
  mass of, 1178
  work done by, *178,* 178–180, *179,* 204
Spring-loaded popgun, speed of projectile from, 213–214, *214*
Stable equilibrium, 218
State function, 565
Statistical mechanics, 596
Steam engine, 623
Stefan's law, 573
Steradian, 696
Stirling engine, 625
Strain, 345, 349
  tensile, 346
Strain gauge, 816
Streamlines, *432,* 432–433, 436, *436,* 441
Stress, 345, 349
  shear, 347
  tensile, 346
  volume, 347
Submerged object, 430, *430*
Sun, mass of, 399
Superconductor(s), 782–783, 888
  critical temperatures for, 782, 782t
  resistance-temperature graph for, 782, *782*
Superposition principle, 459, 460, *461,* 472, 499, 500, 655–656, 659, 672, 1092
Susceptibility(ies), magnetic, 884, 884t
Synchronization, of clocks, 1157–1158

Tacoma Narrows Bridge, collapse of, 380
Tangential acceleration, *87,* 87–89, 280
Telescope(s), 1080–1082
  reflecting, 1080, *1081,* 1081–1082
  refracting, 1080, *1080*
  resolution of, 1126–1127
Temperature, 531–550
  and zeroth law of thermodynamics, 532–533, 543
  average kinetic energy and, 590
  change in, 531–550
  coefficient, resistivity of, 780–781
  conversion of, 536

Temperature *(Continued)*
  critical, 607, 782
    for superconductor, 782, 782t
  Curie, 887, *887,* 887t
  fixed point, 535, 535t
  molecular interpretation of, 589–591
  resistance and, 780–782
  resistivity versus, 781, *781*
  steps for obtaining, 624
Temperature gradient, 569, 576
Temperature scale(s), 533
  absolute, 534, 624
  Celsius, 534, 536
  Fahrenheit, 536
  ideal gas, 624
  Kelvin, 534, 535, *535,* 536, 624
  thermodynamic, 535
Templates, computer spreadsheets and, A.40
Tensile strain, 346
Tensile stress, 346
Tension, 116
  free-body diagram for suspended object and, 118–119, *119*
Terminal speed, 155, *155,* 162
  for various objects falling through air, 157t
Terminal voltage, 798, 798n
Tesla, Nikola, 981
Tesla (unit), 837, 854
Thermal conductivity(ies), 569, 570t, 576
Thermal contact, 532
Thermal efficiency, 617, 624, 637–638
  of heat engines, 617
Thermal energy, 616, 618
  definition of, 552
  heat and, 552–554, 574
  ideal gas and, 591
  temperature and, 558–559, 562–563, *563*
  transfer of, 552, 557, 563, 564, 570, 575, 628, 630–631
Thermal equilibrium, 532, 543
Thermal expansion, of solids and liquids, 536–541, *537,* 537n
Thermistor, 782
Thermodynamic temperature scales, 535
Thermodynamic variables, 542
Thermodynamics, 529–645
  first law of, 552, 563–565, 575, 615, 617
    applications of, 565–568
  second law of, 616, 617, 638
    heat engines and, 616–619
    Kelvin-Planck form of, 617, 618
  third law of, 624n
  work and heat in, *561,* 561–563, *562*
  zeroth law of, 628
    temperature and, 532–533, 543
Thermogram, 572
Thermometer(s), 533
  calibration of, 533
  constant-volume gas, 534, *534*
  gas, 533, 534, *534,* 535
  mercury, 533, *533*

Thermometer(s) *(Continued)*
  platinum resistance, 781
  problems associated with, 533, 533n
Thermos bottle, 574
Thermostats, 539
Thin-film interference, *1103,* 1103–1107
  problems in, 1105
Thompson, Benjamin, 553
Thomson, J.J., 849, *849*
Thrust, 261
Time, 17
  conversion factors for, A.1
  dilation, 1159–1163, *1160, 1162,* 1182
    and atomic clocks, 1166n, 1166–1167
    twins paradox and, *1165,* 1165–1166
  displacement as function of, 38, *38*
  intervals, approximate values for, 8t
  mean free, 605
  proper, 1162
  SI unit of, 6
  simultaneity and relativity of, 1158–1159
  standards of, 4–8
  velocity as function of, 32, 38, *38*
  velocity vector as function of, 74
  versus displacement for simple harmonic motion, 361, *361, 367*
  versus velocity plot for particle, *41,* 41–42, *42*
Time constant, of *RL* circuit, 944, *944*
Top, spinning, motion of, 321–322, *322*
Toroid, iron-filled, 885
  magnetic field created by, 873, *873,* 890
  magnetic field strength in core of, 883
Torque, 287–289, *288,* 296
  and angular acceleration, *289,* 289–292
  and angular momentum, 313
  definition of, 287, 288, 296, 842–843, 854
  net, and angular momentum, 322
  on cylinder, 288–289, *289*
  net external, 290, 296, 313, 325
  on current loop, in uniform magnetic field, *841,* 841–843, *842*
  rolling motion, and angular momentum, 306–336
  vector product and, 309–311, *310,* 325
Torricelli, Evangelista, 427
Torricelli's law, 435
Torsion balance, 1005, *1005*
  Coulomb's, 651, *651*
Trajectory, parabolic, of projectile, 77, *77*
Transducer, 479
Transformation, Galilean, 90–91, 92, 94, 1150, 1167
  of coordinates, *1150,* 1150–1151
  Lorentz, 1182
  velocity, Galilean, 1151, 1169
    Lorentz, 1169–1170, 1172, 1182
Transformation equations, Lorentz, 1167–1171, 1168n
Transformer, 981–983
  ac, 981, *981*
  ideal, 981, *981, 982*

Transformer (Continued)
  step-down, 981
  step-up, 981, 982–983
Transmission axis, 1133, 1139
Triangle method of addition, 57
Trigonometry, A.22–A.24
Tuning fork, frequency of, 512, *512*
Turbulent flow, 431, *432*
Turning points, 218
Turntable, rotating, 281
Twins paradox, time dilation and, *1165*, 1165–1166

Ultraviolet light, 1012–1013
Ultraviolet waves, 1012–1013
Unit vectors, 60–62, *61*
Units, cgs system of, 7
  conversion of, 13
  of acceleration, 112t
  of flux, 877
  of force, 112, 112t
  of heat, 553
  of mass, 112, 112n, 112t
  of physical quantities, A.2–A.3
  SI. *See* SI unit(s).
Universal gas constant, 542
Universe, entropy of, 629, 632, 638
  total energy of, 219

Vacuum, particle falling in, 158, *158*
  speed of light in, 1030n, 1030–1031, 1043
Van Allen, James, 845
Van Allen radiation belts, 845–847, *846*
Van de Graaff, Robert J., 725
Van de Graaff generator, 707, *725*, 725–726
van der Waals, J.D., 606
van der Waals' equation of state, *606*, 606–607, *607*
Vaporization, latent heat of, 558, 558n, 558t
Vascular flutter, 437, *437*
Vector form, 655
Vector product(s), and torque, 309–311, *310*, 325
  definition of, 310
  properties of, 310
  right-hand rule and, 310, *310*, 325
  two vectors of, 311
Vector(s), 53–70
  acceleration, displacement and velocity vectors, 71–74
  addition of, 56–58, *57*, 62, *64*, 64–65
    component method of, *61*, 61–63
    law of, 840
  components of, *59*, 59–60, *60*, 65, *65*
  displacement, 71–72, *72*
    of projectile, 78, *79*
    velocity and acceleration vectors, 71–74
  gravitational field, 401, *401*
  magnetization, 883
  multiplication of, by scalar, 58

Vector(s) (Continued)
  negative, 58
  position, 72, *72*, 74, 93
    as function of time, 75
    magnitude and direction of, 64, *64*
  Poynting, 1002, *1002*, 1006, *1006*, 1007, 1010, 1014
  projections of, 59–60
  properties of, 56–59
  quantities, and scalar quantities, 55–56, 64
  subtraction of, 58, *58*
  two, cross products of, 311
    equality of, 56, *56*
    scalar product of, 174–176, *175*, 175n
    sum of, 62–63
  unit, 60–62, *61*
    acceleration of particle and, 88, *88*
  velocity, 93
    as function of time, 74
Velocity, angular, and angular acceleration, 277–278
  as function of displacement, 33–34
  as function of position for simple harmonic motion, 369
  as function of time, 32, 38, *38*
  as vector quantity, 56
  average, 24, 25, 27, 43, 72, *72*
    and instantaneous, 27
    calculation of, 25
  drift, 774, 784, 784n, 790
  Galilean addition law for, 1151
  instantaneous, 26, 27, 44, 73
    and speed, 26–28
    definition of, 26
  of sinusoidal waves, 466
  position, and acceleration, graphical relations of, 30
  relative, 89–92, *90*
    of spaceship, 1170, *1170*
  tangential, 280
  transformation, Galilean, 1169
    Lorentz, 1169–1170, 1172, 1182
  versus time plot for particle, *41*, 41–42, *42*
Velocity selector, 847–848, *848*
Velocity-time graph, 29, *29, 30*, 32, *33*
Velocity vector, 93
  as function of time, 74
Venturi tube, 435, *435*
Vibration, frequencies of, 367
  modes of, 453, 499
  normal modes of, 513, *514*, 519
    in standing wave, *505*, 505–506
  of string, 470
  resonant, 380
  wave motion and, 453
Viscosity, 431–432, *439*, 439–441
  coefficient(s) of, *440*, 440t, 440–441
Volt, definition of, 709
  electron, 709
Voltage, Hall, 852
  open-circuit, 798, *798*
  root mean square (rms), 967, 983

Voltage (Continued)
  terminal, 798, 798n
  of battery, 799
Voltage amplitude, of ac generator, 965
Voltmeter, 752, 814, *814*
Volume, constant, molar specific heats at, 591, 607
  dimensions of, 11t
  expansion, average coefficient of, 538, 543
Volume elasticity, *347*, 347–348
Volume flux, 433
Volume stress, 347
von Klitzing, Klaus, 853

Wall, R value of, 572
Water, boiling of, 470
  freezing of, 541
  triple point of, 535
Water bed, 426
Water molecule, *756*, 756–757
Water waves, 455, 456, *457*
Watt, 187
Watt, James, 187
Wave equation(s), for electromagnetic waves in free space, 998
  general, 998
Wave front, 485
Wave function, 458, 473, 472
  for sinusoidal waves, 466, 471
  for standing waves, 503
Wave intensity, 1002
Wave number, angular, 466
Wave pulse, 456, *456*, 456n, 458–459, *459*, 472
  speed of, 458, *463*, 463–464
Waveform(s), 516, *516*
  harmonics of, *518*
Wavelength(s), 486
  definition of, 455, *455*
  of normal modes, 506
  "seeing ability of wave and, 1079–1080
Wavelets, 1036, *1038*
Wave(s), amplitude of, 457, *457*
  audible, 479
  complex, *516*, 516–518
  electromagnetic, 455, 1000, *1000*
    detection of, 996, *996*
    energy carried by, 1002–1004
    equations for, 995–996
    frequencies of, 995, 996
    generation of, 995, 996, *996*
    in free space, wave equations for, 998
    momentum and, 1004
    phase changes of, 1102
    plane, *997*, 997–1002, 998n, *999*, *1001*, 1001n
    production of, by antenna, 1008–1011, *1010*, *1011*
    properties of, 1000, 1013–1014
    radiation pressure and, 1004–1005, *1005*, 1005n

Wave(s) *(Continued)*
  sinusoidal plane-polarized, *999*, 999–1000, 1014
    spectrum of, 1011–1013, *1012*, 1014
    speed of, 999
  grain, 455
  harmonic, traveling, 500
  infrared, 1012
  infrasonic, 479
  interference of, 459
  light, fringes of, 1095, 1108
    bright, 1120
    dark, 1120
      positions of, 1120–1121
    distance between, 1096
    order number of, 1095, 1108
  interference of, 1091–1116, 1117–1118, *1118*
    constructive, 1095
    in thin films, *1103*, 1103–1107
  linearly polarized, 1132, *1132*
  path difference of, 1094–1095
  plane of polarization of, 1132
  plane-polarized, 1132
  polarization of, *1132*, 1132–1138, *1133*
    by double refraction, 1135–1137
    by reflection, 1134–1135, *1135*
    by scattering, 1137, *1137*
    by selective absorption, *1133*, 1133–1134, *1134*
  speed of, 1151–1154, *1152*
    constancy of, 1156, 1157
  unpolarized, 1132, *1132*
  linear, 459, 472
  longitudinal, 456, *456*, 472
  mechanical, 453–527
    requirements for, 455
  motion, 454–478, *456*
    and vibration, 453
    stimulator, 460
  nonlinear, 459
  on strings, speed of, 462–464, *463*, 472
  periodic, frequency of, 455
  phasor addition of, *1098*, 1098–1101, *1099*
  physical characteristics of, 455
  plane, 486, *486*
  radio, 1011
  shock, *490*, 491, *491*
  sinusoidal, *465*, 465–469
    frequency of, 466
    general relation for, 467
    interference of, 500
    longitudinal, 481, *481*, 491

Wave(s) *(Continued)*
  on strings, 467–469, *468*, 468n, 473
    energy transmitted by, *469*, 469–470
    power transmitted by, 470, 473
    traveling, 467, *467*
    velocity of, 466
    wave function for, 466, 471
  sound, 455, 456
    amplitude displacement of, 481–482, *482*
    harmonic, intensity of, 491
    intensity of, spherical, 491–492
    intensity variations from source, 486–487
    interference of, 501, *501*
    moving train whistle, frequency of, 489–490
    periodic, 481–482
      intensity of, 482–484, *483*
    pressure of, 481–482, *482*, 491
    siren, frequency of, 490
    speed of, 480–481, 491, 1152–1154
      in various media, 480t, 480–481
  speed of, 455
  spherical, 484–487, *485*
  standing, 499, 502–505, *504*, 519
    formation of, 505
    in air columns, 510–512, *511*, 519
    in fixed string, *505*, 505–508
    in rods and plates, 512–513, *513*
    longitudinal, 510–511, *511*
    motion of, energy and, 503–504, *504*
    wave function for, 503
  stationary, 504
  superposition of, 459
  transmission of, reflection and, *464*, 464–465, *465*
  transverse, 456, 472
  traveling, 456, *456*, 459–460, *461*
    harmonic, 500
    one-dimensional, 457–459
    two, combining of, 1093–1094, *1094*
  types of, 456–457
  ultrasonic, 479
  ultraviolet, 1012–1013
  visible, 1012
  water, 455, 456, *457*
    interference pattern in, *1093*
Weber per square meter, 837, 854
Weber (unit), 877
Weight, 105, 113n, 113–114, 129
  apparent, versus true weight, *123*, 123–124
  gravitational forces and, *123*, 123–124, 394–395
  loss, and exercise, 554

Weight *(Continued)*
  mass and, 111
Wheatstone bridge, 815, *815*
Wheel(s), angular acceleration of, *291*, 291–292
  power delivered to, 190
  rotating, 279
Whistle, moving train, 489–490
Wind, energy from, 437–439, *438*
Wind generators, *438*, 438–439
Windmill, 439
Windshield wipers, intermittent, 811
Work, 171, 173
  and energy, 171–201
  conversion factors for, A.2
  done by central force, 402
  done by conservative forces, 205–206, 222
  done by constant force, *172*, 172–174, 192
  done by gas, 575
  done by spring, *178*, 178–180, *179*, 204
  done by varying force, *176*, 176–180, *178*, 192
  done in isothermal process, 566–567
  done using horizontal force, 174, *174*
  heat and, 615
  in rotational motion, *293*, 293–295, 297
  in thermodynamic processes, 561–563, *562*
  net, done by force, calculation of, from graph, 177, *177*
  units of, in systems of measurement, 173, 173t
Work-energy theorem, for rotational motion, 294
  kinetic energy and, *180*, 180–186, *182*, 192

X-rays, 1013
  diffraction of, by crystals, 1130n, 1130–1132, *1131*
Xerography, 726, *727*

Young, Thomas, 1022, 1024, 1092
Young's double-slit experiment, 1092–1096, *1093*, 1108
Young's modulus, 346, 348, 349

Zeropoint energy, 535
Zero(s), absolute, 535, 624, 624n
  significant figures and, 15
Zeroth law of thermodynamics, 628
  temperature and, 532–533, 543

Standard Abbreviations and Symbols of Units

Abbreviation	Unit	Abbreviation	Unit
A	ampere	in.	inch
Å	angstrom	J	joule
u	atomic mass unit	K	kelvin
atm	atmosphere	kcal	kilocalorie
Btu	British thermal unit	kg	kilogram
C	coulomb	kmol	kilomole
°C	degree Celsius	lb	pound
cal	calorie	m	meter
deg	degree (angle)	min	minute
eV	electron volt	N	newton
°F	degree Fahrenheit	Pa	pascal
F	farad	rev	revolution
ft	foot	s	second
G	gauss	T	tesla
g	gram	V	volt
H	henry	W	watt
h	hour	Wb	weber
hp	horsepower	μm	micrometer
Hz	hertz	Ω	ohm

Mathematical Symbols Used in the Text and Their Meaning

Symbol	Meaning
$=$	is equal to
$\equiv$	is defined as
$\neq$	is not equal to
$\propto$	is proportional to
$>$	is greater than
$<$	is less than
$\gg (\ll)$	is much greater (less) than
$\approx$	is approximately equal to
Δx	the change in x
$\sum_{i=1}^{N} x_i$	the sum of all quantities x_i from $i = 1$ to $i = N$
$\|x\|$	the magnitude of x (always a nonnegative quantity)
$\Delta x \to 0$	Δx approaches zero
$\dfrac{dx}{dt}$	the derivative of x with respect to t
$\dfrac{\partial x}{\partial t}$	the partial derivative of x with respect to t
$\int$	integral